About the Authors

W9-BYI-112
WITHDRAWN

Ian C. Stewart is a chemist at a petrochemical company. He formerly worked with Nobel Prize-winning chemist Robert H. Grubbs and is the recipient of several fellowships. He lives in Houston, Texas.

Justin P. Lomont is a National Science Foundation Graduate Research Fellow whose doctoral research at the University of California Berkeley focuses on organometallic reaction mechanisms using ultra-fast infrared spectroscopy. He lives in Berkeley, California.

Also from Visible Ink Press

The Handy Anatomy Answer Book
by James Bobick and Naomi Balaban
ISBN: 978-1-57859-190-9

The Handy Answer Book for Kids (and Parents),
2nd edition
by Gina Misiroglu
ISBN: 978-1-57859-219-7

The Handy Art History Answer Book
by Madelynn Dickerson
ISBN: 978-1-57859-417-7

The Handy Astronomy Answer Book, 2nd
edition
by Charles Liu
ISBN: 978-1-57859-190-9

The Handy Biology Answer Book
by James Bobick, Naomi Balaban, Sandra
Bobick, and Laurel Roberts
ISBN: 978-1-57859-150-3

The Handy Dinosaur Answer Book, 2nd edition
by Patricia Barnes-Svarney and Thomas E.
Svarney
ISBN: 978-1-57859-218-0

The Handy Geography Answer Book, 2nd
edition
by Paul A. Tucci
ISBN: 978-1-57859-215-9

The Handy Geology Answer Book
by Patricia Barnes-Svarney and Thomas E.
Svarney
ISBN: 978-1-57859-156-5

The Handy History Answer Book, 3rd edition
by David L. Hudson Jr.
ISBN: 978-1-57859-372-9

The Handy Law Answer Book
by David L. Hudson Jr.
ISBN: 978-1-57859-217-3

The Handy Math Answer Book, 2nd edition
by Patricia Barnes-Svarney and Thomas E.
Svarney
ISBN: 978-1-57859-373-6

The Handy Ocean Answer Book
by Patricia Barnes-Svarney and Thomas E.
Svarney
ISBN: 978-1-57859-063-6

The Handy Personal Finance Answer Book
by Paul A. Tucci
ISBN: 978-1-57859-322-4

The Handy Philosophy Answer Book
by Naomi Zack
ISBN: 978-1-57859-226-5

The Handy Physics Answer Book, 2nd edition
By Paul W. Zitzewitz, Ph.D.
ISBN: 978-1-57859-305-7

The Handy Politics Answer Book
by Gina Misiroglu
ISBN: 978-1-57859-139-8

The Handy Presidents Answer Book, 2nd edition
by David L. Hudson
ISB N: 978-1-57859-317-0

The Handy Psychology Answer Book
by Lisa J. Cohen
ISBN: 978-1-57859-223-4

The Handy Religion Answer Book, 2nd edition
by John Renard
ISBN: 978-1-57859-379-8

The Handy Science Answer Book®, 4th edition
by The Science and Technology Department
Carnegie Library of Pittsburgh, James E.
Bobick, and Naomi E. Balaban
ISBN: 978-1-57859-140-4

The Handy Sports Answer Book
by Kevin Hillstrom, Laurie Hillstrom, and Roger
Matuz
ISBN: 978-1-57859-075-9

The Handy Supreme Court Answer Book
by David L Hudson, Jr.
ISBN: 978-1-57859-196-1

The Handy Weather Answer Book, 2nd edition
by Kevin S. Hile
ISBN: 978-1-57859-221-0

Please visit the "Handy" series website at www.handyanswers.com.

LA GRANGE PUBLIC LIBRARY
10 WEST COSSITT
LA GRANGE, ILLINOIS 60525

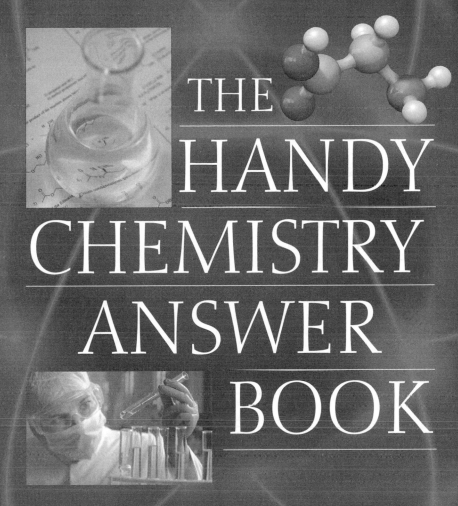

THE
HANDY
CHEMISTRY
ANSWER
BOOK

Ian C. Stewart and Justin P. Lomont

VISIBLE
INK
PRESS

Detroit

THE HANDY CHEMISTRY ANSWER BOOK

Copyright © 2014 by Visible Ink Press®

This publication is a creative work fully protected by all applicable copyright laws, as well as by misappropriation, trade secret, unfair competition, and other applicable laws.

No part of this book may be reproduced in any form without permission in writing from the publisher, except by a reviewer who wishes to quote brief passages in connection with a review written for inclusion in a magazine, newspaper, or Website.

All rights to this publication will be vigorously defended.

Visible Ink Press®
43311 Joy Rd., #414
Canton, MI 48187–2075

Visible Ink Press is a registered trademark of Visible Ink Press LLC.

Most Visible Ink Press books are available at special quantity discounts when purchased in bulk by corporations, organizations, or groups. Customized printings, special imprints, messages, and excerpts can be produced to meet your needs. For more information, contact Special Markets Director, Visible Ink Press, www.visibleink.com, or 734–667–3211.

Managing Editor: Kevin S. Hile
Art Director: Mary Claire Krzewinski
Typesetting: Marco Di Vita
Proofreaders: Shoshana Hurwitz and Crystal Rosza
Indexer: Larry Baker

ISBN 978–1–57859–374–3

Cover images: Background image: Shutterstock. All other images: iStock.

Library of Congress Cataloging–in–Publication Data
Stewart, Ian (Ian Christopher), 1980–
  The handy chemistry answer book / by Ian Stewart and Justin Lomont.
      pages cm – (The handy answer book series)
  ISBN 978–1–57859–374–3 (paperback)
  1. Chemistry–Miscellanea. 2. Chemistry–Examinations, questions, etc. I. Lomont, Justin. II. Title.
  QD39.2.S75 2014
  540–dc23
                                                              2013020166

Printed in the United States of America

10 9 8 7 6 5 4 3 2 1

R
540
STE

Contents

Acknowledgements

We would first like to thank Roger Jänecke and Kevin Hile at Visible Ink Press for their assistance, patience, and for taking a chance on two young science writers.

College students at the University of Michigan and Brandeis University submitted many of the questions in the chapter titled "The World around Us." Our thanks go out to these people for their creative and inspiring questions: Jon Ahearn, Krishna Bathina, Alex Belkin, Matt Benoit, Emma Betzig, Jeetayu Biswas, Ariana Boltax, Nick Carducci, Gina DiCiuccio, Alice Doong, Christina Lee, Shelby Lee, Greg Lorrain, Jake Lurie, Sotirios Malamis, Aysha Malik, Yawar Malik, Katie Marchetti, Nicholas Medina, Leah Naghi, Humaira Nawer, Logan Powell, Nilesh Raval, Alexandra Rzepecki, Minna Schmidt, Leah Simke, Sindhura Sonnathi, Eva Tulchinsky, Afzal Ullah, Anna Yatskar.

Finally, we would like to dedicate this book to the person who initially brought us together, inspired and challenged us with this project and many others, taught us what it really means to learn and to teach, and so much more. This is for you, Brian.

Photo Credits

All line art illustrations by Kevin S. Hile. Photographs in the "Chemistry Experiments You Can Do at Home" chapter by James Fordyce.

All other images courtesy of Shutterstock, with the following exceptions: page 18: Armtuk; page 62: BrokenSphere; pages 67, 79, 129, 156, 186, 193, 222: public domain; pages 106, 141: Wikicommons; page 136: Patrick Edwin Moran; page 163: Стрелец Игорь; page 252: AlexanderAlUS; page 271: Miia Ranta from Finland.

Introduction

What's inside of a zit? Why does eating turkey make you sleepy? How do glow sticks work? What causes a hangover? Chemistry (and this book!) holds answers to all of these questions. There are many wonderful stories about chemistry and the people behind these discoveries. Whether you have studied chemistry in high school or college, or even do chemistry for a living, we think you will enjoy this book. We certainly enjoyed writing it.

As you will see right away, our approach is quite different from a textbook. If you're curious about the answers to hundreds of interesting questions about all the things in the world that you touch, feel, and taste every day, then you've come to the right place. *The Handy Chemistry Answer Book* uses a simple question and answer format to explain the chemistry in our daily lives. There are entire chapters on sustainable chemistry, the chemistry of cooking, and the chemistry of space. Some of these are questions we got from people like you! The questions in the chapter called "The World around Us" were all submitted from college students at the University of Michigan (our alma mater—Go Blue!) and Brandeis University.

We think that you've probably wondered, for just one more example, what sodium laureth sulfate is doing in your shampoo, but maybe never had a chance to ask. We are interested in explaining these things in plain language, and we've kept a conversational tone throughout the entire book, even with some very challenging subject matter. We hope that reading this book feels like you're talking to someone about chemistry, even if you wouldn't be caught dead doing that. There are chemical structures throughout this book, and we've used a simplified drawing system. Take what you can from these abstract drawings, but don't dwell on them. Focus on the stories we're trying to tell about molecules. And if you have a chemistry question you'd like to ask, or a chemistry story you'd like to share with us, please drop us an email.

Finally, we both really enjoy working at our corporations and institutes of higher learning, respectively, and want to continue to do so for years to come. So every fact, implication, mistake, and opinion expressed herein is absolutely ours and ours alone, and do not in any way represent the opinion or position of our employers, or any other person or organization.

Enjoy.

Ian Stewart
Justin Lomont
handy.chemistry.answers@gmail.com

HISTORY OF CHEMISTRY

What is the earliest **historical evidence** of the study of **chemistry**?

Although they didn't call it chemistry, people from ancient civilizations used chemical reactions in many aspects of their lives. Metalworking, including the extraction of pure metals from ores, and then combining metals to make alloys, like bronze, left many artifacts of early man's chemistry experiments. Pottery, including the production and use of various glazes, fermentation to make beer and wine, and pigments and dyes for cloth and cosmetics are all evidence that man has always been fascinated by the ability to change matter.

Where was early **chemistry developed**?

While many civilizations learned how to make dyes and pigments, or ferment fruit into wine, the earliest theories about atoms and what makes up the chemical world came from ancient Greece and India. Leucippus in Greece and Kanada in India both came up with the idea that there must be a small, indivisible part of matter. The Greek word for "uncuttable" is *atomos,* clearly the root of the modern term atom. Kanada's term for this similar concept was "paramanu" or simply "anu," the indivisible element of matter.

What does the city of **Miletus** have to do with **chemistry**?

Miletus, one of the Greeks' greatest cities, was located on the western coast of what is now Turkey and was home to where some of the earliest ideas about chemistry were recorded. During the sixth century B.C.E., the Milesian school of thought was founded, and the musings of three philosophers survived into the modern era: Thales, Anaximander, and Anaximenes. Thales thought the most basic building block of the universe was water and that the Earth floated on top of this celestial water. Anaximander challenged both of these ideas, proposing that the universe was born when fire and water (or hot and cold) separated from one another and that the Earth simply floated on nothing.

1

Anaximenes, who was a friend or perhaps student of Anaximander, countered that air was the most basic substance and that air condensed to form water and evaporated to reverse that process.

Who first proposed the idea of elements?

Plato is often given this accolade as he was the first to use this term for his description of the five basic shapes that he believed made up the entire universe: tetrahedrons, icosahedrons, dodecahedrons, octahedrons, and cubes. He went on to ascribe each shape to a basic element, borrowing from Empedocles (see next question). The tetrahedron was fire; icosahedron, water; dodecahedron, aether; octahedron, air; cube, earth. While this association of basic geometrical shapes to the nature of the Universe obviously didn't work out for him, Plato's ideas did lead Euclid to invent geometry.

What did Empedocles believe were the four basic elements?

A Greek named Empedocles (who was not from Miletus, but rather Sicily) was the first to propose the four basic "elements." These four elements were earth, air, water, and fire. These elements had a much different definition from that which chemists use today (which we'll get to later). Unlike the modern definition of an element, Empedocles' understanding of an element did not require it to be a pure substance. Water, for example, was obviously not the only liquid Empedocles had ever encountered. Earth represented solids, water represented liquids, air represented gases, and fire represented heat.

What fifth element did Aristotle add?

Although Empedocles is understood to have been the first to propose the four basic elements, Aristotle is sometimes given this credit. Aristotle did propose a fifth basic element though, which he called aether. Aether was a divine material that Aristotle said made up the stars and other planets in the sky.

When did the theory of the atom come about?

The idea of the atom was originally proposed by ancient scholars. The philosophers Democritus and Leucippus are often credited with proposing the early notions of the atom, including the ideas that many different kinds of atoms exist, that there is a substantial amount of empty space between atoms, and that their properties are responsible for the properties of materials we see and interact with. For centuries,

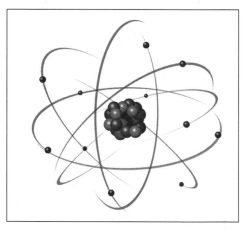

The earliest theories of the atom came about in ancient Greece, where philosophers correctly surmised that there were different kinds of atoms and that they contained mostly empty space—all centuries before the invention of the microscope!

ideas about the structure and properties of the atom were based largely on conjecture and logical arguments, and it wasn't until the 1800s that experiments began to allow atomic theory to advance to where it is today.

What is an **element**?

An element is the most basic form of a chemical substance. If you have an object made of a pure element, all of its atoms have the same number of protons (we'll discuss what this is more a little later) and the same basic chemical properties. There are not many objects that we encounter on a daily basis that are actually composed of only a single element—most things are formed from atoms of several types of elements bonded together.

What separates **ancient** and **modern chemistry**?

While there's not a clear, punctuating distinction between ancient and modern chemistry, there are a few major differences that separate the two. Modern chemists describe the world in terms of atoms, molecules, and electrons and have a relatively complete understanding of the basic particles that make up matter—at least insofar as is necessary to describe chemical transformations. Ancient chemists didn't have this information and relied less on experimental evidence and more on theory and mythology. For example, ancient chemists sought the Philosopher's Stone (see below), for which there was no verifiable evidence, but they were attracted to it for its mythological power to preserve youth.

Who ran the **first chemistry experiment ever**?

Jābir ibn Hayyān, known as "Geber" in Western texts, was probably the world's first alchemist to run actual experiments. Jābir lived during the eighth century in what is now Iran, and like alchemists before and after him, was fascinated by the prospect of changing one metal into another and by creating artificial life. To Aristotle's four elements, Jābir added sulfur and mercury, and proposed that all metals were made of differing ratios of these two elements. He was the first to emphasize the importance of rigorous experimentation and is credited with describing many common lab techniques and equipment.

What is **metallurgy**?

Metallurgy is the branch of science that deals with the properties of metals composed of both pure elements and those of mixtures of metallic elements (which are called alloys). It represents one of the first efforts to manipulate and understand how elemental composition of a substance affects its physical properties.

What is **bronze**?

Bronze is an alloy of copper and tin that may contain up to one-third tin. Early civilizations used bronze because it could make stronger, more durable tools than stone or pure copper.

How did **ancient civilizations** make **bronze**?

Tin had to be mined in the form of an ore and then purified through a process called smelting. Once the tin was pure, it could be added to molten copper, in whatever ratio was desired, to make bronze for use in things like tools and weapons.

What is **iron smelting**?

Smelting is a method for extracting a pure metal from an ore (an ore is a rock made up of metals and other minerals). In smelting, a chemical transformation may be used to purify the metal by changing the oxidation state of metals in the ore (we'll get into more detail on what an oxidation state is later in the book). The smelting of iron dates back to ca. 1000 B.C.E. (or maybe even earlier), and it typically involves first heating the raw material in a furnace called a bloomery. This produces a soft iron material that can be shaped. A hammer is often used to remove other impurities from the soft metal before allowing it to harden to form a relatively pure form of iron.

What is **meteoric iron**?

Meteoric iron is just what the name sounds like: it's iron that comes from meteors. For early civilizations, meteoric iron was one of the few available sources of relatively pure iron (that is, prior to when the extraction of iron from ore was discovered). Meteors containing meteoric iron are composed of mostly nickel–iron alloys. Iron meteorites often have a distinct appearance, and they are typically much easier to recognize than other types of meteorites. For this reason, they are discovered more often than other types of meteorites. Iron meteorites actually account for all of the largest meteorites that have been discovered.

What is **alchemy**?

Alchemy was among the earliest practices of a chemical science, and, in a way, it can be considered a predecessor to the modern science of chemistry. Alchemy is somewhat different than a modern science, though, in that it also has roots in mythology and spiritualism. Practitioners of alchemy were known as alchemists. Among the primary goals of alchemists were to find a method or material that could convert inexpensive metals into precious gold, as well as to find an elixir of life, which could make a person both youthful and immortal. Myths told of the existence of such materials and of the possibility of such feats; the goals of alchemists were based largely on these myths. In medieval times, alchemists could be found in many countries around the world, and those in different regions held somewhat different beliefs. In the western world, people were still considering how to make metals into gold as recently as the late 1700s.

When did **alchemists** finally **abandon** trying to **make gold**?

In the late 1700s, a scientist named James Price was still hard at work trying to "transmute" metals into gold and silver. In 1782, he claimed he could convert mercury into

What is the Philosopher's Stone?

The Philosopher's Stone was an object of legend among alchemists. It was believed that the Philosopher's Stone possessed the power to make gold from other cheap metals. It has also been said to serve as the elixir of life, which was sought by many alchemists. Of course, such a stone has never been discovered (at least not outside of Hogwarts).

silver and gold. At first it appeared that his experiments had worked, but conflict rapidly rose. More and more scientists asked to witness the experiments firsthand, and Price eventually lost confidence in the validity of his own work. After disappearing for a few months, in 1783 he invited scientists to his laboratory to witness his experiments in person, but only a few men showed up. In their presence, Price intentionally ingested a poison, killing himself. He was the last of the modern scientists to claim to have achieved the goals of alchemy, and it is no longer believed that anyone will find a simple way to convert inexpensive metals into gold.

How did **pharmaceutical science** get **started**?

Paracelsus is credited with being the first person to use chemicals in medicine. Before Paracelsus, people believed that illness and disease were caused by an imbalance in the patient. Hippocrates thought it was an imbalance of the four humours (blood, phlegm, black bile, and yellow bile), and Galen furthered these ideas by assigning a symptom to an imbalance of each of Hippocrates' humours. These theories supported the use of medical techniques like bloodletting. Paracelsus believed that illness was the result of something from the outside world attacking inside one's body and that some of the illnesses could be cured by chemicals. He is also known for proposing the basis of toxicology, namely that dosage was critical to whether a substance was poisonous or not.

What was the **first chemistry textbook** published?

Although numerous chemistry texts exist before it, *Alchemia,* published by Andreas Libavius in 1597, is considered to be the first organized chemistry textbook. Libavius, born in Halle, Germany, in 1555, was a chemist and a medical doctor, and also served as a schoolmaster at the end of his life. In addition to his noteworthy textbook, Libavius is significant in the history of chemistry for further advancing the discipline away from the realm of magic, the occult, and alchemy toward a teachable, logical, and scientific discipline.

What's the **difference** between **alchemy** and **chemistry**?

Let's ask Robert Boyle, who in 1661 published *The Sceptical Chymist,* arguing that experiments disproved the idea that the universe was composed solely of Aristotle's four

Herbal medications are natural remedies for treating various ailments. Often these are traditional remedies that can date back hundreds of years and are still used today.

elements. Boyle himself was an alchemist, in that he believed that one metal could be changed into another, but he was a staunch promoter of the scientific method and helped elevate chemistry to a science. So one could simply say that alchemy is a philosophy, while chemistry is a science.

How did **early chemistry** relate to **medicine**?

Early societies all over the world found that certain types of plants could be used for medicinal purposes. Though only relatively recently have people attempted to gain a detailed understanding of the chemistry behind these methods, the overarching reason why these methods work is because a chemical in the plant interacts with the chemicals in your body in a beneficial way.

What is an **herbal medicine**?

Herbal medicines are any plants or plant extracts used for treating ailments, aches, pain, or discomfort. They can range from culinary remedies (like chicken soup for the common cold), to calming extracts (like mint tea), to eating whole herbs. Every ancient civilization seems to have discovered the use of plants as medicines in one form or another. Even as far back as five thousand years ago, humans were using herbal medicines, as evidenced by herbs being found alongside well-preserved, mummified humans like Ötzi the Iceman.

How were **herbal medicines discovered**?

If we had the story of how each and every medicinal herb was discovered, each would likely be an interesting and unique tale. Unfortunately, the use of plants as medicine predates written human history by a few millennia. The earliest written records come from the great ancient civilizations of humankind.

How are **herbal medicines prepared**?

There are many ways of preparing herbal medicines. Tinctures and elixirs are extractions of herbs using some solvent, usually ethanol. If a plant is extracted with acetic acid, the solution is known as a "vinegar," even though the solvent is also vinegar. A tisane uses hot water to extract herbs—like tea.

What **herbal medicines** do people still **use today**?

Aspirin and quinine are probably the most famous herbal medicines that have made the transition to mainstream medicine. Many modern medicines were originally isolated from plants, however, but the commercial sources are now usually man-made. For example, Taxol® (paclitaxel) was originally isolated from the Pacific yew tree. In 1967 this compound was found to be useful as a treatment for various types of cancer. For almost thirty years, most of the paclitaxel that was given to patients was obtained from the yew tree. Alternate supplies of this drug were developed in the 1990s, moving this natural drug into the realm of modern synthetic medicines.

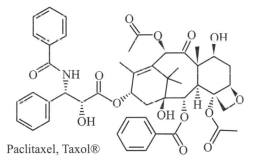

Paclitaxel, Taxol®

How do **herbal medicines differ** from **modern medicine**?

Modern pharmaceutical medicines usually contain only one active ingredient, or a few at most. The rest of the ingredients in a pill are there to aid in its delivery in one way or another. Herbal medicines, because they are made from plants that were once living, can contain many more chemicals, though only one may be the active ingredient in this case as well.

How did **chemistry affect trade** in **ancient times**?

Ancient chemistry was involved in the production of many goods that were important to trade. These included salt, silk, linen dyes, precious metals, wine, and pottery.

What is **fire**?

The chemical description of fire is a combustion reaction. It involves the reaction of oxygen with molecules in some combustible material. The fire itself is caused by energy released by this reaction in the form of heat and light. The fire you see is not only the light that's being released, but also glowing hot gases.

How can a **fire** be started with a **piece of flint**?

Almost everyone has seen a movie character start a fire using a piece of flint, but you may wonder how this is possible. Flint is a hard stone that can produce sparks when it

Before matches and butane lighters were invented, people could use flint to start fires for heat and cooking. Striking a hard piece of metal against flint causes a spark, which can in turn catch tinder on fire.

is struck against a metal, such as steel. The sharp edge of the flint breaks off a small splinter of steel, which is heated significantly by the friction from the strike of the flint. As this splinter of hot steel reacts with oxygen in the air, a spark is produced. The sparks generated in this way can then ignite a piece of dry wood, paper, or other fuel.

Who first **realized** that **air has weight**?

It was actually a mathematician named Evangelista Torricelli who is the first on record to demonstrate that air has weight. His experiment to prove this fact was prompted by the observation that water from a mineshaft could only be pumped upward to reach a certain height. Torricelli thought that the air pushing down on the surface of the water must play a role. To test this theory, in 1643 he placed a sealed tube of mercury upside down in a bowl of mercury. He observed that the weight of the air would keep the mercury in the tube at a certain level, and on different days he observed that the mercury would rise to different levels. We now know this is because the air pressure varies from day to day, and Torricelli's experiment was the first barometer.

Who first realized that **oxygen gas** (O_2) was **required for fire**?

Philo of Byzantium in the second century B.C.E. was the first to observe (or at least the first to record such an observation) that if you placed a jar on top of a candle with water around its base, some water would be drawn up into the jar as the candle burned and eventually went out once all the oxygen was consumed. Although the experiment was well-designed, he ended up with an incorrect conclusion about the process. Robert Boyle repeated the experiment but replaced the candle with a mouse (seriously), and noticed the water also rose up the container. From this experiment he correctly inferred that whatever the component in air was (he called it nitroaerues), it was needed for both combustion and respiration. Robert Hooke, and others, likely produced oxygen gas in the seventeenth century, but didn't realize it was an element as the phlogiston theory (see below) was in vogue at the time. So to really realize that oxygen gas was required for fire, it first had to be, well, discovered.

What is the **theory of phlogiston**?

In 1667, a scientist named Johann Joachim Becher introduced the theory of phlogiston as an explanation for the various observations scientists had made regarding combustion. These observations include the fact that some objects can burn while others cannot, and that a flame in a sealed container can go out before the combustible material is consumed. Becher proposed that a weightless (or almost weightless) substance called phlogiston was present in all materials that could burn and that this phlogiston was the substance being given off during combustion. If a candle placed in a closed container went out, Becher said this was because the phlogiston from the candle was moving into the air and that the air could only absorb a certain concentration of phlogiston before it became saturated and could no longer absorb more phlogiston from the candle. Another tenet of this theory was that the purpose of breathing was to remove phlogiston from the body. Air that had been used for combustion couldn't be used to breathe then because it was already saturated with phlogiston.

How was the **theory of phlogiston disproved**?

Antoine Lavoisier, an eighteenth-century French chemist, disproved the theory of phlogiston by showing that combustion required a gas (oxygen) and that that gas has weight. Lavoisier did this by burning elements in closed containers. These solids gained mass, but the total weight of the containers did not change—what did change was the pressure inside the vessel. When Lavoisier opened the vessel up, air rushed in, and the total weight of the vessel increased. So Becher had it backward: oxygen was being used up by the candle instead of phlogiston being given off by the flame.

How was **oxygen gas** first **discovered**?

Well, to answer that question, you would first want to know *who* first discovered oxygen, and there is no simple answer to that question! There are three people to whom discovery of this can be ascribed: Carl Wilhelm Scheel, Joseph Priestley, and Antoine Lavoisier. Scheele produced O_2 (he called it "fire aire") from mercuric oxide (HgO) in 1772, but the result wasn't published until 1777. Meanwhile, in 1774 Priestley produced O_2 (he called it "dephlogisticated air") using a similar experiment, which was published in 1775. Lavoisier claimed to have independently discovered the gas, and was in fact the first to explain how combustion worked via quantitative experiments, leading to the principle of Conservation of Mass, and ultimately disproving the entire idea of phlogiston. Whew. So Scheel found it first, but didn't report it; Priestley reported it first, but didn't have the explanation correct; and Lavoisier was last, but nailed it. Who would you give credit to?

What is **electrochemistry** and how was it **discovered**?

Modern electrochemistry studies reactions that take place at the interface of an electronic conductor and a source of charged ions (possibly a liquid). The development of electrochemistry began with studies on magnetism, electric charge, and conductivity. The earliest experiments typically focused on questions surrounding properties of materials; for example, which materials can be magnetized and which materials can be charged? As early as the 1750s scientists had discovered that electrical signals were important to human life and were using them to treat medical issues such as muscle spasms. In the late 1700s, Charles Coulomb developed laws describing the interactions of charged bodies, which are still used widely today and taught in any introductory course on electricity and magnetism.

The first electrochemical cells were developed during the 1800s. Electrochemical cells are arrangements of electrodes and sources of ions that either generate electric current from a chemical reaction, or alternatively, use electricity to drive a chemical reaction. Today these cells find applications in daily life, such as in the batteries that power your car or cell phone. Today electrochemistry still constitutes an important field of research and is one that will likely continue to lead to the development of new products and technologies.

What is the **law of definite proportions**?

The law of definite proportions says that a substance always contains the same proportions of each element of which it's composed. For example, a molecule of water (H_2O) always contains two hydrogen atoms for every oxygen atom. This is commonly understood among modern chemists, but it was an important step in working toward a microscopic understanding of the composition of matter. The first to make such claims, in the early 1800s, was the French chemist Joseph Proust. It was a controversial idea at that time, and other chemists believed that elements could be combined in any proportion.

What is **Avogadro's constant**?

Avogadro's constant is a large number used to discuss large quantities of atoms or molecules, usually when chemists talk about quantities they can actually see or measure out. The number itself (rounded at three decimal places) is 6.022×10^{23}. It's just a big number that relates an atomic or molecular mass to the mass of a collection of many atoms or molecules. Avogadro's number of atoms of an element is called a mole of that element, and, similarly, Avogadro's number of molecules of a compound is a mole of that compound. For example, the atomic mass of oxygen is about 16 grams per mole, and 6.022×10^{23} atoms (1 mole) of oxygen weigh(s) about 16 grams. The most recent (and accurate) definition of this constant was $6.02214078(18) \times 10^{23}$, which was calculated by careful measurements of the mass and volume of 1-kilogram (about 2.2 lbs.) spheres of silicon-28, a particular isotope of silicon (see next chapter concerning isotopes).

When was **Avogadro's constant discovered**?

Amedeo Carlo Avogadro published a paper in 1811 describing his theory that a volume of gas (at a given temperature and pressure) contains a certain number of atoms or molecules regardless of what gas it is. Avogadro didn't actually determine what that number was, however. It took just over fifty years for someone to make progress on that: Johann Josef Loschmidt, in 1865, estimated the average size of molecules in air. It's nothing short of amazing that he ended up being off by only a factor of two. Jean Perrin, a French physicist, accurately determined the constant using a few different techniques. He was awarded the Nobel Prize in Physics in 1926 for the work, but Perrin proposed that the constant be named for Avogadro—and the name stuck. (For more on the use of the constant, see "Atoms and Molecules.")

Why is chemistry **"the central science"**?

Chemistry is called the central science because it's related to everything! It connects and draws from topics in biology, physics, materials science, mathematics, engineering, and other fields. Chemistry is important to how our body functions, to the food we eat, to how our medicines work, and to pretty much everything else in our lives. After reading this book, we hope you'll agree!

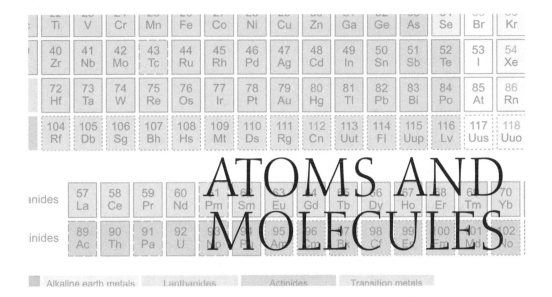

Ti	V	Cr	Mn	Fe	Co	Ni	Cu	Zn	Ga	Ge	As	Se		Br	Kr
40 Zr	41 Nb	42 Mo	43 Tc	44 Ru	45 Rh	46 Pd	47 Ag	48 Cd	49 In	50 Sn	51 Sb	52 Te	53 I	54 Xe
72 Hf	73 Ta	74 W	75 Re	76 Os	77 Ir	78 Pt	79 Au	80 Hg	81 Tl	82 Pb	83 Bi	84 Po	85 At	86 Rn
104 Rf	105 Db	106 Sg	107 Bh	108 Hs	109 Mt	110 Ds	111 Rg	112 Cn	113 Uut	114 Fl	115 Uup	116 Lv	117 Uus	118 Uuo

ATOMS AND MOLECULES

| anides | 57 La | 58 Ce | 59 Pr | 60 Nd | | Pm | Sm | Eu | Gd | Tb | Dy | Ho | Er | Tm | 70 Yb |
| inides | 89 Ac | 90 Th | 91 Pa | 92 U | | | Am | | | | | Cf | | | No |

Alkaline earth metals   Lanthanides   Actinides   Transition metals

STRUCTURE OF THE ATOM

What is an **atom**?

Atoms are among the most basic building blocks, making up all matter. The word atom derives from the Greek word *atomos*, which means "that which cannot be split." The existence of atoms, or a fundamental, indivisible unit of matter, was proposed long before modern chemistry and physics came about. It turns out that atoms are actually made up of even smaller particles, but the atom is the smallest unit of matter that defines an element. The smaller particles that make up an atom are positively charged protons, charge-neutral neutrons, and negatively charged particles called electrons.

What is an **electron**?

The electron is a negatively charged subatomic particle, and it is one of three main subatomic particles (the others being the proton and the neutron) that make up atoms. Electrons are responsible for bonding atoms together to make molecules, and they are also the carriers of electric charges in the conducting materials found in the electronic devices you use every day. While protons and neutrons are both found in the center, or nucleus, of an atom, electrons are located apart from the nucleus and are best described as a cloud of electron density. Most reactions in chemistry deal with changes to the arrangement of electrons in some form.

What is a **proton**?

Protons are subatomic particles that carry a positive charge. They are substantially heavier than electrons (roughly 1,836 times heavier), and carry a positive charge equal in magnitude to that carried by the electron. Protons are found in the nucleus of every atom, and the number of protons present in an atom determines its chemical proper-

13

ties (or, in other words, determines what element it is).

What is a **neutron**?

Neutrons are the other principal component of the nucleus of an atom (along with protons). The neutron is neutral in charge and has a mass roughly similar to that of a proton. Atoms of the same element that contain different numbers of neutrons will generally still have the same behavior as one another in terms of chemical reactivity properties. Both protons and neutrons are, in fact, made up of even smaller particles, but chemistry doesn't usually deal with these even smaller bits.

What were some **early models** for the **atom**?

Experiments suggested that atoms were actually made up of smaller particles,

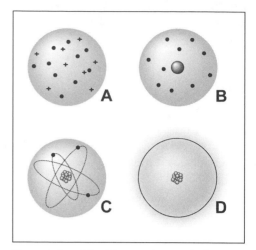

Theoretical models of the atom have evolved over time, including A—the Thomson model (a mix of particles with positive and negative charges), B—the Rutherford model (a positive nucleus surrounded by electrons), C—the Bohr model (stating that electrons follow defined orbits around a nucleus), and D—the quantum mechanical model, based on the idea that you can only determine the probability of an electron's location.

which motivated the development of new models involving protons, neutrons, and electrons. One was Thomson's Plum Pudding Model, which described the atom as a positively charged "pudding" filled with negatively charged electrons. Rutherford later proposed the idea of a positively charged nucleus, but couldn't explain why electrons didn't just fall into it. A Danish physicist named Niels Bohr proposed the idea that electrons travel around the nucleus in specific orbits and advanced the atomic theory to a point very close to where it is today.

How did **scientists determine** that **atoms consist** of **electrons, neutrons, and protons**?

Originally atoms were thought to be the smallest unit of matter, but in the late nineteenth century experiments allowed scientists to finally probe inside atoms. Some of these first experiments were carried out by the British physicist J. J. Thomson, who discovered the electron. He noticed that the rays (actually rays of electrons, though he didn't know it at the time) were deflected by electrically charged plates and concluded that these rays must consist of charged particles that were much smaller than atoms themselves.

Thomson's first graduate student, Ernest Rutherford, continued to investigate the nature of the atom. In the early twentieth century, Rutherford carried out a now-famous experiment in which radioactive particles were shot through extremely thin gold foil. While some bounced off of the nuclei in different directions, most of the particles actually passed through the foil undeflected. Rutherford interpreted this as an indication that the atoms making up the foil must consist of mostly empty space. Over his career, he developed the picture of the atom as a positively charged center surrounded by electrons, and he also proposed that there must be neutral particles (neutrons) to explain the different isotopes of a given element.

What is the **current model** for the **atom**?

The current model for the atom consists of negatively charged electrons orbiting a positively charged nucleus. The nucleus consists of neutrons and protons that are very tightly bound to each other by a strong force. Orbiting electrons behave something like a cloud surrounding the nucleus, and we can't be sure quite where they are at any given time. The electrons are also very lightweight compared to the nucleus, and they move much, much faster.

What **fraction of atoms** are **empty space**?

The fraction of an atom that is occupied by empty space is very large. In fact, over 99.9% of atoms are empty space! The protons, neutrons, and electrons are incredibly small, and the atom occupies such a relatively large effective volume because of the delocalized electron cloud around the nucleus.

What is an **atomic mass unit**?

One atomic mass unit has a mass of 1.66×10^{-27} kg, which is about the mass of a single proton or neutron. These units are convenient, since when the mass of atoms is expressed in atomic mass units, their masses come out to values that are very close to integers. The values of mass given on the periodic table are expressed in atomic mass units.

What is an **isotope**?

Isotopes are atoms with the same number of protons and electrons, but with different numbers of neutrons. Since the numbers of protons and electrons effectively determine the reactivity of an atom, isotopes have the same basic chemical properties and are the same element. They have different masses, though, since they have different numbers of neutrons.

The various isotopes of an element are usually present in relatively fixed ratios throughout nature, but in some cases the ratio can depend on the environment or molecule in which they are found. For example, the element carbon most commonly exists with 6 protons, 6 neutrons, and 6 electrons (referred to as Carbon–12, the total number of particles in the nucleus). A small fraction, however, have 6 protons, 7 neutrons, and 6 electrons (Carbon–13). Roughly 99% of carbon atoms have 6 neutrons, but most of the remaining 1% have 7 neutrons.

Below is a table of the breakdown of the isotopic abundance of several common elements:

Element	Symbol	Nominal Mass	Exact Mass	Abundance %
Hydrogen	H	1	1.00783	99.99
	D or ^{2}H	2	2.0141	.01
Carbon	C	12	12	98.91
		13	13.0034	1.09
Nitrogen	N	14	14.0031	99.6
		15	15.0001	0.37
Oxygen	O	16	15.9949	99.76
		17	16.9991	0.037
		18	17.9992	0.2
Fluorine	F	19	18.9984	100
Silicon	Si	28	27.9769	92.28
		29	28.9765	4.7
		30	29.9738	3.02
Phosphorous	P	31	30.9738	100
Sulphur	S	32	31.9721	95.02
		33	32.9715	0.74
		34	33.9679	4.22
Chlorine	Cl	35	34.9689	75.77
		37	36.9659	24.23
Bromine	Br	79	78.9183	50.5
		81	80.9163	49.5
Iodine	I	127	126.9045	100

TRENDS IN REACTIVITY AND THE PERIODIC TABLE

What was the **law of triads**?

A scientist named Johann Döbereiner discovered trends in the reactivity of groups of elements. Certain sets of three elements, such as lithium, sodium, and potassium, showed similar chemical properties, and Döbereiner noticed that the average of the atomic masses of the atoms of the heaviest and lightest elements in the triad gave the atomic mass of the midweight element. For example, the mean of the atomic masses of lithium and potassium is $(3 + 19)/2 = 11$, which is the atomic mass of sodium. Due to differing numbers of neutrons in each element and the existence of different isotopes, this law isn't always strictly true, but it does tend to work well, especially for lighter elements. For reasons we hope to explain in the coming questions, this trend plays an important role in the structure of the periodic table.

What was the **law of octaves**?

The law of octaves was put forth by the British chemist John Newlands. He noticed that, when the elements were listed in order of increasing atomic weight, elements with similar properties occurred every eight elements. The trend was named "the law of octaves" by analogy to musical scales, and it was the first realization of the relationship between atomic masses and a repeating pattern of elemental properties. This periodicity has since been explained in detail as chemists have gained a better understanding of atomic structure, and the law of octaves played a crucial role in the development of the periodic table chemists use today.

How was the **modern periodic table developed**?

A French geologist named Alexandre Béguyer de Chancourtois is actually the first on record to list all of the elements in order of increasing atomic mass. His first version contained sixty-two elements and they were placed in columns that wrapped around a cylinder; however, there were a variety of issues with this first attempt that were later improved upon. Newlands made the next significant advance, publishing the elements in columns of those with similar properties, which brought the description close to the version chemists use today.

The modern version of the periodic table was proposed by a Russian scientist named Dmitri Mendeleev in 1869. His table was the first to lay out the elements in order of increasing atomic mass in columns of elements with similar reactivity. Elements on the table appeared periodically, essentially in accordance with the law of octaves, hence the name "the periodic table." Mendeleev's table had to include some blank spaces so that the elements were each in the proper column according to reactivity. As more elements were discovered, the blank spaces in the table were eventually filled in, validating Mendeleev's table.

Which **elements** are most **abundant**?

Element	Abundance (ppm atom fraction)
Hydrogen	909,964
Helium	88,714
Oxygen	477
Carbon	326
Nitrogen	102
Neon	100
Silicon	30
Magnesium	28
Iron	27
Sulfur	16

What are the **different groupings for elements** on the **periodic table**?

There are several ways of classifying the elements on the periodic table. One is by the periods, or the horizontal rows, in each of which the properties of the elements change going from left to right. Another common classification is by groups, or vertical columns on the

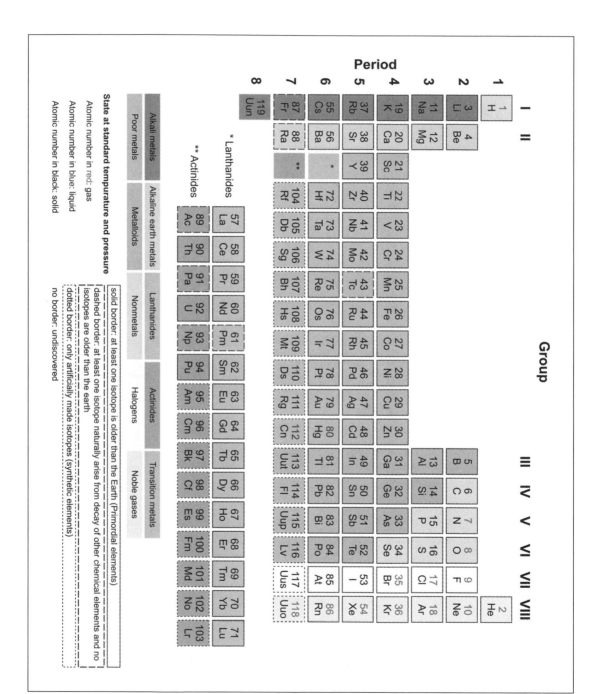

The Periodic Table of Elements.

table. All elements in the same group are expected to have similar properties, which is the periodic property, originally noted in the law of the octaves, after which the table is named.

Yet another classification is by blocks, meaning the elements are classified by the type of orbital in which the highest energy electrons reside (see below). The logic behind this type of classification is that the type of orbital in which the highest energy electron resides strongly influences the reactivity of the element, thus elements in the same block usually have similar properties. There are even more ways still of clustering the elements on the table, but these three are the most commonly used.

What is **scientific notation**?

Scientific notation is a commonly used method for expressing large numbers. The number is written as a product of a decimal number and a power of 10. See the following question for an example of where this is useful.

What do the **numbers** on the **periodic table mean**?

The periodic table lists elements in boxes containing their name, atomic number, chemical symbol, and atomic mass (averaged over the natural abundances of the various isotopes). A typical arrangement for a given element looks something like this:

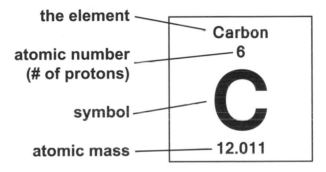

How many elements are there, and will **more be discovered**?

As of the writing of this book, 118 elements have been discovered. The lightest element, with only one proton and an atomic mass of 1.00794 g/mol, is hydrogen. The heaviest is ununoctium with 118 protons and an atomic mass of 294 for the only detected isotope. Considering that five new elements have been discovered since the year 2000, it seems very likely that more elements will be discovered. It is getting harder and harder for scientists to discover, or synthetically create, new elements, though, because the heaviest elements that have been observed to date are usually unstable and decay extremely quickly.

How are **elements named**?

The names of elements often have interesting origins. They have been named after people, places, colors, mythological creatures, or for a variety of other reasons. Some are named after scientists, such as Curium (after Marie and Pierre Curie), Lawrencium (Ernest Lawrence), Seaborgium (Glenn Seaborg), Mendelevium (Dmitri Mendeleev), Einsteinium (Albert Einstein), and Bohrium (Niels Bohr). Others are named after places, such as Lutetium (Lutetia means Paris in Greek), Californium, Berkelium (Berkeley, California), Americium, Dubnium (Dubna, Russia), Hassium (Hessen, Germany), Yttrium, Ytterbium, Terbium, and Erbium (these last four being named after Ytterby, Sweden).

Tantalum (Tantalus), Niobium (Niobe), Promethium (Promethius), Uranium (Uranus), Neptunium (Neptune), Plutonium (Pluto), Palladium (Pallas), and Cerium (Ceres) are all named after mythological creatures.

Though elements can take on different names in different countries, the commonly accepted names are those agreed upon and assigned by the International Union of Pure and Applied Chemistry (IUPAC).

PROPERTIES OF ATOMS AND ELECTRONS IN ATOMS

How large is an atom relative to things we can see?

The smallest object a human eye can see is approximately 0.1 mm, or 10^{-4} m. Atoms have sizes on the order of 10^{-10} m, or approximately one million times smaller than something the human eye can possibly see.

Is it possible to split an atom?

It is possible to split an atom. When people refer to splitting an atom, it's the nucleus of the atom that is being split. One process that splits the nucleus of an atom is called fission, which can happen spontaneously in heavier elements. Spontaneous fission basically involves a nucleus emitting a particle containing one or more protons or neutrons. One of the most commonly emitted particles is called an alpha particle, which consists of two neutrons and two protons. Whenever the number of protons in a nucleus changes, it becomes a different element.

Nuclei can also be split intentionally in laboratories. The nucleus is held together very tightly, so it usually takes a high-energy particle colliding with an atom to break it apart. Typically a high-energy neutron is used to initiate the process of splitting a nucleus. This process results in an overall release of energy so that once one nucleus is split, its products can cause the reaction to happen again. This is called a chain reaction, and it can be used to produce energy in a nuclear reactor (if it happens somewhat slowly), or an explosion (if it happens quickly).

Can elements be converted into one another?

It is possible for atoms of one element to become atoms of another element. One way this can happen is any fission process that results in the loss of one or more protons from a nucleus. The joining of two nuclei to form a single, heavier nucleus is also possible, and this process is known as fusion. Both fission and fusion can result in the creation of new atoms with different numbers of protons than were present before the reaction. These processes are often difficult to control in a laboratory, however, so it's mostly only in specific cases, such as energy production, that chemists and scientists in related fields devote a lot of time to these nuclear reactions.

What is an **atomic orbital**?

Atomic orbitals are mathematical or pictorial descriptions of the locations of electrons in an atom. Electrons are tricky particles to understand because their location isn't easy to define. They can be thought of as clouds of negative charge surrounding a nucleus, and atomic orbitals describe the shapes of these clouds. Atomic orbitals can take on different shapes and sizes, but are essentially very similar from one element to the next. The number and type of orbitals that contain electrons play a central role in determining the properties of that atom.

How many electrons can fit in each orbital?

Each atomic orbital can contain up to two electrons. Electrons have a property called spin angular momentum, which can take on two different values of opposite sign. It

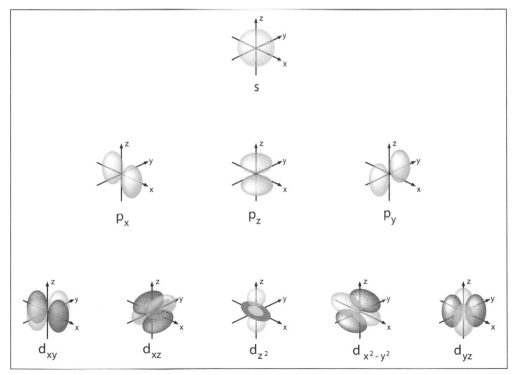

The four main types of atomic orbitals (s, p, and d) and their variants.

turns out that electrons residing in the same atomic orbital must have opposite spin angular momenta. This is a consequence of a physical principle known as the Pauli Exclusion Principle.

What do **atomic orbitals look like**?

There are three main shapes of orbitals relevant to most of chemistry, and these are referred to as s, p, and d orbitals. The designations s, p, and d are abbreviations for sharp, principle, diffuse, and fundamental, which have historical significance describing results of early experiments to probe the electronic structure of atoms. You can see what these orbitals look like in the graphic on the preceding page.

The shapes of these orbitals are determined by their orbital angular momentum, which is a property that describes the motion of the electron around the nucleus.

What is the **valence shell** of electrons?

Electrons fill up orbitals in "shells." The innermost shell consists of just one s-type orbital and can hold just two electrons. The next shell consists of one s-type and three p-type orbitals, and can hold eight electrons. Higher shells consist of more and more orbitals and can thus hold more and more electrons. The valence shell of electrons is the highest occupied, or partially occupied, set of orbitals.

What is the **atomic radius** of an atom?

The atomic radius of an atom is defined as half of the distance between two atoms of the same element held together in a chemical bond. Not surprisingly, these are very small distances! For hydrogen, the smallest atom, the atomic radius is 0.37 Ångströms, or 3.7×10^{-11} meters.

How do the **atomic radii** of atoms **change across the table**?

The atomic radii of atoms generally decrease going from left to right across a period, and increase going top to bottom down a group (see graphic on next page).

The increase in atomic radius going down a group is fairly straightforward to understand: additional shells of electrons are added and they must surround the inner shells, resulting in an increased atomic radius. Though the number of protons in the nucleus increases going down a group, the inner shells of electrons serve to shield the valence shell from the attractive force of the nucleus, resulting in an overall increase in atomic radius.

Moving to the right across a period, the number of protons increases, increasing the attractive force on electrons in the valence shell. Within a period, additional electrons go into the same valence shell, and an increasing attractive pull from the nucleus results in a more contracted valence shell, resulting in a smaller atomic radius. The situation is complicated by the rightmost group (known as the Noble gases), but the atomic radius of these elements is typically not important as they are rarely involved in chemical bonds to other atoms.

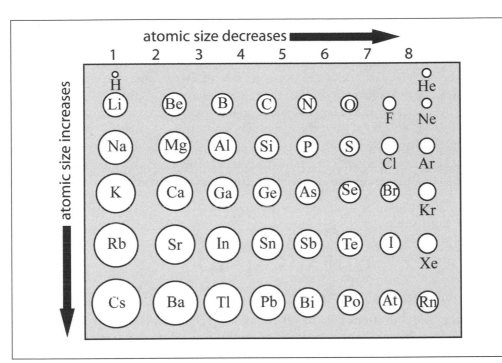

The atomic radii of atoms generally decrease going from left to right across a period, and increase going top to bottom down a group. (Atomic sizes are not to precise ratios and are for illustrative purposes only.)

What is the **ionization energy** of an atom?

The ionization energy of an atom is the amount of energy it takes to remove an electron from the atom. The process of removing an electron leaves the atom with an extra proton, relative to the number of electrons, and thus creates a positively charged ion, known as a cation. The ionization energy can be thought of as a measure of how strongly an atom holds on to its electrons. In general, ionization energies increase from left to right across a period (though there are exceptions) due to an increasing number of protons to attract electrons in the valence shell. Ionization energies decrease going down a group in the periodic table, due to the valence electrons being farther from the nucleus, and thus more shielded from its positive charge. Note that the trends in atomic radii and ionization energy go in the same direction—larger atoms tend to have lower ionization energies.

What **keeps an electron** from **crashing** into the **nucleus**?

Opposites attract, so electrons and protons are attracted to each other, making it somewhat difficult to understand why an electron wouldn't just get as close as possible to the nucleus and crash into it. The key to answering this question has to do with the fact that electrons are very, very small particles, so they are governed by rules that don't apply to larger objects. As we've talked about a little already, electrons are best thought

23

of as clouds of negative charge surrounding the nucleus. Their properties are governed by rules that describe the cloud as a whole, rather than as a single particle. It turns out that there is something favorable about the electron being spread out, or delocalized, around the nucleus. For reasons we won't go into in detail, when the electron's cloud gets packed closer to the nucleus, the energy associated with its motion (its kinetic energy) begins to rise, which makes the situation unstable. There's a balance between the stability associated with placing the electron close to the nucleus (the favorable positive–negative charge attraction) and that associated with spreading out the electron's cloud (to keep its kinetic energy low). This prevents the electron's cloud from getting too close to the nucleus or the electron just crashing into the nucleus.

MOLECULES AND CHEMICAL BONDS

What is a **molecule**?

A molecule is a set of atoms held together by chemical bonds. Molecules are the smallest units of a substance that behave as that substance, and separating the atoms of a molecule will change its properties.

What is a **substituent**?

A substituent is an atom, or group of atoms, attached to a specific position in a molecule. For example, in the molecule 3-bromopentane (see drawing below), we could refer to the bromine as a substituent on the third carbon atom.

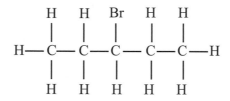

What is a **chemical bond**?

A chemical bond is an attractive interaction that binds atoms together through a sharing of electron density. The simplest bonding arrangement involves just two electrons shared between nuclei such that each effectively has a stable octet of eight valence electrons (or just two in the case of H–H). When two atoms are sharing a total of two electrons between them, the atoms are referred to as singly bonded to each other.

Bonds are what hold atoms together in molecules, and they are usually not easily broken. The arrangement of atoms in a molecule determines the identity of a chemical compound. The making or breaking of bonds is a chemical reaction, which converts one chemical compound into another.

Can I think of **chemical bonds** as **springs between atoms**?

Chemical bonds can be thought of as springs holding together the atoms in a bond. When atoms in a bond are stretched or compressed from their equilibrium separation, the bond provides a force to pull the atoms back together or to keep them from getting too close. For relatively small displacements, the bond actually provides a force that is physically very similar to that of a spring connecting two objects. This model of a spring as a chemical bond can be very useful for getting an intuitive idea of how a chemical bond connects atoms in a molecule.

What is a **Lewis structure**?

Lewis structures are a simple way of depicting the electronic structure of atoms and molecules. They show us which atoms are bonded to each other in a molecule and also show how many nonbonded electrons are present in the valence electron shell of each atom. The easiest way to understand them is probably to just take a look at a few.

The simplest Lewis structure is that for a single hydrogen atom. It has just one electron, and its Lewis structure looks like this:

H •

The letter H lets us know that it's a hydrogen atom, and the one dot represents its one electron.

Moving on to the Lewis structure for a molecule, let's look at the Lewis structure for F_2:

Here the two Fs let us know there are two fluorine atoms. The line connecting them shows that they are bonded with a single bond (containing two electrons). Each has six more electrons surrounding it, and these electrons are nonbonding.

And finally for a molecule with more than one bond, CH_2O:

This molecule is called formaldehyde. The Lewis structure shows us that the carbon is involved in a single bond (sharing two electrons) with each hydrogen atom, and a double bond (sharing four electrons) with the oxygen atom. The oxygen atom also has four nonbonding electrons.

What is a **"stable octet"**?

The term "stable octet" describes the fact that many atoms in molecules are most stable when the valence shell contains effectively eight electrons. This counts both nonbonding electrons and electrons in chemical bonds between atoms. Molecules tend to be most stable when the valence shells of each atom in the molecule contain eight electrons. In the Lewis structures for F_2 and CH_2O (see the previous question), we see that the fluorine, carbon, and oxygen atoms are each surrounded by eight electrons. We get this total by adding both the nonbonding and bonding electrons. Since hydrogen atoms are in the first row and have just a single orbital in their valence shell, they only need two electrons (a single bond) to fulfill their analogue of a stable octet.

What is **electronegativity**?

Electronegativity is a property that describes the tendency of an atom to attract electrons in a chemical bond. The most electronegative atoms are those which "pull" hardest on the electron density they share in a bond with another atom. There is more than one scale and definition for electronegativity, and our description here follows that given by Linus Pauling, which is the most commonly used scale in chemistry courses. Electronegativity can most readily be described in terms of the number of protons in the nucleus of the atom and the distance to which its valence electron cloud extends away from the nucleus. As a general trend, the most electronegative atoms are those with the shortest distance between the valence electrons and the nucleus. Electronegativity isn't a physical quantity that can be directly measured, but several scales have been developed that derive values for this property based on other measurable physical quantities.

What is **polarity** and how is it **related** to **molecular structure**?

Polarity is related to the symmetry of the arrangement of electron density in a molecule. Polar molecules are those which possess a net dipole moment, which means that the electron density is not symmetrically distributed in all directions. Nonpolar molecules have the electron density distributed in such a way that there is no net dipole moment. Typically this doesn't mean that nonpolar molecules have their electron density distributed evenly over every part of the molecule, but rather that the dipole moments created by an unequal sharing of electrons in each individual bond cancel each other out, so that there is no net direction in which an asymmetry of electron density exists.

What is the **charge of a molecule**?

The overall charge of a molecule is determined by the number of protons and electrons in the whole molecule. If there are more protons than electrons, the molecule will possess an overall positive charge. If there are more electrons than protons, the molecule will similarly possess an overall negative charge. A molecule with the same number of electrons and protons is neutral and has no net charge.

How are **formal charges different**?

Formal charges are given for individual atoms within molecules. These are determined by dividing the electrons in every bond equally between the atoms that share them, regardless of the elements involved. Textbooks typically follow this somewhat obtuse statement with an equation (which *always* helps, right?) like this:

Formal Charge = Group Number – Nonbonding Electrons – ½ Bonding Electrons

Let's work through this with an example, starting with carbon monoxide:

$$:C \equiv O:$$

Carbon is in Group 4 of the periodic table; it has two nonbonding electrons (the two dots shown), and since there are three bonds to oxygen, there are six bonding electrons. So the formal charge is $4 - 2 - \frac{1}{2}$ (6), or –1. Oxygen is in Group 6 and has the same number of nonbonding and bonding electrons as carbon does in this example. The formal charge on oxygen is therefore $6 - 2 - \frac{1}{2}$ (6), or +1. Carbon monoxide has no net (or total) charge (because $1 + -1 = 0$), but the individual atoms do have formal charges.

What is **Coulomb's law**?

Coulomb's Law tells us the force experienced by a pair of separated charges. It's a fundamental equation in the study of electrostatics, which is a broad area of physics concerned with the interactions between stationary charges. The equation for this force can be written:

$$F = \frac{q_1 q_2}{r_{12}^{\,2}}$$

where charges q_1 and q_2 are separated by a distance r_{12} and have a "unit of charge" defined by:

$$q_i = \frac{z}{\sqrt{4\pi\varepsilon_0}}$$

in which z is the charge in Coulomb's and ε_0 is the permittivity of free space, a fundamental physical constant.

The key features of Coulomb's Law are that it predicts an attractive force between particles of opposite charge and that this force decreases with the square of the distance between the particles. For chemistry, it's relevant to point out that the force between charges falls off rather slowly with the distance between them, so where charges are present in relatively dense materials (like liquids and solids), they have a significant effect on their environment.

What is a **dielectric constant**?

The dielectric constant of a material characterizes the extent to which it insulates against the flow of charge or against the effects of an electric field. Materials with a high dielectric constant screen the effects of charges within the material, while materials with a low dielectric constant allow the effects of a charge to be felt more strongly. In solutions containing ions, the dielectric constant of the solution will determine the extent to which the other molecules in the solution feel the effects of the charges present.

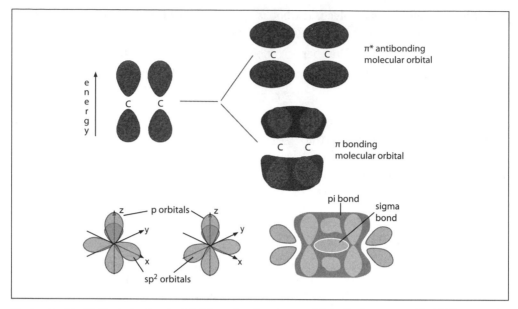

The top graphic (A) illustrates the pi-orbital formation from two p-orbitals; the bottom graphic (B) illustrates the formation of sigma- and pi-molecular orbitals from two sp^2 hybridized carbon atoms.

The lowest possible dielectric constant exists in a vacuum in which there is no material present to screen the charge of a field.

What is **valence bond theory**?

Valence bond theory is one of two main theories (the other being molecular orbital theory) that is used to explain bonding in molecules. Valence bond theory explains bonding by describing the interactions of atomic orbitals on individual atoms as they come together to form chemical bonds. The basic idea is that orbitals with the right shapes to overlap strongly with each other will form the strongest chemical bonds. Today, valence bond theory's description of chemical bonding based on atomic orbitals has become less popular in favor of molecular orbital theory.

What are **molecular orbitals**?

Molecular orbitals are different from atomic orbitals in that they cover several atoms and possibly even a whole molecule. While atomic orbitals originate from a single atom, molecular orbitals are formed from combinations of the atomic orbitals. Because they allow electrons to occupy the space between the atoms in a molecule, they can provide a very useful description of chemical bonds holding atoms together.

What is **molecular orbital theory**?

Molecular orbital theory is the other main theory (the first was valence bond theory) used to explain and predict bonding properties in molecules. Molecular orbital theory describes bonding interactions by using molecular orbitals that are spread out over multiple atoms, and this allows an electron's location to be described by an orbital that bonds atoms together in a more realistic way than valence bond theory.

What are some **common structures/geometries** for **molecules**?

The study of chemistry has benefited greatly from knowledge of properties relating to the geometries and, especially, the symmetries of molecules. To get a sense of what shapes molecules adopt, it's worth taking a look at a few of the geometries that come up often in the study of chemistry.

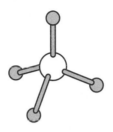

One commonly encountered geometry is that of a tetrahedron. Methane has the molecular formula CH_4 and exists in a tetrahedral geometry with angles of approximately 109 degrees between each pair of C–H bonds.

Linear geometries are also relatively common. Carbon dioxide has the molecular formula CO_2 and exists in a linear geometry with a 180-degree angle between the CO bonds.

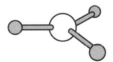

One last geometry we'll look at here is a planar geometry. The molecule BH_3 provides one example of a planar geometry, and in this case the BH bonds are separated by angles of 120 degrees. There are also planar molecules with four bonds in a plane, and in those cases the bonds are separated by angles of 90 degrees.

How **large** are **molecules**?

Molecules span a wide range of sizes. The smallest molecules contain only two atoms, and these diatomic molecules have length scales that are approximately the sum of the atomic radii of the constituent atoms. The smallest molecule, H–H, has a length of only 0.74 Ångströms (7.4×10^{-11} m). Larger molecules can be comparatively quite large. Biologically important molecules, like proteins, often contain thousands of atoms. Polymers, which are highly linked networks of covalently bonded atoms, can be even larger still, sometimes becoming so large they are visible to the naked eye.

Is it **possible** to **see** a single **molecule**?

With some of the largest single molecules, like polymers, they can actually be seen by the naked eye or through a microscope. Most molecules, however, are so small that a single isolated molecule cannot be seen with even the best microscopes. There is a physical limitation that prevents their observation with light, which has to do with the size of small molecules (lengths of ca. 0.1 to 1.0 nm) being significantly smaller than the wavelengths of visible light (400 to 700 nm). Other techniques based on diffracting electrons off of molecules, measuring the force molecules exert against a very small metal tip, and other methods have been developed to image small molecules, but it's impossible to see most small molecules with light in the way that we conventionally see things.

Is **everything made** of **molecules** or **atoms**?

Basically, yes! The only material things that aren't made up of atoms or molecules are the subatomic particles that make up atoms. Anything you find around your house, office, or anywhere else is made of some combination of atoms that are on the periodic table.

How do **molecules interact**?

The forces molecules exert on each other fall into a few main categories:

Van der Waal's interactions—Van der Waal's interactions are the broadest group of intermolecular interactions. This includes basically all attractive and repulsive forces that don't involve ions (charged atoms or molecules) or the rather unique situation of hydrogen bonding. Van der Waal's interactions include forces due to the dipole moments of polar molecules as well as interactions due to induced dipoles that can form even in nonpolar molecules.

Ionic interactions—Another class of intermolecular attractions involves attractive and repulsive forces between pairs of ions, or between ions and neutrally charged atoms or molecules. These interactions are typically stronger than those in the Van der Waal's category. Interactions between pairs of ions are governed by Coulomb's law, while interactions between ions and neutral molecules are either ion-dipole or ion-induced dipole interactions.

Hydrogen bonding—A hydrogen bond is a strong interaction between a hydrogen atom and another electronegative atom (usually fluorine, oxygen, or nitrogen) that are not covalently bonded to one another. The hydrogen atom also must typically be bonded to an electronegative atom (usually oxygen or nitrogen). The origin of this strong attractive interaction is that a hydrogen atom bonded to an electronegative atom has a partial positive charge due to its lack of electron density. This allows the hydrogen atom to have a strong attractive interaction with electronegative atoms (or ions), which have partial negative charges due to their extra electron density.

How strong are intermolecular interactions relative to a covalent bond?

Most intermolecular interactions are fairly weak relative to a covalent bond. Covalent bonds typically involve energies on the order of 100 kilocalories per mole (a unit of energy commonly used in chemistry). Van der Waal's interactions are the weakest type of intermolecular interaction with typical energies of roughly 0.01 to 1 kilocalories per mole (or 0.01% to 1% the strength of a covalent chemical bond) for a pair of interacting atoms. The strengths of ion–ion and ion–dipole interactions can vary widely, particularly in solutions, because the ions and/or dipoles can be separated by very different distances. The charges of ions can also be significantly shielded by solvent molecules around them. If the ions are very close together (like in a solid), their interaction energy can approach (or even exceed) that of a covalent bond. Hydrogen bonds are usually the strongest type of intermolecular interaction with energies of about 2–5 kilocalories per mole (or roughly 2% to 5% the strength of a covalent bond). Because they are such strong interactions, hydrogen bonds can play a dominant role in determining the structures of liquids, solids, and single molecules.

What's a solvent?

In chemistry, a solvent is a liquid (though it can be a gas or a solid, but forget about that for now) that other chemicals are dissolved in. These other chemicals can be called the solute. The solute and the solvent can together be called a solution. Take salt water: water is the solvent, salt is the solute, and we can refer to the salt water as a solution.

What makes something magnetic?

As we mentioned just briefly above, electrons have a property called a spin, or spin angular momentum, which can take on two possible values. This property, combined with the fact that electrons are charged particles, dictates that each electron has an associated magnetic moment, called the spin magnetic moment. In a macroscopic object, magnetism, or lack thereof, is determined by whether these spin magnetic moments are all aligned in the same direction. If all of the spin magnetic moments line up in the same orientation, the object will behave as a magnet. If the spin magnetic moments are oriented randomly, the object won't be magnetic. The trick is that only certain materials have the potential to exhibit magnetism, and we'll get to that next.

What determines which metals can be magnetized?

Chemists discuss magnetism in terms of three basic categories: diamagnetism, paramagnetism, and ferromagnetism.

Diamagnetic materials have all of their electrons arranged in pairs, which, by definition, means that their spin magnetic moments must all be arranged in pairs and thus cancel each other out. For this reason, diamagnetic materials cannot be magnetic, and won't be influenced by magnetic fields.

Paramagnetic materials have unpaired electrons, but in these materials the electrons' spin magnetic moments cannot all be lined up in the same direction, which means that they cannot be strongly magnetic. Since they do have unpaired electrons, they can be influenced by applied magnetic fields, but not to the same extent as the third class, ferromagnetic materials.

Ferromagnetic materials are the materials that can give rise to the magnets we're all familiar with. All of the materials that magnets can interact strongly with are ferromagnetic. In these materials, there are unpaired electrons whose magnetic spin can all be aligned in the same direction. Note that just because a material is ferromagnetic doesn't mean it must be a magnet, but rather just that it has the potential to be magnetized. Take a paper clip, for example; when you first pick it up it's not a magnet, but you can turn it into a weak magnet by holding a magnet next to it for a short length of time. Some of the most common ferromagnetic substances are those made of iron, nickel, or cobalt.

What is an **ideal gas**?

An ideal gas is a collection of atoms or molecules that do not interact with one another and occupy essentially no volume. While this is an idealized model, it turns out to describe many gases very well. The reason it works so well is that the atoms or molecules making up a gas are spread out far from one another so that the intermolecular forces between them are extremely weak (they don't "feel" each other). This description leads to the ideal gas law, which is a relationship between the pressure, volume, and temperature of a gas. The ideal gas law allows chemists to predict how, for example, the volume of a gas will change as its temperature is increased. The equation for the ideal gas law is

$$PV = Nk_bT$$

where P is pressure, V is volume, N is the number of particles (atoms or molecules), T is temperature, and k_b is Boltzmann's constant (a fundamental physical constant).

How many different **chemical substances** have been **discovered**?

According to the Chemical Abstracts Service (CAS), which is the world's largest authority for chemical information, the seventy millionth chemical compound was recently registered (announced December 2012). Today, new chemicals are being discovered and registered at a staggering rate: the sixty millionth chemical was registered only eighteen months earlier!

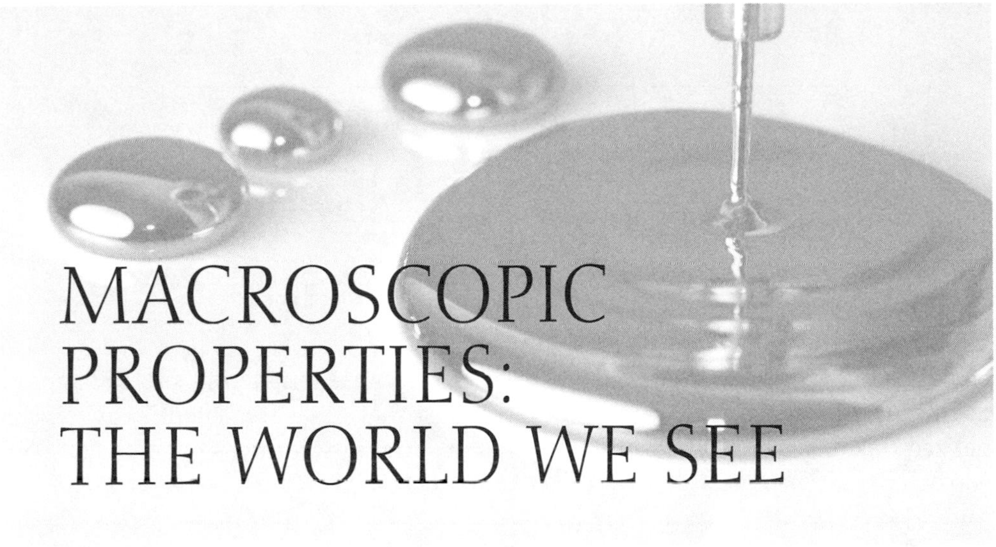

MACROSCOPIC PROPERTIES: THE WORLD WE SEE

PHASES OF MATTER AND INTENSIVE PROPERTIES

What are the **different phases** of **matter**?

There are three phases, or states, of matter that you come across every day—solids, liquids, and gases. There is a fourth phase of matter, plasma, which is only naturally found in stars and elsewhere in outer space. The distinctions between the first three phases are usually made using bulk properties. Solids have a defined shape and volume, while liquids easily change their shape but not their volume. Gases have neither a defined shape nor volume.

What is **plasma**?

Plasma, the fourth state of matter, is a gas where some amount of the particles have been ionized. As little as 1% ionization leads to very different properties for plasmas than gases, including increased conductivity (like lightning) and magnetization.

Are there **plasmas** in our **daily lives**?

Plasmas are found in fluorescent lights and neon lights. If you've ever seen a Tesla coil at a science museum, the arcs of light that they produce are plasma, as is lightning. Plasma TVs and plasma lamps are correctly named—both generate light using plasma, similar to fluorescent lights.

What is a **phase diagram**?

A phase diagram shows the phases of a particular substance as a function of temperature and pressure. An example for a single component phase diagram (as in not a mix-

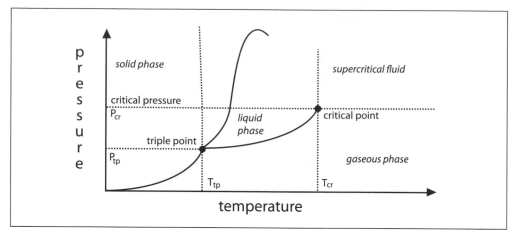

A phase diagram shows the phases of a particular substance as a function of temperature and pressure.

ture) is shown above. Phase diagrams also exist for mixtures, but these get very complicated very quickly.

What is a **triple point**?

The temperature and pressure at which three phases of a substance are in equilibrium is known as a triple point. The three phases can be solid, liquid, and gas, but can also be two solid phases (different arrangements of the molecules in the solid) and a liquid phase.

What is a **critical point**?

A critical point is a combination of temperature and pressure values above which a phase boundary no longer exists. There are liquid–liquid critical points above which the two liquid phases become miscible, and also liquid–gas critical points above which the boundary between the liquid and gas phases disappears and the substance becomes supercritical.

What is a **supercritical fluid**?

Above its critical point, a given temperature and pressure combination, a substance behaves like both a liquid and a gas and is called supercritical. Supercritical fluids are very good solvents, like liquids, and as a result many modern chemical processes use them.

How is **coffee decaffeinated**?

Consider the chemistry required to decaffeinate coffee. Most techniques use an extraction process to remove the caffeine from the green coffee beans before they are roasted. One method starts by steaming the green beans and then rinsing them with an organic solvent (usually dichloromethane) to pull the caffeine molecules out of the beans. The other method uses supercritical carbon dioxide to extract caffeine. The latter obviously avoids the use of toxic solvents, but is energy-intensive. With all extraction techniques, it is challenging to only remove the caffeine and not the flavor compounds that we want to taste in our coffee.

How many phases can be in coexistence?

A rule called the Gibbs phase rule tells us how many phases can be in coexistence for a given substance or mixture. The rule arises from the fact that, to be in coexistence, a constraint exists that the chemical potentials of each component in each phase must be equal. After some math, one can find a relationship between the number of components of a system, the number of free variables (such as temperature, pressure, or the fraction of a given component present in a mixture), and the number of phases that can be in co-existence. This relationship is:

$$F = C - P + 2$$

where F is the number of degrees of freedom, C is the number of independent components, and P is the number of phases.

What is the difference between a homogeneous and heterogeneous mixture?

A homogeneous mixture is one that is uniformly mixed and has the same proportions of components throughout the mixture. An example is a transparent solution of sugar dissolved in water (specifically one in which there is no undissolved sugar floating around). A heterogeneous mixture is one that is not consistent or uniform throughout, such as a glass of sugar water with some chunks of undissolved sugar also floating around.

Can multiple liquid phases exist for a mixture?

Yes. One familiar example is a mixture of oil and water. The immiscible oil and water phases are two different phases of liquid matter.

Can there be more than one solid phase for a given substance?

Yes, solids can adopt different types of microscopic arrangements. If there is a repeating pattern to the atoms in a solid, it is called a crystal. If the ordered structure exists for the entire material, then the phase is known as a single crystal (think of a diamond). If a sample is a bunch of individual crystals, then we refer to the material as polycrystalline. Lots of solids also just have no pattern to the arrangement of their atoms, and this class is known as amorphous.

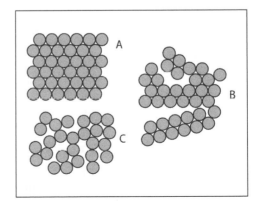

The arrangement of atoms in a solid can take several form, depending on the arrangement of the atoms: A) crystalline, B) polycrystalline, or C) amorphous.

What is the **density** of a material?

Density is the mass (or weight) of a material divided by a unit of volume. Water, for example, has a density of 1.0 g/cm³ (1.0 gram per cubic centimeter).

What **determines the density** of a substance?

At the most basic level, density is determined by how close the atoms or molecules in a substance are packed, as well as the mass of those atoms. While it's not quite as simple as assuming that the heaviest elements on the periodic table have the highest densities, having a high mass does help: heavy metals like iridium and osmium are the densest metals known to date. Remember that the density of a material does not depend on how much material you have; the density of 1 gram of lead is the same as the density of 1 kilogram of lead. Density is an intensive property, so changing the amount of a material you have does not affect its density.

Why does **ice float**?

Ice floats in water because it is less dense than water, though this is actually a very unusual case in terms of comparing the densities of the solid and liquid phases for a given substance. Most substances increase in density when moving from the liquid to the solid phase of matter, but H_2O does the opposite. When water freezes, it forms a network of hydrogen bonds between H_2O molecules, and because of the spacing of the molecules in this lattice, ice is less dense and floats in water.

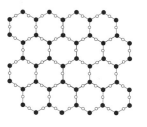

What is **temperature**?

Temperature is a measure of the average kinetic energy of particles in a substance. What does that mean? "Average kinetic energy" is a precise way of saying how fast something is moving, in this case on a molecular level. The faster molecules vibrate, the hotter they feel because heat is being transferred from the object to your hands.

How are the **Fahrenheit, Celsius, and Kelvin** temperature scales related?

The Celsius and Kelvin scales use the same size degree ("incremental scaling" is the technical term), but set their zero values at different absolute numbers. Let's explain that sentence a bit more: If you go up by one degree Celsius or one degree Kelvin, you've raised the temperature the same amount, but 0 °C (the temperature at which water freezes) is 273.15 K. Thus the two scales are offset from one another by 273.15.

Fahrenheit is completely different though. Water freezes at 32 °F, and a change of one degree on the Fahrenheit scale is equal to a change of 0.55 °C.

What makes metals feel colder than air to the touch?

Metals feel colder than air when you touch them because they are good conductors of heat. The cold metal is able to conduct heat away from your hand through the entire object quickly, making it feel colder than the air around you.

What is a boiling point?

The technical definition of boiling point is the temperature at which the vapor pressure of the liquid phase equals the pressure of the surrounding gas (atmospheric pressure usually). This is a precise way of saying the temperature at which a liquid turns to vapor.

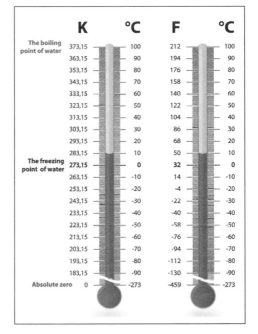

A comparison of the Fahrenheit, Celsius, and Kelvin scales.

What molecular properties lead to higher boiling points?

There are a few factors that play important roles in boiling points of substances. The first is molecular weight: in general, heavier molecules have higher boiling points, which is pretty straightforward considering how we defined boiling point above (heavier molecules take more energy to move from the liquid to the gas phase).

All other properties that affect boiling point deal with intermolecular forces, or interactions between molecules. Think of this like the affinity of one molecule of a substance to be attracted to, or stick to, another molecule. Noncovalent bonds, like ionic or hydrogen bonds, significantly increase boiling point. Why? Because to move to the vapor phase the molecules typically have to break these interactions. Dipole interactions and Van der Waals forces have similar effects (see "Atoms and Molecules"), but these interactions are weaker, so the effect they have on boiling points is smaller. Finally, branching of the carbon backbone of a molecule is also frequently touted as a factor leading to lower boiling points; while this is true, it is really the weakening of Van der Waals forces at work here too.

What is a melting point?

The melting point is the temperature at which a substance changes from the solid state to the liquid state. At this exact temperature, the two phases are in equilibrium, so frac-tions of the sample are constantly moving between the two phases of matter. In practice, it is pretty difficult to observe the exact melting point of a substance.

What **molecular properties** lead to **lower melting points**?

Most of the trends we talked about for raising boiling points hold true for melting points for most of the same reasons. There is one major exception, though. The more branched, and therefore compact, a molecule is, the higher its melting point, because, in general, compact molecules will pack better in a crystal lattice. The better packed a lattice is, the more stable it is, and the more heat (energy) it takes to break up that lattice and melt the solid.

How do the **boiling point** and **melting point** of a substance **change** as a **solute/impurity is added**?

The addition of a solute typically raises the boiling point and lowers the melting point of a substance. These effects are appropriately named "boiling point elevation" and "melting point depression."

Boiling points are raised when a nonvolatile solute (like NaCl) is added to a solution because the solute lowers the vapor pressure of the solution. That is a somewhat circular explanation though. It is important to know that this change in boiling point does not depend on what you add to the liquid, so there are no specific interactions going on here (like forming hydrogen bonds, etc.). As long as the solute has lower vapor pressure (remember we said it was nonvolatile, so its vapor pressure is essentially zero), this effect will be present. It is perfectly correct to think about it as just lowering the vapor pressure of the mixture (if you add something with very low vapor pressure, the average vapor pressure of whatever liquid you're adding it to will go down).

Melting (or freezing) points are usually lowered when a solute is added to a liquid. The best explanation for this effect is based on entropy (see "Physical and Theoretical Chemistry"). When a molecule of solvent moves from the liquid to the solid phase (freezes), the amount of liquid solvent (i.e., its volume) is reduced. This means the same amount of solute is in a smaller space, which reduces their entropy (or raises their energy). This raising of energy means that you have to take even more energy out of the system for each molecule that joins the solid phase. Less energy means lower temperature, so adding a solute lowers the freezing point. An alternative way to look at this is that any impurity will disrupt the crystal lattice, raising its energy, relative to the liquid phase. This also contributes to lowering the freezing point as solutes are added.

How is the **concentration** of a **solution defined**?

The concentration of a substance is the amount of that substance in a solution divided by the volume of the solution. Chemists typically use molar concentration (moles of material/volume).

What **properties influence solubility**?

The most significant properties are intermolecular forces and temperature. If there are favorable interactions between the solute and the solvent, solubility will be higher. This is actually a balance of the interactions of the solute with the solvent and the stability of the solute in the solid phase. Temperature also influences solubility, and for most substances, solubility increases as the temperature of the solvent rises.

Why does putting **salt** on the road **help to melt snow**?

Like we discussed, when a solute is added to a solution, its freezing point is lowered. When salt is placed on snow, it begins to dissolve into any small amount of water on the ice with which it is in immediate contact. This lowers the freezing point of the surrounding water/ice, causing it to melt into the water. This process continues until the salt is completely dissolved.

What are the **basic units of length**?

Almost every length scale used in science is based on the meter. Chemistry frequently deals with very small lengths, so while you're probably familiar with millimeters (10^{-3} m), it's hard to have an intuitive sense about just how small a nanometer (10^{-9} m) is. There's another length scale commonly used when talking about chemical bonds—the Ångström. An Ångström (Å) is one ten-billionth of a meter (10^{-10} m). The lengths of chemical bonds vary depending on the elements and other factors, but are usually around 1–2 Å.

How much **space** does an **atom occupy**?

The atomic radius of the smallest atom, hydrogen, is 53×10^{-12} meters, so it is about 10^{-10} meters in size. The atomic radius of the largest atom, cesium, is about 270×10^{-12} meters, or roughly five times that of hydrogen. These are all very, very small sizes!

How much **space** does the **nucleus take up**?

The nucleus of an atom takes up a very, very small fraction of the total space occupied by the atom. The diameter of a nucleus is on the order of 100,000 times smaller than that of a whole atom.

How **long** are **chemical bonds**?

Chemical bonds are typically about 2 atomic radii in length, since they are formed from two atoms joined together. These distances are on the order of 10^{-10} meters.

What are the basic **units of pressure**?

Unlike units of length and temperature, pressure is reported in at least six common units. The pascal (abbreviated Pa) is the official standard unit, but bar, millimeters of

39

mercury (mmHg), standard atmospheres (atm), torr, and pounds per square inch (psi) are all used in different areas.

How do **planes** stay in the **air**?

Airplanes are very heavy, so the force required to balance gravity and keep them in the air must be large. The engines propel the airplane forward, but we need to understand what gives the upward push, or lift, necessary to keep the plane in the air. This lift comes from the shape of the wings, which are typically curved on the top and flat on the bottom. This design requires air to flow more rapidly over the top of the wings than over the bottom, which creates a lower air pressure above the wing than below. The lower air pressure above the wing is what lifts the plane off the ground and keeps it in the air. This is commonly referred to as the Bernoulli principle. If you blow across the top of a sheet of paper, you will see it lift into the air for the same reason.

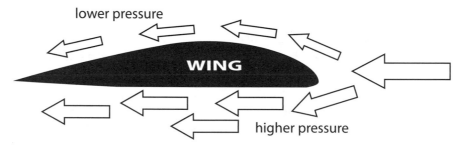

What makes **oil more slippery** than water?

The purpose of lubricants, like motor oil, is to reduce the friction between surfaces so that parts last longer and less energy can be expended in the process of moving them. The key to a good lubricant is that the characteristic length scale for the formation of a thin film of the lubricant must be much smaller than the characteristic length scale of movement in the application. Basically, oils are good lubricants because they can form very thin films that persist even when the parts they serve to lubricate are constantly in motion. This ability to form thin films typically correlates with other properties that are easier to recognize. For example, good lubricants often have a high boiling point, low freezing point, high viscosity, and are stable toward chemical oxidation and changes in temperature.

What **prevents** all of the **air from escaping** Earth's atmosphere?

Gravity! Every molecule on Earth is pulled toward the planet by gravity, even the very lightest gas molecules. To overcome a planet's gravitational pull, an object, be it a space ship or a helium atom, must reach at least the escape velocity. Almost all of the molecules in the atmosphere are below this velocity because of the temperature of the planet Earth (higher temperature = faster molecules).

Almost all? Yes. Earth is very slowly losing its atmospheric gases. The lightest ones go first—about three kilograms of H_2 escape the gravitational pull of Earth per year. A few

> ## What is the composition of air?
>
> **E**arth's atmosphere is made up of 78% nitrogen (N_2) and 21% oxygen (O_2), if you ignore water vapor, which shifts too much to include in averages for the entire planet. The last 1% is made up of mostly argon (Ar), followed by carbon dioxide (CO_2) and other trace gases.

particles can escape because kinetic energy of gases is a Boltzmann distribution, and in a Boltzmann distribution there are always very small probabilities of very high values.

What gives a substance its color?

The color of a substance is the combination of the light that is reflected back at your eye. In other words, you're seeing the light that is not absorbed by the substance. Certain frequencies can be absorbed because of the electronic structure of a substance, while others simply bounce off back into your eye.

What is a glass?

A glass is a noncrystalline solid—it lacks order in the solid state. In polymer chemistry, scientists are concerned with the glass transition temperature (T_g), which is the temperature at which a material changes from a hard to a rubbery state. What's neat about this transition is that the material is not changing phases (i.e., from solid to liquid), but is changing from one type of solid to another.

What makes mercury so dangerous?

Mercury can be absorbed through the skin, making it particularly dangerous to handle. Organometallic mercury compounds, like dimethylmercury (CH_3HgCH_3), are particularly dangerous and have caused the deaths of a number of laboratory research chemists. Most research on this most toxic of mercury compounds has ceased. Be careful cleaning up that broken thermometer!

What is a vacuum?

A vacuum is space without matter. The word vacuum derives from a Latin word that means "empty." A perfect vacuum, or one with absolutely no matter in a given space, is very hard to achieve, but through modern engineering scientists can get pretty close without having to go into deep space to run their experiments.

Can sound move through a vacuum?

No. Sound is a mechanical wave, which means that for the wave to move, actual molecules must bump into one another. In a vacuum, there is no matter, so in space, no one can hear you scream.

How does a vacuum cleaner work?

We commonly use the word vacuum to refer to any area that is of relatively low pressure. The air in areas of higher pressure will spontaneously move into areas of lower pressure, and this is the principle upon which a vacuum cleaner operates. A fan is used to push air out of the vacuum cleaner, creating an area of lower pressure behind the fan. Air from the outside then comes rushing in to reduce the gradient in pressure, carrying dust and dirt along with it. Since the fan is running continuously, the gradient in pressure is constantly being maintained, so the vacuum cleaner is able to keep running even though air is always flowing in.

Can **light move** through a **vacuum**?

Yes. Unlike sound, light is an electromagnetic wave, so no molecules are needed to propagate the wave. But you knew this—the Sun's light crosses the vacuum of space to reach Earth each and every day.

What are some **chemical reactions** that **we can observe** with our naked eye?

There are lots of chemical reactions that we can easily detect by sight. Some examples are the formation of rust on metal, wood burning, fireworks exploding, silver becoming tarnished, or baking soda and vinegar reacting with one another.

What makes **paper towels absorbent**?

Paper towels are composed of cellulose fibers, which are polymers that contain many sugar monomers. These sugar monomers can interact strongly with water molecules, which is what makes paper towels so useful for cleaning up when something gets spilled.

What is an **electric current**?

Electric currents are the flow of electrons through a material. The electricity that comes out of the wall in your home or office is just the flow of electrons through a wire.

What makes a material a **good conductor** of **electricity**?

Materials that are good conductors of electricity have many "free" electrons. Here we mean free in the sense that they are not bound strongly to a given atom or molecule. Metals are often good conductors of

An explosion is a dramatic example of a chemical reaction we can see with our naked eyes.

electricity. Whether or not a material has freely conducting electrons is related to the detailed electronic structure of the material, and a thorough description would be beyond the scope of this book. Basically, the more electrons that are free to move around in a material, the more easily it can carry current, and the higher its conductivity.

What makes a **rubber band stretchy**?

Rubber bands are made of long polymer molecules. These are all tangled together, and you could think of them as being similar to a bunch of interwoven springs. The polymers can be stretched to a more extended state, which is what allows a rubber band to stretch without breaking. More accessible configurations exist, however, when they are in a more contracted state, which means that the more contracted states have higher entropy (see "Physical and Theoretical Chemistry") and for this reason are more favorable. This is what makes the rubber band want to contract and is what gives rise to its elasticity. (Check out "Polymer Chemistry" for more questions on polymers.)

How are **soft drinks carbonated**?

Soft drinks are carbonated using pressurized CO_2 gas and a siphon that introduces the pressurized gas into the water or soda. The CO_2 gas is forced into the liquid at a concentration beyond that which could exist under atmospheric conditions and then the container is sealed, preventing it from being released. That's why if you leave a glass of soda sitting out, it will go flat; the CO_2 escapes into the air.

Why might a **soft drink freeze** when you **open it**?

Remember that the freezing point of a solution is decreased by having anything dissolved in it. When a soft drink is opened and CO_2 rushes out, less CO_2 is present in the solution, causing the freezing point of the solution to increase. So at a constant temperature, the increase in freezing point causes the soda to freeze. If you haven't seen this happen, you can try to make it happen by placing a soda in the freezer for a while, but be careful not to leave it in too long or it might explode!

Why do **helium balloons float**?

Helium balloons float because helium is less dense than air, so gravity pulls down

Helium balloons float because the gas inside the balloon is much less dense than the surrounding atmosphere.

on air more than it does on the helium balloon. The difference in density is enough that the displaced air supports the weight of the balloon and makes it float...up, up, and away.

What is **dry ice**, and why does it **"evaporate"** from the **solid phase**?

Dry ice is solid carbon dioxide (CO_2). At constant atmospheric pressure we see that an increase in temperature will bring it from the solid phase directly to the gas phase. This process is called sublimation, and it takes place at -78.5 °C.

How far away can **one molecule "feel" another**?

Molecules "feel" one another through the intermolecular forces they exert on one another. These typically span lengths that are a little longer than the length of chemical bonds, or something in the neighborhood of 5×10^{-10} meters.

FOOD AND SENSES

What **elements** are in **your body**?

Six elements make up all but 1% of the human body by mass. In decreasing order, they are: oxygen, carbon, hydrogen, nitrogen, calcium, and phosphorus. Oxygen and hydrogen are so prevalent because of most of our cells are over 50% water.

What **elements** are in our **food**?

Our food is made up of pretty much the same elements that we are, which makes sense because at some point, the elements in our food literally become the elements in our bodies. We are what we eat.

What's special or different about **organic foods**?

While the precise definition is still changing, everyone agrees that organic foods are those grown without the use of pesticides or synthetic fertilizers. The "organic" label also frequently excludes the use of irradiation and genetically modified fruits and vegetables. Whether the food tastes better, or is healthier to eat, is for you (or at least someone other than us) to say.

What gives a **food** its **taste**?

Molecules, of course! You probably learned in school that there are four basic tastes: sweet, bitter, sour, and salty. Your science textbook probably also had a diagram like the one on the following page, showing that your bitter tastebuds are located on the back of your tongue, and sweet is tasted on the tip. Wrong. Wrong. Wrong.

Not only are there at least five basic tastes, but they're basically spread out evenly across your tongue. The five basic tastes are the four you know from elementary school,

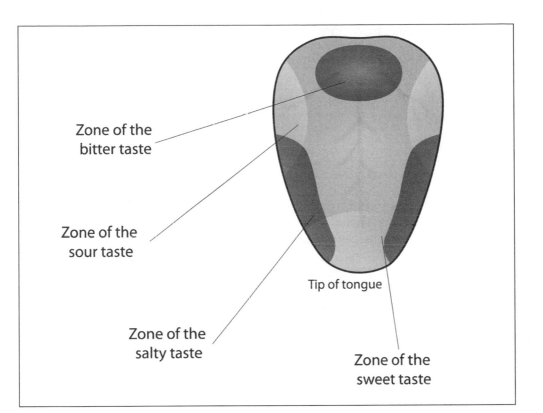

Zone of the
bitter taste

Zone of the
sour taste

Tip of tongue

Zone of the
salty taste

Zone of the
sweet taste

The old notion that different regions of your tongue detect different tastes has been shown to be false. Instead, taste buds all over the tongue's surface can detect all the basic flavors of sweetness, sourness, saltiness, and bitterness.

and umami. If you grew up eating Western cuisine, umami tastes like MSG, or maybe you just think "Asian" food. There is debate as to whether there is a sixth basic taste bud that senses fat, or maybe another sensor for piquance (spiciness).

But not even those five (or six, or seven) basic tastes fully account for all the sensations you get during eating. Wine is probably the best example (if you drink wine, of course). What makes wine taste "dry"? The "dry" taste certainly doesn't fall into one of the five basic groups, but it is known to be related to the presence of tannins. So is there a tannin taste bud? No one knows yet.

What makes something poisonous?

There are many ways to seriously disrupt our biological machine. Carbon monoxide binds to hemoglobin and prevents oxygen from getting to our cells. Cyanide shuts down the production of ATP in mitochondria. Hemlock is a weed that contains a mixture of at least eight rather toxic molecules that target the nervous system. Thallium ions (Tl^+) are particularly toxic because they are highly water soluble and once in the body they bind to ion channels and disrupt other processes that normally function with potassium ions (K^+). 45

Do **pesticides** make food more **dangerous**?

Pesticides certainly don't help make food safer, but their adverse long-term effects are difficult to measure. Limiting exposure as much as possible is certainly a good idea.

What gives a substance its **smell**?

Substances smell because your nose (or olfactory system) is able to detect the volatile molecules that are being released. Your nose has roughly 350 different receptors that detect molecules and then send a signal up through various parts of your olfactory system, ultimately ending up in your brain. A smell doesn't come from a single receptor firing off a signal, but rather a whole array of receptors. Your brain takes the combination of signals it receives and translates that into the perception of the odor.

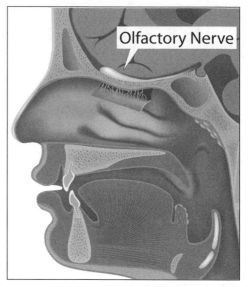

Receptors in your nose detect molecules in the air, sending signals to your olfactory nerve, which then transmits the signals to your brain, which interprets them as smells.

Why do **some substances smell stronger** than others?

There are a couple of reasons that a substance might smell "stronger" than another. The first is volatility, or vapor pressure, of the molecule you smell. Simply, if there is more of it, it smells stronger. Additionally, some molecules interact more strongly with the receptors in your nose than others, which results in a stronger smell sensation.

Which **elements** are **liquids** at **room temperature**?

Actually there are only two elements that are liquids at room temperature: mercury and bromine. Bromine is a diatomic compound, Br_2, while mercury is a liquid metal.

How **dense is mercury** compared to other liquids?

The density of liquid mercury (it is a liquid at room temperature) is about 5.43 g/mL. This is about five and a half times as heavy as water! One of the next most dense liquids is bromine with a density of 3.03 g/mL, but mercury remains almost twice as dense.

How does the **density of snow compare** with that of **liquid water**?

The density of recently fallen, damp snow is approximately one-tenth that of liquid water. This means that one centimeter of rainfall and ten centimeters of snowfall contain approximately the same quantity of water molecules.

What is a **surfactant**?

Surfactants are molecules that reduce the surface tension of a liquid. These are typically amphiphilic organic molecules, which means that they have one hydrophobic and one hydrophilic section. This will cause them to align in a certain way at the surface of a liquid which interrupts the order of the liquid at the surface, lowering the surface tension.

What is **precipitation**?

Precipitation is the formation of a solid in a solution. This happens when a substance has too low of a solubility to remain dissolved in the solution. This process can occur for a variety of reasons, such as when a new product of a chemical reaction has low solubility or when a temperature change takes place. Precipitation begins when a small crystal begins to form, which is part of a process called nucleation.

Mercury is the only metal that is a liquid at room temperature. It is also toxic and should not be handled with bare hands.

What is the **Curie point** of a material?

As the temperature of a ferromagnetic material is increased, it can eventually become paramagnetic (see "Atoms and Molecules" for a discussion of ferromagnetism and paramagnetism). The temperature above which it becomes paramagnetic is called the Curie point. This is actually also a phase transition, though it's not one that's easy to see with our eyes, such as when water freezes or evaporates.

What is a **physical change**?

A physical change is a change involving the macroscopic properties of a substance without any accompanying change in chemical composition. A few examples of physical changes would be evaporating, melting, cutting, slicing, breaking, and grinding. The key similarity is that none of these processes involve changing the chemical makeup of the substance.

What is the **Mohs scale of hardness**?

Substances, mostly minerals, are ranked by their ability to scratch one another. If one mineral can scratch another mineral it gets a higher ranking on the Mohs hardness scale. When Friedrich Mohs devised the scale, diamond was the hardness substance known, so it was given a value of 10. Talc (also known as talcum powder), being very soft, has a hardest rating of only 1. Mohs devised this scale to help sort out the private rock collection of an Austrian banker, and later the Archduke's museum collection. There are more quantitatively accurate measurements of hardness available today, but the simplicity of the Mohs scale keeps it relevant and practical.

CHEMICAL REACTIONS

KINETICS AND THERMODYNAMICS

What is a **chemical reaction**?

A chemical reaction is any process that involves the transformation of one or more molecules—most of the time the reactant(s) and product(s) will be different molecules. A chemical reaction almost always involves the breaking and/or formation of new chemical bonds.

How do we **write an equation** for a **chemical reaction**?

Chemists often write "equations" to describe chemical reactions. It is conventional to list the initial species, or reactants, on the left side of the equation, followed by an arrow, and the final species, or products, on the right side of the equation. The equation below shows the reaction of methane and oxygen to produce water and carbon dioxide.

$$CH_4 + 2\,O_2 \longleftrightarrow CO_2 + 2\,H_2O$$

The arrow to the right indicates that the reactant species are converted to the product species during the course of the reaction. In some cases, reactions are reversible, which will be represented by two arrows, one pointing in either direction ($\leftrightharpoons$). Be careful, because in chemistry $\longleftrightarrow$ is not the same as $\leftrightharpoons$!

What is the **yield** for a **reaction**?

The yield of a chemical reaction is the amount of product that is produced (an example might be two grams). It is often of greater interest to consider the percent yield, which describes the amount of product formed relative to the maximum amount of product one could have expected based on the quantities of reactants used. The percent yield provides a measure of how efficient a process is for producing the target product.

What does **selectivity** mean for chemical reactions?

Selectivity can have several meanings for chemical reactions, but there are two main categories: either a reaction will occur selectively with a particular chemical species or at a particular location in a molecule to avoid unwanted side-reactions, or a reaction will produce a particular product selectively.

How do modern **chemists characterize the products** of chemical reactions?

Chemists need to characterize the products of their chemical reactions so that they can be sure of the structure and composition of the molecule(s) they have made. One common way is by measuring the melting point of a solid substance. This doesn't provide specific information about the arrangement of chemical bonds in a molecule, though, so more advanced techniques are required to completely characterize a molecule. These techniques often involve using electromagnetic radiation to probe the energy levels in the molecules (see "Physical and Theoretical Chemistry" for more on these topics). Knowledge of what energies/wavelengths of light the molecule can absorb can be related directly to structural features of the molecule.

Are chemists **still looking** for **new reactions**?

Yes. Chemistry has hundreds of years of knowledge to build on, but it is in no way a complete field. New ways of making existing molecules, and making chemical structures that are completely new to our planet, are absolutely goals of modern chemistry. Developing new chemical reactions, and understanding old ones, are topics of eternal interest to chemists.

What is the **law of conservation of matter**?

The law of conservation of matter states that matter cannot be created or destroyed. This is relevant to chemical reactions because it tells us that we must have just as many atoms of each element at the beginning of a reaction as we do at the end of a reaction. In the example above, this is reflected by the fact that we use two molecules of oxygen, or four oxygen atoms from our reactants, to produce one molecule of CO_2 and two molecules of H_2O for a total of four oxygen atoms in our products.

What is the **stoichiometry of a reaction**?

The stoichiometry of a reaction is closely tied to the idea of the conservation of matter; it tells us the ratio in which the molecules react. Again using the example at the beginning of this chapter, the reaction stoichiometry 1:2:1:2 describes the ratio in which methane and oxygen react to form carbon dioxide and water.

Why do **some chemical reactions** cause a **color change**?

The colors we see all have to do with what wavelengths of light something absorbs or reflects. For a chemical reaction to cause a change in the color of something, all that has

> ## What are a couple of examples of familiar chemical reactions?
>
> Fire is one example of a chemical reaction that everyone has seen take place. Fire involves a combustion reaction, which is any reaction where a hydrocarbon reacts with oxygen to form carbon dioxide and water. Another example is when your car accumulates rust. This reaction involves oxidation of the iron in the metal. Lots of complicated chemical reactions are taking place all the time in our bodies too. Every movement you make, for example, involves many chemical reactions taking place in your muscles and nerves.

to happen is that the products of the reaction absorb and reflect different wavelengths of light than the reactants do. We'll discuss how light interacts with molecules in more detail in "Physical and Theoretical Chemistry."

What is **enthalpy**?

Enthalpy is a measure of the energy that something contains, and it's defined as the total heat content of a system. In terms of chemical reactions, we are most often interested in the change in enthalpy (denoted H) associated with a reaction. The H for a reaction is defined as the enthalpy of the products minus the enthalpy of the reactants, and this is typically measured via changes in temperature that take place during the reaction.

What is a **calorie**?

A calorie is a unit of heat energy defined as the amount of energy it takes to raise one gram of water by one degree Celsius. Calories are also often used to describe the energy content of foods. When used for foods, a "calorie" actually refers to 1,000 calories, or a kilocalorie of energy (which can be rather confusing).

What is a **bond enthalpy**?

Bond enthalpy refers to the amount of energy it takes to break a chemical bond. This tells us how favorable a chemical bond is relative to the separation of the two fragments on either side of the bond.

What is a **heat of formation**?

The standard heat of formation for a substance is the change in enthalpy associated with its formation of one mole (see "History of Chemistry") of a substance from its elements with the constituent elements in their standard states (see "Analytical Chemistry").

What is **Gibbs free energy**?

Gibbs free energy is a quantity that describes the amount of useful work (see "Physical and Theoretical Chemistry") that can be obtained from a system at a constant temperature and pressure. In the context of chemical reactions, changes in Gibbs free energy will typically dictate whether or not a reaction is favorable.

What makes a **reaction** happen **spontaneously**?

A spontaneous chemical reaction is one for which the associated change in Gibbs free energy is negative. The fact that a reaction is spontaneous actually doesn't tell us anything about how quickly the reaction takes place, though. A spontaneous reaction may happen very quickly or take thousands of years!

What is a **unimolecular reaction**?

A unimolecular reaction involves only a single reactant molecule undergoing a chemical reaction to form products. One possible outcome is that the bonds rearrange within a single molecule to form only one product molecule, while another possibility is that the reactant molecule will fragment, producing multiple product molecules.

What is a **bimolecular reaction**?

As you might be able to guess if you've read the previous question, a bimolecular reaction involves two reactant molecules undergoing a chemical reaction. They may form a single product molecule (if they combine) or multiple product molecules.

What is the **equilibrium constant** for a reaction?

Some reactions can go both in forward and reverse, while others can only go in one direction. For a reaction that can go both ways, the equilibrium constant describes the ratio of products to reactants. For the reaction:

$$A \rightleftharpoons B$$

The equilibrium constant would be:

$$K_{eq} = [B]/[A]$$

For the reaction:

$$A + B \rightleftharpoons C$$

The equilibrium constant would be:

$$K_{eq} = [C]/[A][B]$$

and for the reaction:

$$A + B \rightleftharpoons C + D$$

The equilibrium constant would be:

$$K_{eq} = [C][D]/[A][B]$$

Reactions with a large equilibrium constant ($K_{eq} > 1$) favor formation of the products, while reactions with a small equilibrium constant ($K_{eq} < 1$) favor formation of the reactants.

What is **Le Chatelier's principle**?

Le Chatelier's principle tells us how to predict the effect a change in conditions will have on a chemical equilibrium. It tells us that a system at equilibrium will shift to counteract changes that disturb the equilibrium. These could be changes in concentrations of chemical species, temperature, pressure, or other conditions. The most commonly discussed changes involve changes in concentration of chemical species, so we'll just focus on those here. For this equilibrium:

$$A + B \rightleftharpoons C + D$$

If we decrease the concentration of A, some C and D will react to replenish the A that is depleted, so the concentrations of C and D will decrease. As species A is replenished, more B will be created as well. So the net effect is that decreasing the concentration of A will also decrease the concentrations of C and D, and at the same time increase the concentration of B. More generally, decreasing the concentration of a reactant will cause the equilibrium to shift toward the reactants, increasing the concentrations of other reactants and decreasing the concentrations of products. The converse is also true: Decreasing the concentration of a product will cause the equilibrium to shift toward the products, increasing the concentrations of other products and decreasing the concentrations of reactants.

It is important to keep in mind that Le Chatelier's principle only applies to reversible chemical processes (chemical equilibria), so everything we have said here does not apply to reactions that can only proceed in the forward direction.

What is a **free energy diagram** for a chemical reaction?

A free energy diagram is probably easiest to understand by taking a look at one (see diagram) as we explain the key features.

The y-axis measures the relative free energy of the chemical species we're dealing with, while the x-axis describes the reaction coordinate (it's common that going left to right is forward progress in the reaction, but this isn't necessarily the case 100% of the time). On the left we have our reactants. In general there may be any number of reactants, and here we've just denoted two species, A and B. The "hill" in the middle is the energetic barrier to the chemical reaction, and the quantity Ea denotes the height of this energy barrier. The quantity Ea is commonly referred to as the activation energy for the reaction. On the right-hand side of the diagram we have our

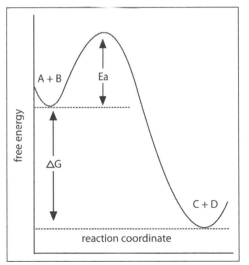

An example of a free energy diagram.

products. Again, there can be any number of products, and here we've denoted them C and D. Finally we have the quantity G, which describes the change in Gibbs free energy associated with the reaction. The fact that the reactants are higher in free energy than the products tells us that this particular example is a spontaneous reaction. If the reactants were lower in free energy than the products, the reaction would not be spontaneous.

Can chemical **reactions** involve **multiple steps**?

Yes, and many do. While some chemical reactions may only involve a single step, others may involve ten or more elementary steps. Of course, chemists working in different subfields may have different definitions of what constitutes a step of a reaction, depending on what aspects of the reaction they focus on.

What is an **example** of a **multistep chemical reaction**?

Many reactions in biological chemistry (see also "Biochemistry") are multistep chemical reactions. Glycolysis, which is the process of breaking down sugar to generate energy, for example, involves ten sequential steps. Each step is carried out by a special type of catalyst, called an enzyme. There are countless multistep processes in biological systems.

What is meant by **"dynamic equilibrium"**?

Equilibrium conditions in a reversible chemical reaction are described as a dynamic equilibrium. This means that even at equilibrium the reaction has not stopped, and the forward and reverse reactions are still taking place. The bulk concentrations of reactants and products don't change, but this is just because the forward and reverse reactions are happening at equal rates. The reaction never stops, it just reaches equilibrium.

What is the **rate-determining step** of a reaction?

In a reaction with multiple steps, the rate-determining step is the slowest step. It's the step that limits the rate of formation of the final products, usually because it has the highest activation energy.

What does it mean when we say a chemical reaction takes one minute?

When we say a reaction takes one minute, what we're really saying is that a certain fraction (specifically that fraction is $1-1/e$, or about 63%) of the reactant molecules have gone on to form products after one minute has passed (e refers to the irrational number $e = 2.7182818\ldots$). If we say a reaction takes one year, or any other amount of time, we're also referring to that same fraction of reactants becoming products. It's not that 100% of the reactants have become products, but rather just a certain fraction.

Why are **chemical reactions important** in **biological systems**?

Chemical reactions make everything in your body work! This is true in all living things (plants, animals, insects, bacteria, etc.). Anytime you make even a slight movement, lots of chemical reactions have to take place. Digesting food, building up or breaking down fat, breathing, cell reproduction, or pretty much any other process that happens in your body involves lots of chemical reactions. Chemistry isn't only important in the lab, it's essential to everything related to life!

ACIDS AND BASES

What are **Lewis acids and bases**?

The definition of a species as a Lewis acid or a Lewis base is based on whether a species has a tendency to donate its electrons to another species or whether it tends to accept electrons from another species in a chemical reaction. Lewis acids are electron acceptors, while Lewis bases are electron donors. As you can probably guess, Lewis acids tend to react with Lewis bases, since they each have what the other one is looking for.

What are **Bronsted acids and bases**?

The Bronsted acid/base definition differs somewhat from the Lewis definition. Rather than focusing on electron acceptors and donors, the Bronsted definition deals with donating or accepting H^+ atoms (protons) in chemical reactions. A Bronsted acid is a species that tends to donate protons in a chemical reaction, while a Bronsted base is a species that tends to accept those protons. To accept protons, a Bronsted base must typically have at least one pair of nonbonding valence electrons.

What is an **amphoteric molecule**?

An amphoteric molecule is one that can act as either an acid or a base. Either the Lewis or Bronsted definition of acid/base behavior can be used to define a species as amphoteric.

What is the **pK$_a$** of a **Bronsted acid**?

The pK$_a$ of a Bronsted acid provides a description of the tendency of the acid to donate a proton in an aqueous (water-based) solution. A lower pK$_a$ is associated with a stronger tendency to donate protons, and thus a stronger Bronsted acid. The pK$_a$ is calculated mathematically as the negative logarithm (base 10) of the equilibrium constant for dissociation of a proton from the acid in water.

What is the **pH** of an **aqueous solution**?

The pH of an aqueous solution provides a description of the concentration of H_3O^+ ions (or water molecules that have accepted an extra proton) in the solution. The pH is calculated as the negative logarithm (base 10) of the activity of H_3O^+ ions in solution. A pH near 7 indicates that the solution is nearly neutral, in the sense that this would be the pH of pure water without any acid or base added. A pH value lower than 7 indicates that the solution is at least somewhat acidic, while a pH above 7 indicates the solution is basic.

The equation describing the pH of a solution is:

$$pH = -\log [H_3O^+]$$

What is a **general chemical equation** for a **combustion reaction**?

A combustion reaction involves the burning of a hydrocarbon in the presence of oxygen to form carbon dioxide and water as products. A specific example would be:

$$CH_4 + 2\,O_2 \; CO_2 + 2\,H_2O$$

In general, the equation for a combustion reaction will look something like:

$$\text{Hydrocarbon} + \text{Oxygen} \; \text{Carbon Dioxide} + \text{Water}$$

Why do **baking soda and vinegar fizz** so much when they react?

The chemical name for baking soda is sodium bicarbonate, and its chemical formula is $NaHCO_3$. Vinegar consists mainly of acetic acid (CH_3COOH) and water. When baking soda is added to vinegar, the following reaction takes place between the sodium bicarbonate and acetic acid:

$$NaHCO_3 \text{ (aq)} + CH_3COOH \text{ (aq)} \; CO_2 \text{ (g)} + H_2O \text{ (l)} + CH_3COONa \text{ (aq)}$$

It's the released CO_2 that creates the bubbles and foaming appearance you see when the two are mixed.

Why does it get so **hot** when you **pour a strong acid in water**?

When a strong acid is poured into water, many proton transfer reactions take place very rapidly, and these are exothermic reactions. The fact that these reactions are exothermic means that they release energy (in the form of heat), increasing the temperature of the solution before the heat has time to dissipate to the surrounding environment.

CATALYSIS AND INDUSTRIAL CHEMISTRY

How can we **make chemical reactions** go **faster**?

Two of the most common ways to make a reaction proceed faster are to increase the temperature at which the reaction is carried out or to use a catalyst to lower the energetic barrier to the reaction. With increased temperature, the amount of thermal energy available to overcome the energetic barrier to the reaction is increased. Of course, you'll also get more products faster if you just increase the concentration(s) of the reactant(s). Light can also serve as the catalyst, breaking a chemical bond to produce a species that is more reactive that the original reactant species.

Plants like this one combine nitrogen and hydrogen using the Haber-Bosch process to manufacture ammonia.

What is a catalyst for a reaction?

A catalyst for a chemical reaction is any chemical species that lowers the amount of free energy necessary to achieve the chemical transformation. This may involve a significant change in the mechanism by which the reaction takes place. Light may also serve as the catalyst.

What are some industrial processes that make use of catalysts?

The chemical industry uses catalysts in a staggering array of chemical reactions that make products that we all use every day. The production of ammonia (NH_3) from nitrogen (N_2) and hydrogen (H_2), known as the Haber-Bosch process, uses iron or ruthenium catalysts; this reaction is used to generate about five hundred million tons of fertilizer per year! Most of the world's plastic is made by one form of catalysis or another, and the refining of crude oil into gasoline and other fuels is also possible because of catalysts. Catalysts are also used in food processing and production—margarine is made by reacting fats with hydrogen gas in the presence of a nickel catalyst.

What chemicals are produced on the largest scale?

In terms of volume (not profit), sulfuric acid is the largest chemical produced worldwide. Global consumption is around two hundred million tons per year—a number that is just

57

too impossibly large to imagine. Nitrogen (N_2), ethylene (CH_2CH_2), oxygen (O_2), lime (mainly CaO), and ammonia (NH_3) round out the top five, though sometimes these change.

What is the **difference** between **heterogeneous catalysis** and **homogeneous catalysis**?

Homogeneous catalysis takes place when the catalyst and the other reactants are all dissolved in the same solution. Heterogeneous catalysis typically involves the use of a catalyst that is insoluble, or perhaps only weakly soluble, in the solution in which the reaction takes place. Thus, in heterogeneous catalysis, the catalyst and solution may form a suspension, or the catalyst may simply be a solid that is placed in the solution.

OTHER KINDS OF CHEMICAL REACTIONS

Why does **iron rust**?

Rust is iron that has been oxidized. The oxidation of iron, or any iron alloy (See "History of Chemistry"), can occur whenever iron is in the presence of oxygen and water. The chemical reaction involves electrons from the iron being transferred to oxygen atoms, which react with water molecules to eventually form iron oxides. The presence of ions, like salt or H_3O^+ in acidic solutions, can accelerate the rate of these reactions. Preventing rust usually requires a protective coating that prevents the iron from reacting with oxygen and water.

Is **melting** a **chemical reaction**?

No, melting is not a chemical reaction. Melting is a change between phases of matter, and it does not involve the breaking or formation of any new chemical bonds. As the temperature increases to cause a substance to melt, the arrangement of the molecules in the solid/liquid changes, but no chemical reaction takes place. The same is true for liquid to gas phase changes as well as for the reverse processes (freezing and condensing).

What is **thermite**?

Thermite is a mixture of a metal powder and a metal oxide that is capable of causing a very strong exothermic reaction. Different metals can be used, but the most com-

Everyone is familiar with rust, which is what happens when iron oxidizes—the result of iron being exposed to water and oxygen.

> ### What causes lightning?
>
> During a storm, collisions of water and ice particles in the clouds can result in a charge separation within the clouds, resulting in the buildup of significant electric fields. If the fields become large enough, lightning may occur between the cloud(s) and the ground, or between different clouds, which reduces the charge separation. The lightning in the sky is just a very large spark, not all that different from how you might occasionally get a small shock from static electricity when you touch a doorknob.

mon mixture is iron oxide (Fe_2O_3) and aluminum (Al) powder. The mixture is stable at room temperature, but when it is ignited, it burns in an incredibly exothermic reaction, releasing a lot of heat.

How can **light** be used to **initiate a chemical reaction**?

When light strikes a molecule, it has the potential to excite an electron to a higher energy level. When this happens, the molecule becomes less stable and more reactive. This can even cause the molecule to fragment, generating a more reactive species or catalyst, which can then go on to react with another molecule.

What is **diffusion**?

Diffusion is essentially the random motion of molecules (or atoms) through a medium. It causes the concentrations of species in a liquid or gas to even out across the accessible volume. Diffusive motions of molecules in a liquid or gas are typically responsible for the chance encounters that lead to bimolecular reactions.

Can **reactions** take place in **gas, liquid, and solid phases**?

Absolutely. Bimolecular (two species) reactions taking place in the gas phase and in solutions are quite similar in that they involve random diffusion of reactant molecules until they collide, at which point a reaction may take place. For reactions involving solids, the surface of a solid is often reacting with another species that is present at the interface between the solid and a gas or liquid with which it is in contact. An example would be your car rusting to form oxides.

What is the **partial pressure** of a **species** in a **gas**?

If you have a rigid container containing a mixture of gases, the partial pressure of a species is simply the pressure the gas would have if that one species occupied the entire volume of the container.

How does **pressure affect** the **rate** of a **chemical reaction** in the **gas phase**?

In a gas phase chemical reaction, the pressure (or partial pressure) of a chemical species is directly related to its concentration. In just the same way we could write rate equations for solution phase reactions involving a constant multiplied by the concentrations of species, we can write rate equations for gas phase reactions that involve a constant multiplied by the partial pressures of species in the gas phase. So just like in a solution phase reaction, increasing the pressure of a reactant in the gas phase will increase the rate at which products are formed.

What is an **electron transfer reaction**?

It's just what it sounds like: a reaction that involves movement of an electron from one species to another. This most often means the electron is moving between two different molecules, although intramolecular (one molecule) electron transfer reactions are also encountered commonly in the study of chemistry.

Does **reaction stoichiometry** also apply to **electron transfers**?

Yes, for just the same reasons it applies to chemical reactions. The conservation of matter tells us that we must have the same amount of stuff before and after a reaction has taken place, so the number of electrons on each side of a chemical reaction equation need to be balanced for the equation to be accurate.

Where are **electron transfer reactions important**?

Electron transfer reactions are important in many areas. In biological systems, processes like photosynthesis, nitrogen fixation, and aerobic respiration (the process your body uses to make energy using oxygen) all rely heavily on electron transfer reactions. Electron transfer reactions are also frequently used to obtain pure metals from ore. Electrochemical cells (see "Analytical Chemistry") also rely on electron transfer reactions; the batteries that power your cellular phone and other devices use electron transfer reactions to do so.

What is a **pyrophoric reagent**?

Pyrophoric reagents are substances that will ignite spontaneously when exposed to air. Very often, this is due to a reaction with the water in the air. Thus pyrophoric reagents should be used only under inert atmospheres, such as in a glovebox filled with an inert gas such as argon or nitrogen. Often, pyrophorics will be sold as solutions already dissolved in a solvent so that they do not tend to create fires so easily. Some of the more mild pyrophoric substances can be handled in the air, but caution needs to be exercised to flush air out of the container before storing them for an extended period of time. They also must be disposed of with caution, or they might accidentally set the wastebucket on fire!

Perhaps not surprisingly, pyrophoric materials can be useful for starting fires in a controlled manner. Pyrophoric materials are present in the spark-generating mechanisms in lighters and some firearms.

What is the **flash point** of a substance?

The flash point of a substance is the temperature above which it can form an ignitable vapor in the air. This requires an ignition source—the vapor may no longer burn if a source of ignition is not present.

What is the **autoignition temperature** of a substance?

The autoignition temperature is similar to the flash point, except that it does not require an ignition source to begin or to continue burning. Above the autoignition temperature, the vapor will start to burn, and continue to burn, even in the absence of any ignition source.

ORGANIC CHEMISTRY

STRUCTURES AND NOMENCLATURE

What is an **organic compound**?

Any molecule, or compound, that contains carbon atoms is referred to as "organic." The usage of this term is a bit arbitrary, though, as some forms of carbon (like graphite and diamond) and carbon-containing ions (like formate and carbonate) are not thought of by chemists as "organic molecules."

Where do I run into **organic chemicals** in my **life**?

The food you eat, the clothes you wear, the gasoline in your car, the plastic bags you may (or may not!) get at the grocery store…. The list could go on forever.

Why have **chemists devoted** so much **focus** to the chemistry of **carbon**?

Because it's everywhere! Carbon is the fourth most abundant element in the universe (fifteenth on our planet), and the building blocks of life (DNA, amino acids) all contain many carbon atoms. Many biologically active molecules and medicines rely on carbon to define their overall shape.

What was the **first organic chemical** ever **synthesized in a lab**?

Urea. In 1828 Friedrich Wöhler was trying to make ammonium cyanate ($NH_4^+CNO^-$), but this salt turned out to be unstable. It reacted to form urea, demonstrating for the first time (arguable according to some) that an organic chemical could be made from inorganic starting materials.

What else did Friedrich Wöhler discover that helped make him famous?

Not only is Wöhler known for making the first organic chemical outside of a living cell, he also discovered the elements beryllium (independently discovered by Antoine Bussy), silicon, aluminum, yttrium, and titanium. In case that wasn't enough to cement his position in the annals of chemistry, he also discovered that meteorites contained organic compounds and developed a process to purify nickel.

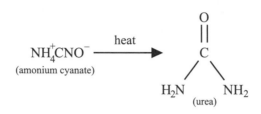

How **many bonds** can **carbon form**?

Carbon has four electrons available for bonding with other atoms. When carbon is bonded to four other atoms, they are arranged in a tetrahedral geometry. These two simple bonding rules have important consequences, as we'll see in this chapter.

What **types of bonds** can **carbon form**?

Carbon can form single (σ) or double (π) bonds to other elements. Double bonds use two of carbon's four available electrons, so carbon can form two double bonds (like carbon dioxide, CO_2), or one double bond and two single bonds (like formaldehyde, H_2CO), or four single bonds (like methane, CH_4).

Can **carbon form more** than **one π-bond**?

Yes, if two carbon atoms form one σ and two π bonds (for a total of three bonds, known as a triple bond), the group is called an alkyne. The simplest alkyne is acetylene (C_2H_2). Welding torches use a combination of oxygen and acetylene to reach temperatures of over 6000 °F (3300 °C).

$$H-C\equiv C-H$$

What is the **shape** of a **carbon–carbon double bond**?

The geometry of the carbon atoms in double bonds is planar. This shape comes from the hybridization of the carbon atom, which is sp^2 (one p orbital is not involved in forming single bonds). To get a bonding interaction between these two remaining p-orbitals, they have to overlap in space. So in a molecule like ethylene (C_2H_4), all of the hydrogen atoms are located in the same plane.

64

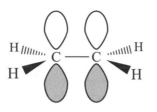

What are **hydrocarbons**, and **how many different ones** are there?

Hydrocarbons, as you might have figured out from the name, are molecules that contain only hydrogen and carbon atoms. There are literally an infinite number of ways to arrange these two elements together, especially if you include polymers (see "Polymer Chemistry"). Hydrocarbons are important molecules. Different sizes and types of hydrocarbons are known as natural gas, gasoline, waxes (like candles), and plastics.

How do **chemists name** so many **different hydrocarbons**?

With a bunch of rules! Let's start with just straight chains of carbon atoms. Here we just need to define how many carbon atoms there are in the molecule. If the molecule doesn't have any double bonds, we use the suffix "-ane." The prefix indicates how many carbon atoms there are. Most of these prefixes are based on Greek numbers (one is Latin, and a few are just weird). Collectively, these molecules are called alkanes.

Number of "C" Atoms	Word Root	IUPAC Name	Structure	Molecular Formula
1	Meth	Methane	CH_4	CH_4
2	Eth	Ethane	$CH_3—CH_3$	C_2H_6
3	Prop	Propane	$CH_3—CH_2—CH_3$	C_3H_8
4	But	Butane	$CH_3—(CH_2)_2—CH_3$	C_4H_{10}
5	Pent	Pentane	$CH_3—(CH_2)_3—CH_3$	C_5H_{12}
6	Hex	Hexane	$CH_3—(CH_2)_4—CH_3$	C_6H_{14}
7	Hept	Heptane	$CH_3—(CH_2)_5—CH_3$	C_7H_{16}
8	Oct	Octane	$CH_3—(CH_2)_6—CH_3$	C_8H_{18}
9	Non	Nonane	$CH_3—(CH_2)_7—CH_3$	C_9H_{20}
10	Dec	Decane	$CH_3—(CH_2)_8—CH_3$	$C_{10}H_{22}$

Are **hydrocarbons always straight chains** of carbon atoms?

No. There could be carbon atoms attached to the linear chains we talked about in the previous question. Let's learn the next step in naming alkanes and have a look.

First, we need to define the names of branches (see graphic, next page). Chemists use the same prefixes to indicate the length of the branch, but now the suffix is "-yl" instead of "-ane." So methane (CH_4), when it's a branch off of a chain of carbon atoms, becomes methyl (-CH_3), ethane (CH_3CH_3) becomes ethyl (-CH_2CH_3), and so on.

Next we have to indicate where along the main carbon chain the branch point is. This part is pretty simple—just number the carbon atoms and put this number before the

65

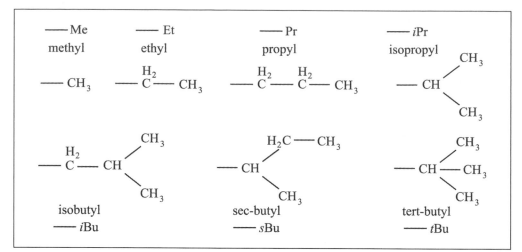

Hydrocarbons with additional atoms attached to linear chains have suffixes ending in "-yl" instead of "-ane." Here are some examples.

name of the branch. So if you had an eight-carbon chain (octane) with a two-carbon branch (ethyloctane) on the third carbon from the end, it would be called 3-ethyloctane and look like this:

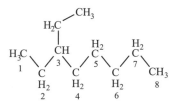

There are a lot more rules to naming organic compounds, but that's enough for now.

Can **carbon chains form rings**, too?

Yes—chains of carbon atoms can connect back to themselves, forming rings of atoms. The prefix *cyclo-* is added to the name of the linear carbon chain to indicate that a ring is present (so hexane becomes cyclohexane). The chemistry ring structures can be different than their linear cousins because of the added energy that some rings contain. We know that sp^3-hybridized atoms like to form bonds that are separated by 109.5°. The more that a ring forces those bonds to deviate from that ideal angle, the more energy (called ring strain) that is released when that ring is opened during a chemical reaction.

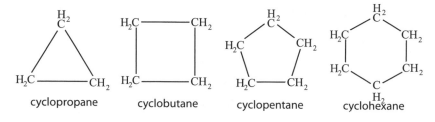

What is the **structure of diamond**?

Diamond has a repeating structure of carbon atoms in which all the atoms are bonded to four others in a tetrahedral geometry. It's easiest to see if we first look at the structure of cyclohexane, a ring (cyclo-) of six carbon atoms (-hex-), with no double bonds (-ane).

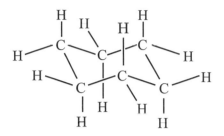

If we repeat the structure of cyclohexane over and over, we arrive at the structure for diamond.

What is **charcoal**?

Charcoal is made of carbon and ash and is formed when water and other substances are removed from animals or plants. It can be produced by heating wood or other biologically derived materials in the absence of oxygen.

What is a **heteroatom**?

A heteroatom is any atom that is not a carbon or hydrogen atom. Some examples of typical heteroatoms include oxygen, sulfur, nitrogen, phosphorus, chlorine, bromine, and iodine, though anything other than carbon or hydrogen fits the definition.

What is a **chalcogen**?

A chalcogen is an element in group 16 on the periodic table. This includes oxygen, sulfur, selenium, tellurium, polonium, and livermorium. This name comes from the Greek word meaning "copper-former," and has its origins in the fact that some of these elements tend to coordinate to metals to form compounds with metals in ores.

What is a **cation**?

A cation is a positively charged atom or molecule. Cations have a larger number of protons than electrons, such that they have a net positive charge.

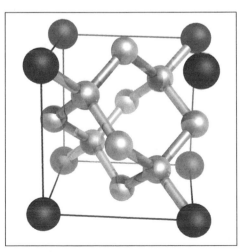

The structure of a diamond crystal.

What is an **anion**?

An anion is a negatively charged atom or molecule. Anions have a larger number of electrons than protons such that they have a net negative charge.

What is a **free radical**?

A free radical is an atom or molecule that contains unpaired electrons in one (or more) of its orbitals. Typically these species will be highly reactive as the unpaired electron(s) can pair with other electrons in a favorable manner. Radical species can have any charge or be neutral.

What are **isomers**?

Isomers are chemical compounds with the same molecular formula, but which are different in some way. The major types are constitutional isomers, stereoisomers, and enantiomers (the last one is actually a subset of the second-to-last one, but we'll get there in a minute).

Constitutional, or structural, isomers have the same number of atoms, but they are arranged in a different order. For example, four carbon atoms and ten hydrogen atoms can be arranged in two different ways.

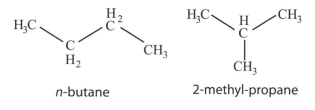

n-butane   2-methyl-propane

What is a **geometric isomer**?

Geometric isomers are molecules containing the same set of atoms and bonding arrangements, but with a different spatial arrangement of the atoms or groups. For example, cis and trans isomers are an example of geometric isomers.

What are **stereoisomers**?

Stereoisomers have the same number of atoms connected in the same order, but differ in their arrangement in space. There are two major types of stereoisomers: enantiomers and diastereomers.

What is **chirality**?

Chiral objects have nonsuperimposable mirror images. What does that mean? Superimposable means one object can be placed over another, or less technically, that they're identical. So enantiomers are not identical, but they are mirror images. Take a look at your hands— they are enantiomers. If you put one hand up to a mirror, it looks like your other hand (so they are mirror images). But if you try to put one hand on top of your other (no, not palm to palm, that's cheating), you see they're not identical (therefore nonsuperimposable).

What are **enantiomers**?

Enantiomers are molecules that are chiral. In organic chemistry, if a carbon atom is bonded to four different atoms (or groups of elements), then we can draw two enantiomers of the molecule. Remember that the connectivity does not change, just the arrangement of the atoms in space.

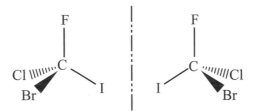

Wait—what do those dashed and wedged bonds mean?

Up to this point we've mostly been representing molecules as flat objects, where chemical bonds are just shown as straight lines. But molecules are not flat. In the previous question, the four halogen atoms around the central carbon form a tetrahedron. Chemists use dashed bonds to indicate that they are behind the plane of the paper, and wedged bonds come toward you, above the plane of the paper.

What are **diastereomers**?

This is going to sound like a cop-out, but diastereomers are stereoisomers that are not enantiomers. That's the real, technical definition. One type of diastereomers show up when carbon forms a double bond. Recall from previous chapters that when there are three groups bonded to a carbon atom, it will be planar (sp^2 hybridized). If the double bond is in the middle of a carbon chain, there are two possible isomers.

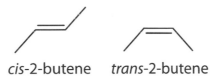

cis-2-butene   *trans*-2-butene

These two molecules are not superimposable, but they're also not mirror images, so they are called diastereomers. There are many other forms of diastereoisomers, but this form is the easiest to understand.

What is a **racemic mixture**?

A racemic mixture contains equal amounts of both enantiomers of a molecule.

What does **enantiomeric excess measure**?

The enantiomeric excess is a measure of how much more of one enantiomer is present in a mixture. It's often reported as a percentage. Racemic mixtures have an enantiomeric excess value of 0% because both enantiomers are present in equal amounts. For a solution composed of 75% of one enantiomer, the enantiomeric excess would be 50% (75% − 25% = 50%).

How and when was **molecular chirality discovered**?

Nonracemic mixtures rotate the plane of a beam of polarized light in either a clockwise or counterclockwise direction. Jean-Baptiste Biot, a French physicist, observed this effect in 1815 with quartz crystals, turpentine, and sugar solutions. These were important results in understanding the nature of light, but it was Louis Pasteur in 1848 that figured out that the effect was based on molecular properties. Pasteur painstakingly separated enantiomerically pure crystals from a racemic mixture of tartaric acid and showed that the two enantiomers rotated light in opposite directions.

Are all the **carbon–carbon bonds** in **benzene** the **same length**?

Yes, but you might not think so by looking at a single line structure of benzene. The actual structure of benzene is a combination of two structures, shown below. In technical terms, the electrons in the π bonds are delocalized (spread out) by resonance. The drawing convention that chemists use to represent molecular structure just can't display this properly in a single structure. The electrons do not move from one place to another, and the carbon–carbon bonds do not oscillate between long and short—the structure is an average of these two drawings. After all, a molecule of benzene doesn't really care that we can't properly draw it.

Sometimes, you might see benzene drawn with a single circle in the center, representing the delocalization of the π-electrons.

What is **resonance**?

Resonance is a way that chemists represent delocalized electronic structure. Let's take that statement apart to understand what it means. "Delocalized" means that an electron, or a pair of electrons, is not located entirely around a single atom or bond. Take a look at the two structures of nitrogen dioxide (NO_2) on the following page. The negative charge is located

---

When was the term "aromatic" first seen?

It was first seen in a chemistry publication in 1855. The paper was written by August Wilhelm von Hofmann, but he gives no indication why he chose this term. It's an unusual case since only a few aromatic compounds are actually smelly, while many nonaromatic compounds in chemistry labs do truly stink.

on one oxygen atom in one resonance structure, but can be found on the other oxygen in the second resonance structure. Notice that we said "electronic structure" and haven't said anything about atoms moving here—that's because they don't. Resonance only deals with electrons, and the atoms are located in the same arrangement in every resonance contributor. This is important, and makes sense if you remember that the electrons aren't "moving" from one resonance structure to another. These structures are needed because the actual molecule is more complicated than our simple drawing system can represent.

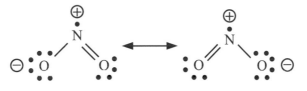

What is **aromaticity**?

Aromaticity is a special kind of resonance delocalization, one we've actually already seen. Delocalization of electrons always makes molecules more stable (compared to imaginary molecules where the electrons cannot spread out). If this delocalization takes place in a flat ring of n carbons, and the number of electrons involved is 4n + 2 (i.e., 2, 6, 10, 14, etc.), then the system is called aromatic.

What is a **functional group** in organic chemistry?

Functional groups are groups of atoms that tend to show similar reactivity in different molecules. Chemists use these groupings to help understand and predict how different molecules react with each other, and they're also used for naming compounds.

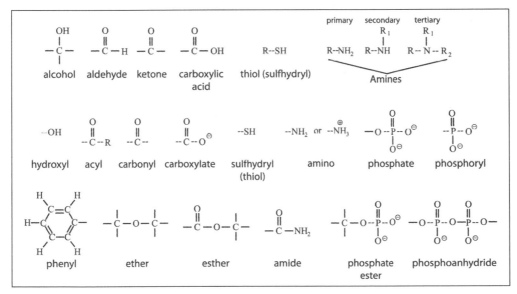

Some examples of functional groups.

What is a **carbocation** and what is a **carbanion**?

A carbocation is a positively charged compound in which a carbon atom takes on a significant fraction of the positive charge. The definition for a carbanion is probably easy to guess, then—it's a negatively charged compound in which a carbon atom bears a significant fraction of the negative charge.

REACTIONS OF ORGANIC COMPOUNDS

What are **"curved arrows"** in organic chemistry?

Chemists used curved arrows to depict the flow of electrons in a chemical reaction. The arrow starts at the nucleophile (a lone pair of electrons, a π bond, or a σ bond), and points toward the electrophile (an atom, or bond, with a full or partial positive charge). Here's an example of a transesterification reaction (i.e., changing one ester for another).

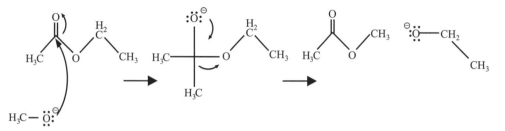

What is a **nucleophile**?

A nucleophile is a molecule that donates electrons (to an electrophile) in a chemical reaction. These are typically functional groups with lone pairs of electrons, but can also be π bonds, or, in some rare cases, σ bonds.

What is an **electrophile**?

Electrophiles are species that accept electrons (from a nucleophile) in a chemical reaction. Usually, electrophiles have either a full- or partial-positive charge, or in some cases (like BH_3) have an unfilled octet of electrons.

What is a **substitution reaction**?

The exchange of one functional group, or atom, for another is called a substitution reaction. The transesterification reaction above is an example of substituting a $-OCH_2CH_3$ group for a $-OCH_3$ group. Another simple substitution reaction is the exchange of one halogen for another on a methyl group:

What is a **unimolecular substitution reaction**?

We've already talked about substitution reactions, so what makes one "unimolecular"? If the transition state (remember, this is the highest energy state of an individual chemical reaction) involves one (uni-) molecule (-molecular), then it is referred to as a unimolecular reaction. This might seem like an odd thing to distinguish, but there are many differences between uni- and bimolecular substitution reactions. These differences all result from how many species are involved in the transition state.

For an example, here's the reaction of tert–butyl chloride with hydroxide ion:

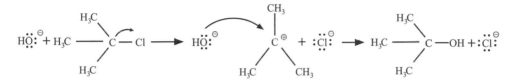

The first step is breaking the C–Cl bond, and this involves only the $(CH_3)_3C$–Cl molecule. Then the hydroxide ion reacts with the tert–butyl carbocation in the second step. Since there is only one molecule in the slower first step (just assume that's true so we can illustrate this point), this is a unimolecular substitution reaction.

What is a **bimolecular substitution reaction**?

From the last answer, you can probably guess that a bimolecular substitution reaction has two (bi-) molecules (-molecular) in the transition state. This means that the slowest step involves two molecules interacting. In the previous example, there was only one.

Let's look at a similar substitution reaction, but instead of tBuCl, we'll react hydroxide with MeCl:

Here the nucleophile (OH⁻) directly displaces the leaving group (Cl⁻), without forming a carbocation intermediate. This is because the methyl cation (CH_3^+) is much less stable than the tert–butyl carbocation formed in the previous question.

Why are more **substituted carbocations more stable**?

Great question! First, let's review what led you to this question. In the unimolecular substitution reaction we looked at, an intermediate tert–butyl cation was formed, but in the bimolecular substitution reaction, the cation (this time a methyl cation, CH_3^+) was too high in energy. In this case the hydroxide ion directly displaced the chloride ion.

Generally, the more highly substituted a carbocation is, the more stable it is. There are a number of ways to explain why this is true. The first is that carbon substituents are more electron-donating than hydrogen atoms. Electrons on neighboring carbon atoms can help stabilize the cationic center. Simply put, the order of carbocation stability matches the number of carbons bonded to the cationic carbon, with higher numbers leading to greater stability.

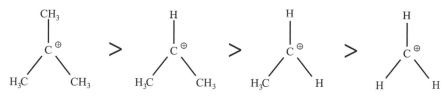

What is **hyperconjugation**?

Hyperconjugation is another way to explain why substituted carbocations are more stable. The electrons in either the C–C or C–H bonds that are near the cationic center can interact with the empty p-orbital. It's not the bonds that are directly connected, but rather one bond removed from the center carbon, that are the most important (see the illustration below). At the simplest level, this provides an explanation for the statement in the previous answer that neighboring carbon atoms are more electron-donating than hydrogen atoms: it's really the neighboring C–H bonds that help stabilize the empty p-orbital.

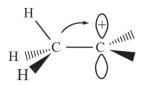

Note: The arrow above is just to illustrate the overlap of the C–H σ bond with the empty p orbital. It's not to suggest that the hydrogen atom actually moves…though this does happen sometimes!

What is an **addition reaction**?

In an addition reaction, two or more molecules combine to make one molecule. This is different than the substitution reactions we have been looking at, where two molecules combine to make two different molecules.

The addition of an acid to an alkene is the simplest example. Here, an acid protonates a carbon–carbon double bond. The carbocation that is formed will be the more

What is Markovnikov's rule?

We actually just looked at an example of Markovnikov's rule! This rule states that when you add a protic acid (like HBr) to an alkene, the proton ends up attached to the carbon with fewer alkyl substituents, while the conjugate base ends up bonded to the carbon with more alkyl substituents. This is because of the stability of the carbocation intermediate that is formed (remember, more substituted means more stable).

substituted one of the possible products that could form. The conjugate base of the acid then reacts with this carbocation.

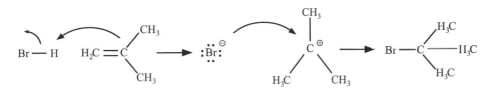

What is an **elimination reaction**?

If an addition reaction adds two groups to a molecule, then an elimination reaction takes them away. The textbook example again involves carbon–carbon double bonds, but this time we're making that double bond.

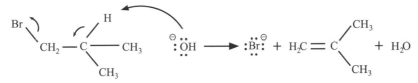

A molecule of base (hydroxide ion, ⁻OH) causes the elimination of HBr. In this case, the process happens in one step, as shown.

Are there **uni- and bimolecular elimination reaction mechanisms,** like there were for **substitution reactions**?

Yes! And what do you think controls whether an elimination reaction happens in one step or two? Right—the stability of the carbocation. In the previous example, if bromide ion had dissociated first, a primary carbocation would have formed:

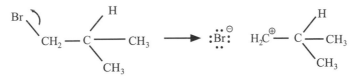

This is a much more difficult reaction than the bimolecular process where the elimination of HBr takes place in a single step.

But if an alkyl halide can form a stable carbocation, the unimolecular elimination reaction is faster. It's referred to as "unimolecular" because the slow step has only one molecule in the transition state, just like substitution reactions.

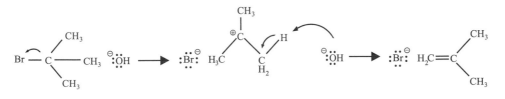

Who were Diels and Alder?

The Diels-Alder reaction, like many reactions in organic chemistry, is named after the chemists who discovered it—in this case Otto Paul Hermann Diels (1876–1954) and Kurt Alder (1902–1958). Kurt Alder was actually Diels' student at the University of Kiel, and Alder was awarded a PhD in 1926. Alder and his advisor Diels jointly received the Nobel Prize in Chemistry in 1950.

What's a **cycloaddition reaction**?

In the most general terms, a cycloaddition reaction is one that forms a ring from multiple π-bonds. To illustrate some of the concepts and ways that these reactions are classified, let's look at the cycloaddition reaction shown below. This reaction, known as the Diels-Alder reaction, is between a conjugated diene (double alkene) and a another alkene that likes to react with dienes (hence dienophile, or diene lover). It is classified as a [4 + 2] cycloaddition, referring to the number of atoms directly involved in bond-forming events.

(There's a second system for classifying cycloaddition reactions that uses the number of π-electrons involved in ring formation, but let's just leave that aside.)

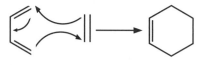

diene + dienophile

What's a **pericyclic reaction**?

A cycloaddition reaction is actually a type of pericyclic reaction, but the term "pericyclic" includes other types of reactions. The textbook definition of pericyclic is a reaction whose transition state has a cyclic structure (i.e., the electrons are flowing in a closed loop). In addition to cycloaddition reactions (which exchange two π-bonds for two σ-bonds, or vice versa), pericyclic reactions include sigmatropic reactions, electrocyclic reactions, and cheletropic reactions (and a few others which we'll ignore).

What's a **sigmatropic reaction**?

While a cycloaddition reaction uses two π-bonds to make two new σ-bonds, a sigmatropic reaction takes one σ-bond and makes…well…one σ-bond. One of the most well-known sigmatropic rearrangements is known as the Cope rearrangement. Using the system we talked about earlier, this reaction is classified as a [3,3]-sigmatropic reaction (the methyl substituent is not directly involved, so we don't count it).

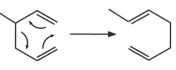

What's an **electrocyclic reaction**?

Okay, so a cycloaddition reaction was $2\pi \rightarrow 2\sigma$, and we just saw that sigmatropic reactions are $1\sigma \rightarrow 1\sigma$, so what about a pericyclic reaction that is $1\pi \rightarrow 1\sigma$? That's an electrocyclic reaction. Like most pericyclic reactions, electrocyclic reactions can either make or break a σ-bond. If a σ-bond is made in the process, it's an electrocyclic ring-closing reaction. If the σ-bond is consumed in the process, the process is named an electrocyclic ring-opening reaction, but it's the same process.

One noteworthy example of an electrocyclic ring-closing reaction is the Nazarov cyclization, which converts divinyl ketones into cyclopentenes, usually in the presence of an acid.

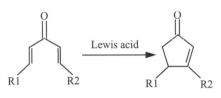

What's a **cheletropic reaction**?

Finally, there's a second type of reaction that involves two π-bonds being transformed into two σ-bonds. The difference here is that two bonds are formed (or broken) at a single atom, while in cycloaddition reactions, only one bond is made at each reactive atom. An example involving SO_2 is shown below (surprisingly this reaction isn't named after anyone!). In this reaction both σ-bonds are made (or broken) at the sulfur atom. Compare this with the Diels-Alder reaction mentioned above, where the two new σ-bonds are connected to the two different ends of the dienophile.

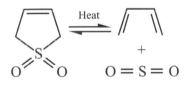

What is a **tautomerization reaction**?

A tautomerization reaction interconverts two constitutional isomers of an organic molecule, referred to as tautomers, which differ only by the positions of hydrogen atoms. The scheme below shows how the reaction typically takes place at the ketone functional group of an organic molecule. The ketone form of the molecule is commonly referred to as the "keto" form of the molecule, while the alcohol form is commonly referred to as the "enol" form of the molecule.

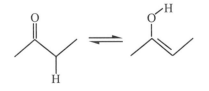

77

INORGANIC CHEMISTRY

STRUCTURE AND BONDING

Which **molecules** are considered **inorganic**?

Any molecule without a carbon atom is technically considered inorganic, but in practice, there are some exceptions. Many salts (like carbonates, CO_3^{2-}, or cyanides, CN^-) are thought of as inorganic, even though they do contain carbon.

What **elements** are **metals**?

Metals are elements that easily conduct electricity. There are a number of sets of metals in the periodic table. Instead of just listing the metallic elements, let's take a look at those sets one at a time in the next few questions.

What are the **alkali metals** and **alkaline earth metals**?

The first two columns of the periodic table are known as the alkali (Group 1) and alkaline earth (Group 2) metals. The elements in both groups have very low ionization energies, which means they readily give up one or two electrons to reach a Noble gas electron configuration (e.g., Na^+, Mg^{2+}). In their elemental forms, these elements are soft, silver-colored substances. These elements are almost never found in their pure elemental state in nature because they quickly react with air or moisture, sometimes violently.

What are the **transition metals**?

Elements in the d-block (Groups 3–12) of the periodic table are referred to as transition metals. (The lanthanides and actinides are usually excluded.) With the exception of Group 12, the transition metal elements have an incomplete d-shell of electrons. The chemistry of these elements depends strongly on these d-electrons, and most transition metals are stable in multiple different oxidation states. If the metal has any unpaired d-electrons, it can have magnetic properties. Many transition metals have uses as catalysts in bond-forming reactions, and we'll talk about a few later in this chapter.

What is a **metalloid**?

Metalloid is a term for elements that are sort-of metals, and sort-of not metals. Sometimes this group of elements is referred to as semimetals. To be more precise, these elements exhibit some of the physical and chemical properties of metals. Generally metalloids have some electrical conductivity, but not nearly as much as true metals. Because of these ambiguous definitions, even which elements are called metalloids can vary. Usually boron, silicon, germanium, arsenic, antimony, and tellurium are included as metalloids; sometimes polonium and astatine; rarely selenium.

What are the **valence orbitals** of **transition metals**?

The transition metals are also referred to as the d block because their valence orbitals are the d shell (see illustration on this page). There are five different d orbitals. Three (d_{xy}, d_{xz}, and d_{yz}) of these five look similar, just with different orientations in space. The fourth ($d_{x^2-y^2}$) has the same shape as the previous three, but the loops point in along the axes as opposed to between them. Finally the d_{z^2} looks like a p-orbital pointed along the z-axis with a ring (technically a torus) around it.

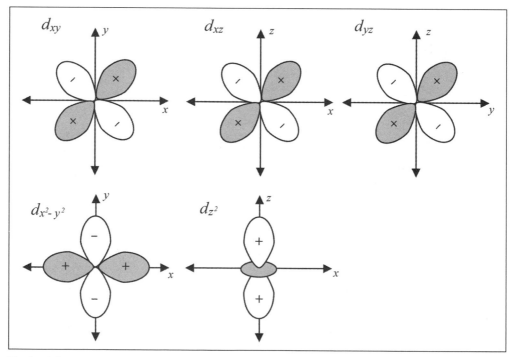

The five different valence orbitals of transition metals.

What are the **lanthanides** and **actinides**?

The actinides and lanthanides make up the f-block of the periodic table. These two rows are frequently separated from the main table, but that's just so the periodic table isn't so wide on a piece of paper (seriously). Many of these elements have radioactive isotopes, but the rate at which they decay can vary immensely. For example, ^{238}U has a half-life of 4.5 billion years, but ^{234}Pa has a half-life of only seventy-two seconds!

What is **crystallography**?

Crystallography is the study of the arrangement of the atoms in a solid material. Today, this term generally refers to methods that rely on the patterns of photons (commonly, X-rays), neutrons, or electrons that are diffracted after impacting a sample. The patterns of the diffracted radiation or particles can be interpreted to determine the structure inside the crystal. The interpretation of the diffraction patterns to yield a chemical structure is by no means a simple task, but crystallographers have been doing this for a long time and it has become a commonplace technique. Crystallographic methods have been used for decades to study the structures of inorganic solids and organometallic complexes.

We point out that, while crystallographic methods are commonly used to study inorganic compounds, they have also frequently been applied for studying other types of molecules as well, including biomolecules. While it can often be difficult to obtain a crystalline sample of a biomolecule, such as a protein, crystallography can be extremely useful in deducing protein structures.

What is a **crystal lattice**?

Crystalline solids have regular, repeating arrangements of atoms or molecules. In order to classify these arrangements, chemists use a three-dimensional lattice to encompass the smallest repeat unit (known as the unit cell) of the crystal structure. There is really a lot of math involved, but we can explain it visually. A unit cell is really just a tiny box containing some atoms or molecules. We draw the edges of this small box so that if you were to line up lots of identical copies of this box in all directions, you would get the structure of the entire crystal lattice.

Are all **crystal lattices cubic**?

In other words, are all the three axes of the unit cell always the same length? No. In fact, cubic is only one of the seven "crystal systems." These systems are based on the lengths and angles of the unit cell. The cubic group has three sides of equal length, and all the internal angles are 90°. If one side is longer, we get a tetragonal lattice. If all three sides are different lengths, the lattice system is called orthorhombic. If one angle is not 90°, it's monoclinic. If the angles are all equal but they're not 90°, it's a rhombohedral lattice. If the angles are all different and not 90°, the lattice is triclinic.

Yes, you're right, that's only six. The seventh isn't based on a cube at all. The hexagonal lattice system is (you guessed it) based on a hexagon.

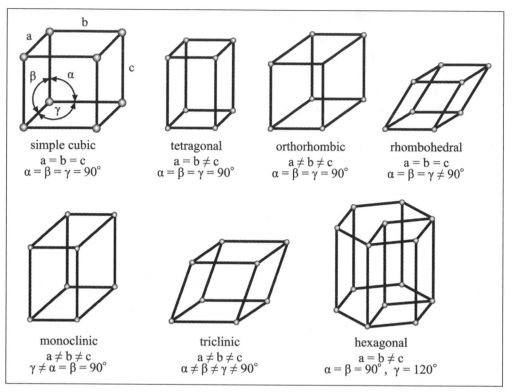

The seven crystal systems.

What **packing arrangements** are possible for a **crystal lattice**?

There are three basic packing arrangements, which we can describe by again imagining a tiny box. If we place an atom at each of the eight vertices of this box, the arrangement is referred to as simple cubic. If we take a simple cubic unit cell and add an atom to the center of each face, we have a face-centered cubic arrangement. If we instead add an atom to the center of the cube, it's called a body-centered cubic unit cell. There are more possibilities, but these three simple ones described here cover a lot of the crystals that chemists encounter.

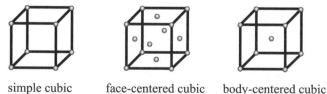

simple cubic   face-centered cubic   body-centered cubic

What determines the **most favorable packing arrangement**?

Thermodynamics! Okay, that's a cop-out answer. Attractive interactions (like Van der Waals forces or hydrogen bonding) can play a large role in determining the stability of a crystal lattice. The most stable packing arrangement can also depend on the pressure and temperature at which the crystal forms.

Can **more than one packing arrangement** exist for a **given chemical molecule**?

Yes—this is referred to as polymorphism, and examples are known for most types of crystalline materials—organic and inorganic molecules, polymers, and metals. The way in which molecules pack in the solid state can actually alter some of its properties. Some pharmaceutically active molecules have more than one solid-state structure, or polymorphs. Sometimes certain polymorphs of drugs can be more useful. For example, a specific arrangement could be more soluble, making it more active in the human body. A second polymorph of aspirin (acetylsalicylic acid) was discovered in 2005, but it's only stable at −180 °C.

ELECTRICITY AND MAGNETISM

What is the **difference** between **paramagnetic** and **diamagnetic complexes**?

In chemistry, atoms or molecules that have at least one unpaired electron (so there is a net spin to the molecule) are known as paramagnetic. If all electrons are paired, chemists refer to the compound as diamagnetic. When a magnetic field is applied a paramagnetic substance will be attracted to the field, while diamagnetic molecules will be repelled from the field.

What gives rise to **magnetism**?

The type of magnetism you're most familiar with (the kind that keeps magnets on your fridge) is technically known as ferromagnetism. Ferromagnets are permanent magnets—they generate their own magnetic field. Ferromagnets have unpaired electrons (so from the information in the previous question we can say that ferromagnets are paramagnetic, not diamagnetic). But ferromagnets have one additional key trait—the unpaired electron spins are all aligned in the same direction, which generates a permanent magnetic field.

Let's go through this again from the beginning: Electrons have spin (a quantum mechanical property, but we don't need to go *that* far back), and this spin

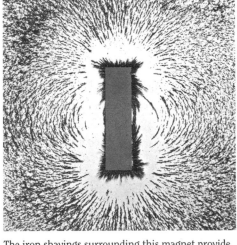

The iron shavings surrounding this magnet provide a good idea of the shape of the magnetic field surrounding this bar magnet, including its north and south poles.

Electric magnets provide both power and a little levitation to maglev trains, resulting in an efficient, smooth, and very fast mode of transportation.

generates a very, very tiny magnetic field. If all the electrons in a substance are paired, it is diamagnetic. If there are unpaired electrons, the substance is paramagnetic. If there are unpaired electrons *and* those unpaired electrons are aligned so that there is a "net spin" for the (macroscopic) substance, it is a ferromagnetic. Ferromagnets are the magnets you know—they stick to your fridge.

Why are **metals attracted to metal**?

Or *why* do magnets stick to your fridge, right? Ferromagnets generate their own magnetic field. This means that other paramagnetic substances will be attracted to ferromagnets, just like they are attracted to magnetic fields. Most metals are paramagnetic, so magnets stick to your metal fridge, but your fridge itself is not magnetic.

So why do **opposite ends of magnets attract** or **repel** one another?

Remember that "real world magnets" are ferromagnetic substances—they have a net overall spin. Again this is a quantum mechanical spin, but it's okay to think of it as spin "up" or "down." In a ferromagnet all of these spins are pointing in the same direction. If you put the "up" ends of a magnet together, they will repel—the magnetic fields are pushing in opposite directions. If you put an "up" end next to a "down" end, they will attract—the magnetic fields are working in the same direction, just like the electron spins are aligned within the magnet.

Are the **North** and **South Poles** of the Earth **magnetic poles**?

Without getting too technical on the physics here (it gets way more complicated with a whole planet versus a tiny piece of iron on your fridge), yes, the North and South Poles are magnetic poles. But compasses point north, so what we call "The North Pole" is actually the south magnetic pole and vice versa.

What is **magnetic levitation**?

Magnets apply forces on one another, and these forces can be either attractive or repulsive. If they are repulsive, and if the forces balance against the force of gravity in a carefully designed way, then the magnetic forces can be used to make an object levitate. This is put to practical use in high-speed trains, which can hover above the track while moving quickly on a magnetic cushion.

What makes **metals good conductors**?

An electrical current is the movement of electrons, so a conductive material allows the free movement of electrons. Metals are good conductors of electricity because of their electronic structure. There are basically two big groups of orbitals in metals—the valence band and the conduction band. The valence band is the group of orbitals that are normally filled in a metal, while the conduction band is empty. These are called bands because they are made up of sets of closely spaced energy levels. Metals, and other good conductors of electricity, can have very small or no band gaps at all. Semiconductors have small band gaps, and electrons can be promoted from the valence band to the conduction band either by heat or light. Insulators, materials that do not conduct electricity, have large band gaps.

What is a **band gap**?

The band gap for a material is the difference in energy between the valence band and the conduction band. This number tells you how good of a conductor a material is. The smaller the band gap (or the less energy) it takes to promote an electron from the filled to the empty orbitals, the better the material can conduct electrons.

How is a **band gap relevant** to the design of **photovoltaic materials**?

The band gap of a material determines what wavelengths of light a photovoltaic (or any other) material can absorb. Photovoltaic materials need to be capable of absorbing the wavelengths of light that the Sun emits, and this requires that they have a band gap that spans the appropriate range of energies.

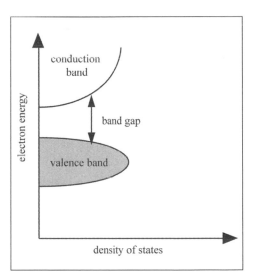

A sample graph showing the band gap of a material between the conduction band and the valence band. The smaller the gap, the better the material is for conducting electricity.

ORGANOMETALLIC CHEMISTRY

What is an **organometallic complex**?

These are molecules with bonds between carbon and a metal. This includes alkali and alkali earth metals (Groups 1 and 2 of the periodic table), the transition metals (Groups 3–12, including the *f* block), and sometimes Group 13 metals are also included. The bonds between metal and carbon can vary widely in their ionic or covalent character. The bonding between a metal and carbon is mostly ionic in two situations: (1) if the metal is very electropositive, as in Groups 1 and 2; or (2) if the carbon group is a stable anion (by being delocalized via resonance as in cyclopentadienyl anion). There are also organometallic molecules with more covalent bonds between the metal atom and the carbon atom. This is usually seen with the transition metals or elements like aluminum. The nature of the metal–carbon bond plays an important role in how organometallic complexes react. The resonance structures of the cyclopentadienyl anion are below:

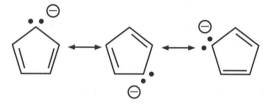

What is the **18-electron rule**?

Earlier we talked about the octet rule, which says that elements like carbon and oxygen are most stable when they have eight valence electrons. For transition metals, as we've mentioned, the d-orbitals start to become important, so eight electrons just aren't enough anymore. Since there are five d-orbitals, we need an additional ten electrons to fill the valence shell. And $8 + 10 = 18$, hence the 18-electron rule for transition metals.

How does one **determine** the **oxidation state** of an organometallic complex?

There are a few ways to approach this problem (and, of course, every chemist thinks the way they use is the most correct), but let's look at what is hopefully the simplest way to think about this. We're going to only talk about complexes with metal–carbon bonds and no other types of ligands. If you're curious enough to want to know how nitroxide binds a transition metal, chances are you have an inorganic chemistry textbook at home anyway.

Okay, let's look at tetra(methyl) zirconium, which has no net charge. In this counting technique, all of the electrons in the metal–carbon bonds are placed on the carbon atoms. So we get four methyl anions and one zirconium ion. Since this complex has no net charge, the zirconium center must balance out the net four negative charges from the methyl groups. Zirconium must be in the +4 oxidation state. Easy, right?

One more example: $K_3[Fe(CN)_6]$ or potassium hexacyanoferrate. We again move the electrons in the metal–carbon bonds to the carbon groups, making six cyanide ions (CN^-), an iron ion, and three potassium ions. We have six negative charges (6 CN^-) and three positive charges (3 K^+ ions), so to balance out the charges, the iron center must be in the +3 oxidation state.

$$ZrMe_4 \quad 4\ Me^{1-} + Zr^{4+}$$
$$K_3[Fe(CN)_6] \quad 3\ K^{1+} + 6\ CN^{1-} + Fe^{3+}$$

It's important to remember that when we count electrons in this way, we're doing only that—counting. Don't get the idea that all of these metal–carbon bonds are the same, or that they're all ionic bonds—this is just for electron bookkeeping!

Why do **organometallic complexes** make **good catalysts**?

Catalysts, by definition, provide a lower-energy route to the product. Organometallic complexes react with organic molecules in ways that are very different than how organic molecules react with themselves. At the most basic level, this is because transition metals have d-orbitals that are involved in making and breaking chemical bonds. With additional orbitals that are of different symmetry than the s- and p-orbitals that are available to elements like carbon, nitrogen, and oxygen, organometallic complexes can accomplish feats that are just impossible (forbidden!) for other elements.

Are **inorganic elements important** in **biological systems**?

Absolutely! Not only do ions like sodium and potassium play a huge role in nervous systems, but many enzymes (nature's catalysts) have metal ions in their active sites. Go find a bottle of vitamins and see just how many metals you need daily!

What is the **hapticity** of an **organometallic ligand**?

The hapticity of an organometallic ligand is a pretty simple concept—it just tells how many atoms from a given ligand are coordinated to the metal center. This is typically denoted in chemistry literature with the Greek letter "eta." A η^2 thus involves a ligand with two atoms coordinated to the metal center, a η^1 one atom, and so on. The most common hapticity for a ligand is η^1, but it is not uncommon to find ligands, such as the cyclopentadienyl anion ligand (usually η^5), with higher coordination numbers.

What is a **carbene ligand**?

A carbene is a molecule that contains a carbon atom with two bonds and two unpaired valence electrons. This leaves the carbon atom neutral in terms of formal charge (see "Atoms and Molecules"), but still typically much more reactive than a typical carbon atom with four bonds. Carbenes are often found coordinated to metal centers in organometallic complexes. These carbene ligands are less reactive than a free carbene species, and actually you might be surprised to learn that organometallic carbene complexes aren't always prepared from the reactions of free carbenes with metal centers. As an organometallic ligand, a carbene may be be either electrophilic or nucleophilic at the carbon atom (see "Organic Chemistry"). Carbene ligands that are electrophilic at carbon are termed Fischer carbenes, and those that are nucleophilic at carbon are called Schrock carbenes. A third class of particularly unreactive carbene ligands are termed persistent (or Arduengo) carbenes.

What is a **monodentate ligand**?

A monodentate ligand is a ligand that coordinates to a metal center via a single atom on the ligand.

What is a **polydentate ligand**?

A polydentate ligand is a ligand that coordinates to a metal center via two or more atoms, forming two or more bonds. This resulting complex is known as a chelate complex.

What are some of the **first organometallic complexes**, and when were they **discovered**?

Among the earliest organometallic complexes was a compound known as cacodyl, or tetramethyldiarsine. This compound, with the chemical formula $C_4H_{12}As_2$ (shown below at left), was first discovered in 1760 and was known for its toxicity and terrible odor. The first platinum olefin complex, $C_2H_4Cl_3KPt$ (also known as Zeise's salt; see below, right), was discovered in 1829. This early organometallic complex was influential in establishing some of the essential underlying concepts that have since become crucial to organometallic chemistry. These examples represent just a couple of the first organometallic species, and to date, thousands of organometallic complexes are known.

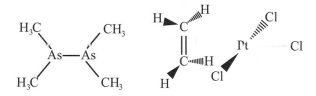

What is a **carbon–hydrogen bond activation reaction**?

Carbon–hydrogen activation reactions are very much what the name would imply—they break typically unreactive carbon-hydrogen bonds. Considering the prevalence of carbon–hydrogen bonds in organic compounds, this is another extremely important type of reaction to be able to carry out both effectively and with selectivity. Successful examples of carbon–hydrogen bond activation reactions have only come about relatively recently (with the first genuine example reported ca. 1965), and organometallic reagents have played a key role in the development of carbon–hydrogen bond reactions.

What is the **Grignard reaction**?

The Grignard reaction is one of the most well-known and powerful reactions in organo-metallic synthesis, primarily due to its ability to readily form new carbon–carbon bonds. In this reaction, a Grignard reagent reacts by adding to a carbonyl functional group of an aldehyde or ketone and forms a new carbon–carbon bond at the carbonyl carbon. Grignard reagents, which are organometallic species that carry out the Grignard reaction, are typically formed by adding magnesium metal to an alkyl or aryl halide (R^1-Br in the scheme below). The discovery of this important reaction was awarded with a Nobel Prize in 1912, and the reaction is actually also named after the French chemist François Auguste Victor Grignard, who discovered it.

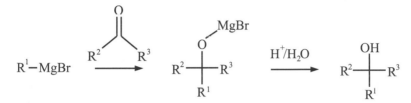

Do **reactions** always take place at the **metal center** of an **organometallic complex**?

Not always. Reactions can also take place at the ligands of a complex! For example, a nucleophile can add to an alkene ligand while it is coordinated to a metal center.

What is **cisplatin**, and how does it help to **fight cancer**?

Cisplatin is a platinum-based compound (see structure below) that reacts with DNA causing it to crosslink, eventually leading to programmed cell death, also known as apoptosis. Cisplatin was the first in a class of platinum containing anticancer drugs. It has been used to fight several types of cancer and is particularly effective against testicular cancer.

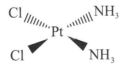

What is the **valence shell electron pair repulsion model**?

The valence shell electron pair repulsion model (VSEPR) is used to predict the bonding geometry about a central atom based on a set of rules that predicts the repulsive forces between the electrons in chemical bonds and in lone pairs. In this model, two factors are used to predict the bonding geometry: the steric number and the number of lone pairs of nonbonding electrons. The steric number is defined as the number of atoms bonded to the central atom plus the number of lone pairs of nonbonding electrons. Based on these two numbers, the table below predicts the bonding geometry that will be observed about a central atom. It should be noted that this is just a predictive model, and it will not be correct 100% of the time.

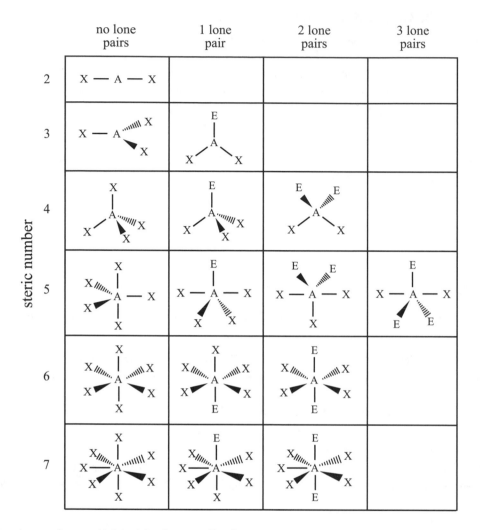

	no lone pairs	1 lone pair	2 lone pairs	3 lone pairs
2	X — A — X			
3				
4				
5				
6				
7				

From top row down, and left to right, the types of bonding geometries are: Steric Number 2) linear; Steric Number 3) trigonal planar, bent; Steric Number 4) tetrahedral, trigonal pyramidal, bent; Steric Number 5) trigonal bipyramidal, seesaw, T-shaped, linear; Steric Number 6) octahedral, square pyramidal, square planar; Steric Number 7) pentagonal bipyramidal, pentagonal pyramidal, and pentagonal planar.

What is **group theory** and how is it useful in chemistry?

Group theory, in the chemistry sense, involves making use of the symmetry of a molecule to better understand its physical properties. By examining the symmetry properties of a molecule, it can be placed into a "point group" that describes the symmetry operations (rotations of reflections) that characterize the molecule's symmetry. Group theory allows chemists to understand an impressive number of molecular properties including the spacing of energy levels of the orbitals in a molecule, the symmetry of the molecular orbitals, the types of transitions that can occur (like vibrational and elec-

> ### When doctors give people "lithium," is that the same lithium as in lithium-ion batteries?
>
> It's the same element, yes. Lithium has quite a few interesting applications! It is used to treat bipolar disorder and in lithium and lithium-ion batteries. It is a metal that can be cut using a spoon, is used to make thermonuclear igniters, and also as a component of armor for army vehicles like tanks. All of this from the tiny element lithium.

tronic excitations), and what the delocalized vibrational motions of a molecule look like. It is very impressive that symmetry can tell us so much about molecules without the need for any complex calculations!

What are "hard" and "soft" Lewis acids and bases?

The words "hard" and "soft" are commonly used to describe two broad classes of Lewis acids and bases (See "Chemical Reactions"). Hard acids and bases typically have small atomic (or ionic) radii, high oxidation states, high electronegativity (for bases), and are not very polarizable. Soft acids and bases tend to be the opposite in that they have relatively large atomic (or ionic) radii, low oxidation states, low electronegativity, and are highly polarizable. As it turns out, hard acids tend to react more rapidly and form stronger bonds with hard bases, and the same is true for soft acids paired with soft bases. This pattern is what makes the theory of hard and soft acids and bases useful for predicting and understanding reactivity in inorganic complexes.

What happens when you add sodium metal to water?

Sodium metal reacting with water can produce a violent response! Sodium and water react according to the equation:

$$2\ Na\ (s) + 2\ H_2O\ (l) \rightarrow 2\ NaOH\ (aq) + H_2\ (g) + heat$$

As you can see, this reaction produces heat, and it turns out that the amount of heat produced can be pretty substantial. The heat can even cause the hydrogen gas (produced in the reaction) to ignite, thus reacting according to this equation:

$$H_2\ (g) + O_2\ (g) \rightarrow H_2O\ (g) + heat$$

The heat produced by both of these reactions can even cause any currently unreacted sodium metal to also ignite and burn according to the equation:

$$Na\ (s) + O_2\ (g) \rightarrow Na_2O_2\ (s)$$

From these reactions, it is probably clear that the reaction of sodium with water can produce quite a lot of heat!

What is **electron paramagnetic resonance spectroscopy**?

Electron paramagnetic resonance (EPR) spectroscopy, also known as electron spin resonance (ESR) spectroscopy, is a method used to probe unpaired electrons in a molecule. This method is fairly similar to nuclear magnetic resonance (NMR) (see "The Modern Chemical Lab"), but it involves exciting electronic spin states instead of nuclear ones. One downside to this technique, relative to NMR, is that the majority of molecules do not contain unpaired electrons and thus cannot be studied using EPR. On the other hand, the lack of interfering signals from most solvents and other molecules can very often be an advantage for the same reason.

How is **inorganic chemistry relevant** to **biological chemistry**?

Because metals are important to many biological processes! In the active sites of enzymes (See "Biochemistry"), where the important chemical reactions take place, metal centers are frequently crucial to the catalytic bond-forming and bond-breaking events. Metals are also important for maintaining gradients in ion concentrations, allowing muscles to move, and a variety of other biological processes. Bioinorganic chemistry is thus a huge field, and these two topics are often taught in conjunction with one another.

Why do **metals** make for such effective **active sites** in **enzyme catalysis**?

Metals are crucial components in enzymatic catalysis for largely the same reasons that organometallic complexes make good catalysts. These include the fact that many metal centers can complex to a variety of substrates, undergo facile changes in oxidation state, and provide good electron donors or acceptors. These characteristics can, and often do, work together to accomplish some remarkable chemical feats.

What are **MRI contrast agents** and what are they used for in medicine?

Magnetic resonance imaging (MRI) (see "The Modern Chemical Lab") contrast agents are used to make tissues and other structures inside the body easier to view in an MRI scan. Many of these contrast agents are based on gadolinium with various ligands attached. The contrast agents are injected or ingested into the body, and they function by changing the relaxation time of the atoms observed in an MRI scan. The overall result is that these complexes improve the ability of MRI to see what is going on in your body.

How are **transition metals** important for **nitrogen fixation**?

Nitrogen fixation is the term for a process that converts diatomic nitrogen gas (N_2) in the atmosphere into ammonia (NH_3). This is a very important biological process because it converts N_2, which is very unreactive, into a form that can more readily be incorporated into amino acids and other molecules. Transition metals (such as vanadium or molybdenum) are found in the active sites of the enzymes that carry out this important reaction.

Why is calcium important for strong bones?

Your bones need calcium to stay strong because they are made of a calcium-based substance called calcium hydroxylapatite, with the chemical formula $Ca_5(PO_4)_3(OH)$. It is recommended that people between the ages of 18–50 consume about 1000 milligrams of calcium per day, while older people should consume 1200–1500 milligrams of calcium each day.

How are **metals** important in maintaining **osmotic balance** in **cells**?

The metals sodium and potassium (really their ions, Na^+ and K^+) are responsible for moderating ion and concentration gradients across cellular membranes. These ions can be selectively allowed to pass, or be pumped, across cellular membranes.

How are **transition metals** important in **photosynthesis**?

Chlorophyll, the green pigment that plays a crucial role in photosynthesis, contains an atom of magnesium at the center of its porphyrin ring (a porphyrin ring is an organic ring molecule found in many biochemical systems). The Mg atom plays a crucial role in the absorption of light from the Sun, the energy from which is then channeled through the molecule and is eventually put to work by the plant's cells.

What **metals** are **naturally present** in **biological systems**?

Sodium, magnesium, vanadium, chromium, manganese, iron, copper, nickel, cobalt, zinc, molybdenum, and tungsten are all naturally present in varying quantities in biological systems.

What **metals** are used as **probe agents** in **biological systems**?

Yttrium, technetium, gold, silver, platinum, mercury, and gadolinium are used as probe agents in biological systems.

What **metal** is at work in **alcohol dehydrogenase** in your **liver**?

The active site of alcohol dehydrogenase contains a zinc center that is responsible for catalyzing the reaction of ethanol to acetaldehyde.

ANALYTICAL CHEMISTRY

A LITTLE MATH

What is the **difference** between a **qualitative** and a **quantitative observation**?

Quantitative observations are, as the name suggests, observations that attempt to quantify how much of something there is. Chemists are often interested in measuring concentrations of chemical species, the amount of energy released in a reaction, the lengths of chemical bonds, and numerical values associated with a multitude of other chemical properties. A few examples of measurements an analytical chemist might make are determining how much fat is in a cookie (about 10 grams), how much CO_2 is in the air (about 390 ppm), or how much lead is in your drinking water (hopefully very little!).

Qualitative observations are observations that tell about a general property of an object, but don't quantify exactly how much of something there is. Typically this means the observation isn't giving a number of something, but rather it is describing a quality of an object. Examples might be that a piece of candy tastes sweet, that the sky is blue, or that a circle is round.

Some observations can be tough to classify as strictly qualitative or strictly quantitative. For example, a judge at the Olympic games may assign a score of 9.5

Chemists use quantitative measurements to discover, for example, the concentration of lead or other substances in our drinking water.

to a gymnastics performance, indicating that she really liked it. Quantitatively this allows for a comparison among the different competitors, but it still feels somewhat qualitative since it reflects a judge's opinion and could be influenced by factors like how carefully the judge observed the performance, what elements of a gymnastics routine the judge values the most, or even just by the judge's mood.

What is an **analyte**?

An analyte is simply the chemical species whose properties we are trying to measure in an analytical chemistry experiment. Most often, analytical chemists are interested in measuring the concentration of an analyte present in a sample. The measurement of an analyte concentration is a quantitative observation.

What are **interferences**?

Interferences are substances that interfere with the measurement of the analyte we are interested in. They give a similar response to our analyte in whatever analytical measurement we are making. Typically analytical chemists want to remove any interferences from a sample prior to making a measurement, or, alternatively, choose the method of analysis carefully, so as to avoid any signals caused by interferences.

What is **experimental error**?

Experimental error describes the uncertainty or variation in a measurement that is repeated several times. This variation could come from many sources, such as limitations in the quality of the device being used to make the measurement, the ability of the experimenter to accurately repeat the experiment or measurement, or something intrinsic about the quantity being measured.

What are some **examples** of **experimental error**?

Say, for example, we use a stopwatch to record the lap times for a racecar driving several laps around a track. Our measurements of the lap times will be affected by the quality of the stopwatch, our ability to accurately operate the stopwatch during each lap, as well as variations in the ability of the racecar driver and car to maneuver around the track.

What is **standard deviation**?

One common way of reporting the error in a measurement is to report the standard deviation for the set of measurements. Standard deviation is calculated from the formula:

$$std.\,dev = \sqrt{\tfrac{1}{N}\sum_{i=1}^{N}\left(x_i - \mu\right)^2}$$

where N is the number of trials, x_i is the value of the measurement of the ith trial, and μ is the arithmetic mean value of the N measurements.

> ## What is the smallest quantity of lead that can be detected in drinking water?
>
> Lead is something we definitely don't want in our drinking water. The maximum contaminant level goal set by the U.S. Environmental Protection Agency (as of 2011) is zero, indicating that there is no amount of "safe" exposure to lead in drinking water. The EPA's maximum contaminant level, which is based on more practical considerations, is 15 parts per billion. So how little can we detect? Thankfully, way less than that. Even in the 1990s it was possible to detect lead concentrations as low as 0.1 parts per billion. So analytical chemists are on top of keeping us safe from lead in our drinking water (as long as someone is checking on it regularly).

What is the **signal-to-noise ratio** in a measurement?

The signal-to-noise ratio is a comparison between the strength of a measured signal to the background variation naturally present in the experiment. There are different ways of reporting the signal-to-noise ratio, but one commonly used measure is to compare the mean value of a measurement to the standard deviation of the background region.

For example, we might want to measure the intensity of a laser beam using a detector. The signal would be the measured intensity when we shine the laser beam on the detector. The noise would be measured as the standard deviation of the intensity measured due to the ambient light when the laser was not shined onto the detector.

Such a definition is commonly used when looking at weak signals in imaging or microscopy. Choosing an appropriate way of reporting the signal-to-noise ratio can depend on details of the situation in question, such as whether the data can take on both positive and negative values, or whether the data can only have positive values.

What is a **detection limit**?

The detection limit for a measurement is the lowest amount of signal that can be distinguished from the background level of noise. The detection limit is commonly reported with a confidence level describing the percent confidence with which the measured signal at the detection limit can be stated to be above the noise intrinsic to the experiment.

What is **random error**?

Random errors are variations that increase the standard deviation of a measurement randomly about its mean value. Since random errors should be distributed in a random fashion, both above and below the true value of the quantity being measured, taking repeated measurements can reduce their effect on the measured experimental value. Random errors could be caused by things like noise in an electrical circuit or random fluctuations in the temperature or humidity of the laboratory environment. Random errors are characterized by their Gaussian distribution of values about the mean measured value.

What is a **Gaussian distribution**?

A Gaussian distribution is a type of probability distribution that characterizes the distribution of a randomly distributed variable about a single mean (average) value. This is the most commonly encountered type of distribution in statistical analysis, so it is good to be at least a little familiar with it. Provided one knows that the probability distribution for a variable is Gaussian, the distribution for that variable can be completely characterized by its mean and standard deviation, which has the shape of a bell curve:

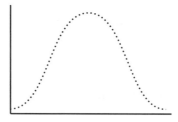

What is the **area under the curve** for a **normalized Gaussian distribution**?

As should be the case for any normalized probability distribution, the area under the curve of a Gaussian distribution totals to 1. This is necessary for the probability of a set of events to have a physically meaningful value since the sum of the probabilities of all possible outcomes must total to 1 (or 100%). Probability distributions that satisfy this property are said to be normalized, and the Gaussian distribution is actually also commonly referred to as the normal distribution. With any Gaussian distribution, 68% of the possible outcomes will lie within 1 standard deviation of the mean value, 95% within two standard deviations, and 99.7% within three standard deviations.

What is **systematic error**?

Systematic errors are unlike random errors in that they change the observed value from the true value in a consistent direction. One example would be if you read a thermometer from the wrong angle; you might consistently read the level of the liquid inside to be a few degrees higher than it actually reads. This would systematically bias the observed value toward a higher temperature than the true value. Another example would be if a scale were calibrated incorrectly, such that it read 5 grams even when no weight was placed on it. This could result in an observed mass that is systematically higher than the actual mass.

MEASURE TWICE

What is **accuracy**?

Accuracy is used to refer to how close a measured value is to the true value of the quantity. This is pretty simple to understand. For example, if the true temperature in the room is 20 °C and a thermometer reads 20 °C, the measurement is pretty accurate. If the thermometer read 0 °C, the measurement is not very accurate.

What is **precision**?

Precision describes how reproducible a measurement is, regardless of whether the observed value matches the true value. Measurements can be very precise even if they are not perfectly accurate. For example, if you weighed yourself on a scale that was offset by 5 kg, you would always be off by the same amount, but you would not be measuring your real weight. So despite not being accurate, the value you obtain could still be very precise. This will generally be true in situations involving a systematic error that is simply a constant offset from the actual value.

What are some **common units of concentration** for a **species in solution**?

The most common unit of concentration in solution–phase chemistry is molarity, or the number of moles of a species present per liter of solution. Recall that the number of moles of a species is equal to the number of molecules of that species divided by Avogadro's number (see "History of Chemistry" for information on Avogadro's number).

In analytical chemistry, it can often be more convenient to discuss concentrations in terms of other units. This is particularly true when the species being studied is present in very small concentrations. Some other commonly used units are parts per million (ppm), parts per billion, or parts per trillion. These units refer to the mass fraction of the analyte in question to total mass of the sample. One part per million means 1 millionth (10^{-6}) of a gram of the species in question, or analyte, per gram of the sample present. Similarly, one part per billion means 1 billionth (10^{-9}) of a gram of analyte per gram of sample, and one part per trillion means 1 trillionth (10^{-12}) of a gram of analyte per gram of sample. It is also possible to use each of these defini-

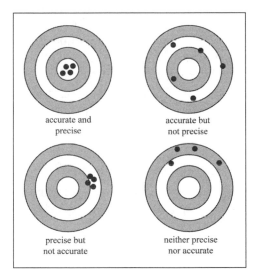

accurate and precise

accurate but not precise

precise but not accurate

neither precise nor accurate

There is a difference between being precise and being accurate, as the bullet holes in these targets demonstrate.

tions based on a per unit volume basis, rather than a per unit mass basis (though this is more common in the gas phase). For example, 1 ppm by volume of an analyte means the analyte occupies 1 millionth (10^{-6}) of the total volume of a sample.

A meniscus is the curve that forms on the surface of a liquid in a container. It can make measuring liquids by eye somewhat of a challenge.

What is a **meniscus**, and **how** should it **be read**?

A meniscus is the curve in the surface of a liquid contained within a (usually narrow) container, such as a graduated cylinder. Take a look at the graphic to the right.

The numerical value we read for the volume of liquid contained in this cylinder will depend on what part of the meniscus we choose to look at. Graduated cylinders and other containers are calibrated such that we should get the most accurate reading by looking at the center of the meniscus from eye level. If we don't look at the meniscus from eye level, something called parallax error will occur that will distort our perception of level of the meniscus.

What is a **titration**?

A titration is a laboratory technique used to determine the concentration of a species in a solution. This is achieved by adding a second species of a known concentration (the titrant) that will react with the species whose concentration is to be determined (the analyte). There must be some indication of when the titration is complete, or, in other words, there must be some indication of when the unknown species has reacted completely with the titrant. A common example is the titration of a base by an acid. In these cases, a small amount of a pH-sensitive colored indicator can be added, and the end of the titration can be determined by observing a sudden large change in pH via a change in the color of the solution.

What is an **aqueous solution**?

An aqueous solution is any solution in which water is the solvent. This term is used pretty frequently in chemistry, so it's good to be familiar with it.

What is a **"standard state"**?

A standard state is a set of reference conditions with respect to which properties can be described/calculated under other conditions. For example, the properties at standard-state conditions of a gas may be defined by its properties at 293.15 K and 1 bar of pressure. Knowledge of the properties of a gas under these standard conditions can be used to calculate the properties under other conditions.

What is a **chemical indicator**?

As we mentioned, for a titration to be successful there must be some indication of when it has finished. Chemical indicators are a common way of determining whether a titration is complete. In acid–base titrations, chemical indicators are typically small molecules whose colors change depending on the pH of the solution (or, in other words, on their protonation state). In other cases, the indicator may change color based on whether it is coordinated to another species in a solution. It is also possible to use a device like a pH-sensitive electrode to determine when the titration has ended; this doesn't rely on you (or another person) being able to notice a visible change in the color of the solution. Such a device can be more accurate, since it doesn't rely on a qualitative interpretation of a color change.

How do you use **pH paper**?

Another way of measuring the pH of a solution is with pH paper. This is a paper strip that contains a chemical indicator that changes color depending on the pH of a solution, and it can be capable of determining a very wide range of pH values. To properly use pH paper, it should not be dipped directly into the solution, but rather a drop of the solution should be placed on the pH paper. It can then be held next to a gradient color scale to obtain a pH value. Since this relies on visible color changes, it isn't the most accurate method of determining pH, but it can be a useful tool in the lab for getting a quick estimate of the pH of a solution.

How long does it take a **chemical reaction** to reach **completion**?

The length of time it takes a reaction to run to completion can range from a fraction of a second to thousands of years. It all depends on how large the energy barrier is for reactants to convert to products. If it's a simple acid-base reaction, like the addition of an HCl solution to water, the reaction will proceed almost instantly. Other reactions can be slower, such as the rusting of a car door. What you first notice as a small spot of rust could take years to spread over the rest of the door. Other reactions can be even slower yet. For example, the slowest biologically relevant reaction known in humans is estimated to take about one trillion years in the absence of an enzyme to catalyze it. That's longer than scientists believe the universe has even existed! Fortunately, enzymes have evolved that allow this reaction to take place in only a few milliseconds.

What are some different **types of titrations**?

We already considered the titration of a base by an acid. Certainly acids can also be titrated by bases, but there are also many other types of titrations, so let's look at just two more.

101

Complexometric titrations—This type of titration involves using a titrant that forms a complex with the analyte. In this case, the species of unknown concentration is often a metal ion, and the indicator may be a dye molecule that forms a weak complex with the metal ions. The indicator-ion complex is bound more weakly than the titrant-ion complex, such that the titrant can displace the indicator from the metal ions. Thus when the titrant has bound to a significant fraction of analyte molecules, the indicator molecules are displaced and the solution will change color.

Redox titrations—A redox titration is a titration that involves the use of a reducing agent or an oxidizing agent to determine the concentration of the unknown species. In this case, the indicator can again be sensitive to the presence of excess titrant, giving a color change when the titrant becomes present in excess.

In a titration, two solutions—one of a known concentration and one of an unknown concentration—are slowly combined until a reaction occurs. When that happens, the concentration of the second solution can be determined.

What is a **buffer solution**?

A buffer solution is a solution that has a tendency to resist changes in pH, even when an acid or base is added to the solution. Buffer solutions consist of a mixture of either a weak acid and its conjugate base, or a weak base and its conjugate acid, dissolved in water. When an acid is added, the base present in the solution tends to bind to the H^+ from the acid, serving to "buffer" the change in pH and preventing the concentration of H_3O^+ from changing very much. Similarly, when a base is added, the acid in the solution tends to neutralize the base, again serving to buffer the change in pH by preventing the concentration of H_3O^+ from changing significantly.

How are **buffer solutions** important in the **human body**?

One very important buffer solution is human blood! An equilibrium between carbonic acid (H_2CO_3) and its conjugate base bicarbonate (HCO_3^-) helps blood to maintain a relatively constant pH of around 7.4. The carbonic acid buffer system is created by carbon dioxide (CO_2) dissolved in blood; carbon dioxide reacts with water (H_2O) to form carbonic acid. Since the amount of carbon dioxide in the blood depends on the rate at which you breathe, your blood pH is influenced by your breathing rate. Your body can actually tell when your blood pH gets too high or too low and adjusts your respiratory rate accordingly!

What is **precipitation**?

In chemistry, precipitation means something different from when the weather forecaster on the news predicts rain. A precipitate is a solid substance that does not dissolve in a solution, and precipitation describes the process of the precipitate forming. This generally takes place when one or more soluble reactants in a solution undergo a reaction to form one or more insoluble products. The result is that a solid substance will form in a solution, either depositing on the bottom of the container or sometimes floating around in the solution. Depending on the circumstances, precipitation can be a desirable or undesirable result.

What is **gravimetric analysis**?

Gravimetric analysis refers to any method for quantifying the amount of an analyte based on mass. For analytes dissolved in a solution, this may require first reacting the analyte with another species to cause it to precipitate out of the solution. The solid product can then be filtered and weighed, and the mass of the analyte initially dissolved in the solution can be determined. Other situations may require different methods of preparing the analyte in an appropriate form for weighing, but the general idea is always that the composition of the weighed species is known, so it's possible to figure out the mass of the analyte originally present.

What is the **difference** between **mass** and **weight**?

Mass and weight are similar terms, but they have slightly different meanings. Weight tells us how much of something there is by telling us how much gravity is pulling down on it. Mass, on the other hand, just tells us how much of something there is, independent of the force of gravity on the object. One might reasonably ask, why does this distinction even matter? The reason it matters is that if we went to the Moon, for example, the weight of an object would change (because the force of gravity on the Moon is different from that on Earth), but the mass would be the same, since the object is made of the same amount of stuff; we just need to make sure everyone is talking about the same thing.

We should point out that people commonly use scales to measure either weight or mass, but when we're using them to measure mass, the measurement relies on the fact that the scale has been calibrated based on the strength of gravity on the Earth, so it "knows" how to figure out the mass of an object from measuring its weight.

What is **spectrophotometry**?

Spectrophotometry is a technique that involves passing light through a sample to measure the fraction of light transmitted or reflected by the sample. This can be useful for determining the concentration of an analyte in a solution or for determining the identity of an analyte present.

A law known as Beer's law (see also "Physical and Theoretical Chemistry") is particularly important in spectrophotometry. Beer's law tells us that the concentration of a species is directly proportional to its absorbance in a solution. This means that if we know the absorbance of a solution of our analyte at a known concentration, we can determine the concentration of that analyte in any other solution.

What is an **electrolyte**?

An electrolyte is a substance that contains ions (charged particles) and is capable of conducting electricity. Most often this is a solution containing dissolved ions, like an aqueous solution of sodium chloride (which contains Na^+ and Cl^- ions). The presence of ions in the solution is what makes it capable of conducting electricity. It's for this reason that it can be dangerous to get the electrical outlet wet when drying your hair in the bathroom. Pure water itself doesn't conduct electricity, but in practice, there are almost always some ions present, making tap water an electrolyte.

Another place electrolytes are important is in living things. Fluids in the human body, and in other organisms, contain free ions and are thus electrolytes. Gradients in ion concentrations in cellular fluids are often important for regulating cellular processes and bodily functions such as muscle and nerve activity.

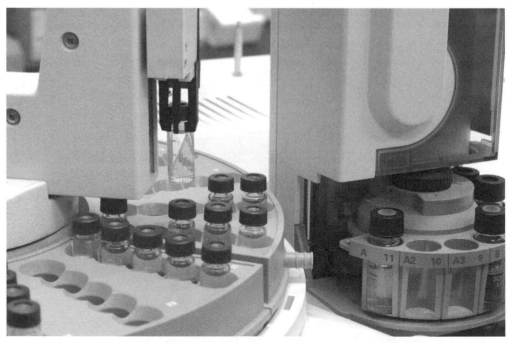

An autosampler selects a sample to be separated by gas chromatography, one of several types of chromatography techniques.

What is **chromatography**?

Chromatography describes a series of techniques used for separating the components of a mixture of chemicals. The most common type of chromatography is column chromatography. This involves dissolving the mixture to be separated into a solution and flowing it over a column of stationary material that interacts to a different extent with each component of the mixture. As the liquid flows over the stationary phase, the components that interact most strongly will be held behind longer while those that interact weakly will be eluted from the column first. The liquid is collected into a series of small vials or test tubes, which will, hopefully, each only contain one component of the mixture. The solvent can then be evaporated from each vial to recover the individual components of the mixture. A wide variety of chromatographic techniques exist, and each generally follows the same basic principles as the column chromatography described here.

What do the terms **miscible** and **immiscible** mean?

Miscible and immiscible are terms that refer to whether two liquids that come into contact will mix to form a single uniform solution. Two miscible liquids are able to mix together to form a uniform solution, while two immiscible liquids will separate into two layers. If the liquids are immiscible, the liquid with the lower density will stay on top and the liquid with the higher density will be on the bottom.

What is a **liquid–liquid extraction**?

Extractions are techniques used to separate the components of a mixture based on their solubilities in different media. A liquid–liquid extraction specifically uses two immiscible liquids to separate the components of a mixture. Two liquids are placed in the same container and shaken vigorously to allow the dissolved components to equilibrate between the two liquid phases. One liquid can then be drained from the container, and the solvents can then be evaporated to recover the separated components of the mixture.

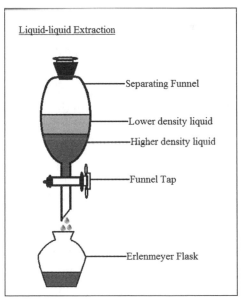

What is **aqua regia** and what is it used for?

Aqua regia translates to "royal water" in Latin, and it consists of a mixture of concentrated nitric and hydrochloric acids, typically in a 1 to 3 ratio, respectively. Since these are both strong acids, aqua regia can be quite dangerous and must be handled

An apparatus for liquid–liquid extractions.

with extreme caution. This mixture is capable of dissolving several types of metals and has found many applications in chemistry, including the refinement of very high purity gold!

What is an **oxidation/reduction reaction**?

Oxidation and reduction reactions are chemical processes that involve the transfer of an electron between two species. Oxidation is the process of taking an electron away from a species, so if a molecule loses an electron, it has been oxidized. Reduction is the process of giving an additional electron to a species, so if a molecule gains an electron, it has been reduced. Oxidation and reduction reactions are particularly important in the field of electrochemistry.

ELECTROCHEMISTRY

What is **electrochemistry**?

Electrochemistry is a branch of chemistry that deals with electrons transferring between an interface and a molecule in an electrolyte solution. The interface is typically a conducting metallic material. Electrochemistry is a rich field that has led to the development of batteries, a widely used chemical separation technique called electrophoresis, a process for plating metals known as electroplating, and an immense body of knowledge surrounding oxidation–reduction chemistry, among other achievements.

What is an **electrochemical cell**?

One common example of a redox reaction in electrochemistry involves the transfer of electrons from zinc (Zn) metal to copper (Cu) ions in an electrochemical cell.

Zinc (Zn) gives up electrons more readily than copper (Cu), so electrons spontaneously transfer from the Zn metal to the Cu metal, depositing Zn^{2+} ions into the solution and causing Cu metal to come out of the solution and onto the solid. This process releases energy, which can be used to drive external processes (such as powering a lightbulb, for example).

In this example of an electrochemical cell, a piece of zinc (Zn) and a piece of copper (Cu) are placed in solutions of zinc sulfate and copper sulfate, respectively, connected by a salt bridge of potassium nitrate. Electrons are given up by the zinc anode and transferred to the copper cathode, generating an electrical current.

How can **metals be plated** onto a surface?

Metals can be plated onto surfaces using a technique from electrochemistry called electroplating. Positively charged metal ions dissolved in a solution are attracted to a negatively charged electrode. This electrode reduces (gives electrons to) the positively charged metal ions, making them into a neutral metal species. The neutral metal is no longer soluble, so it forms a coating on the surface of the electrode.

How does a **battery work**?

Batteries operate based on chemical redox reactions that are set up to cause the spontaneous flow of electrons through the object you want to power.

Rechargable batteries work because the chemical reactions that occur inside them can be more easily reversed than in non-rechargable batteries.

Why can some **batteries be recharged** and others cannot?

Batteries produce electricity from a chemical reaction. So to recharge a battery, we would have to be able to reverse the chemical reaction by applying an electric current. In truth most batteries use chemical reactions that could be reversed, but there are efficiency considerations that come into play. Rechargeable batteries make use of reactions that can be reversed easily, many times, without too much degradation of the battery materials. One-time-use batteries typically use reactions that could only be reversed and reused a couple of times before they would not work very well anymore.

What is the **reduction potential** of an **electrochemical cell**?

The reduction potential of a species describes its tendency to accept electrons, or, in other words, to be reduced. A full electrochemical cell typically involves two separate half-reactions, each of which is associated with its own reduction potential. The overall reduction potential for an electrochemical cell is the difference of the reduction potentials for the two half-reactions.

What is the **Nernst equation**?

The Nernst equation relates the reduction potential of an electrochemical cell to a standard-state reduction potential, along with the temperature, reaction quotient, and number of electrons transferred during the reaction in question. It was first developed by Walther Nernst, who won the Nobel Prize in Chemistry in 1920 for his influential work in physical chemistry.

What is **calorimetry** and what is it used for?

Calorimetry involves measuring the amount of energy released during a chemical reaction in the form of heat. This is accomplished using a device called a calorimeter. A calorimeter is a container inside which a chemical reaction can be carried out while thermally insulated from the surroundings. This can even be something as simple as a sealed Styrofoam® coffee cup with a thermometer inserted through the lid (though more sophisticated devices certainly do exist). The temperature change inside the calorimeter can be monitored to determine the amount of energy given off during the chemical reaction, which can then be used to determine the enthalpy change associated with the reaction.

What is **flame ionization detection**?

Flame ionization detection (FID) is a method of detecting organic analytes in gas-phase chromatography experiments. A very hot flame is used to burn the analytes emerging from the chromatography instrument, which produces positively charged ions. These ions are then attracted to a negatively charged electrode, which detects them by generating an amount of current proportional to the number of positively charged carbon atoms reaching the electrode. FID can be particularly useful due to the fact that it detects organic analytes without any interference from a wide variety of other gases, which can be present in the sample or in the carrier gas used in the gas chromatograph. There are some downsides to FID, one of which is that the sample is destroyed as it is analyzed.

ANALYTICAL CHEMISTRY IN OUR LIVES

How is **analytical chemistry** used in **medicine**?

Analytical chemistry techniques are routinely used in many areas of medicine. When your blood is drawn, for example, the doctors use techniques borrowed from analytical chemistry to determine levels of several analytes in your body, including cholesterol, vitamins, glucose, and white blood cells. Analysis of tissue samples and other biological fluids is also based on analytical techniques. Other examples of medically related applications include tests for illegal drugs and steroids, glucose sensors used by diabetic patients, or to check for toxic substances in your body if you are exposed to hazardous chemicals.

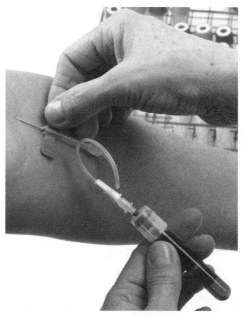

Analytical chemistry is commonly used in medicine, such as when your blood is drawn for tests to see how much glucose, cholesterol, or other chemicals are present.

What **analytical techniques** are used in **quality control** for **pharmaceuticals**?

Some of the most commonly used analytical techniques in pharmaceutical quality

The numbers on gas pump indicate octane ratings, which relate to the amount of compression a fuel can undergo before combusting.

control include infrared spectroscopy, UV/Visible spectrophotometry, melting point determination, reactions based on color changes, and various types of chromatography, including thin-layer chromatography, high-performance liquid chromatography, and gas chromatography.

What do the **different octane ratings** mean at the gas pump?

The octane number provides a measure of the amount of compression a fuel can undergo before it combusts. The name "octane rating" comes from the fact that isooctane is used as the standard against which a fuel's compression characteristics are measured. A fuel with an octane rating of 89 means that it has the same ability to withstand compression as a mixture of 89% isooctane and 11% heptane. In reality, gasoline contains many other components. Fuels with higher octane ratings are typically used in higher performance automobiles. Since fuels may be able to withstand compression better than isooctane, it's entirely possible to achieve octane ratings of over 100.

How is the presence of **illegal narcotics identified**?

Analytical chemists have developed simple chemical tests that can identify the presence of illegal drugs within seconds. Law enforcement officers routinely use these when they are unsure whether someone is in possession of an illegal narcotic. A common field test for cocaine, for example, uses a cobalt thiocyanate complex that turns blue in the presence of cocaine.

109

How can the **presence of blood** be detected by **forensic experts**?

An ultraviolet light source can be used to detect a blood stain that has been cleaned up. It's also possible to use a chemical substance like luminol or phenolphthalein to detect the presence of hemoglobin, revealing its presence even after the visible stain has been washed away.

How does a **breathalyzer work**?

When a person blows into a breathalyzer device, the air they exhale enters a chamber containing two glass vials. One is a mixture of aqueous sulfuric acid, silver nitrate (a catalyst), and potassium dichromate. The air is bubbled through this mixture and the sulfuric acid and potassium dichromate react with the alcohol, producing chromium sulfate and other products. The chromium ion of chromium sulfate is green in the solution, whereas the initially present dichromate ion was red/orange in color. The second vial contains the same reactants but doesn't interact with the air sample, thus serving as a standard for comparison. The color change in the sample exposed to the air is monitored by a photocell (light sensor) system that is then used to determine the alcohol content of the air sample.

Nutrition Facts

Serving Size 1 Cup (53g/1.9 oz.)
Servings Per Container About 9

Amount Per Serving

Calories 188   Calories from Fat 25

	% Daily Value*
Total Fat 3g	5%
Saturated Fat 0g	0%
Trans Fat 0g	
Cholesterol 0mg	0%
Sodium 80mg	3%
Potassium 300mg	9%
Total Carbohydrate 37g	12%
Dietary Fiber 8g	32%
Soluble Fiber	
Insoluble Fibe	4%
Sugars 13g	
Protein 9g	14%
Vitamin A 0%	C 0%
Calcium 4%	10%
Phosphorus 10%	0%

* Percent Daily Values are based o
Your daily values may be higher o

Calories in a serving of food can be calculated by the amount of protein, fat, and carbohydrates in the food item.

How are the quantities on **food nutrition labels** determined?

The calorie content of food can be determined using a simple calorimetry experiment. The food just needs to be placed inside a calorimeter and burned to determine its calorie content. These days, however, the calorie content of a food can also be determined by simply using standard values of calories in each gram of protein (4 calories/gram), fat (9 calories/gram), and carbohydrates (4 calories/gram).

The trick is just to accurately determine the quantities of protein, fat, and carbohydrates in the food, and that's where analytical chemistry comes in particularly handy. The basic idea is that the fat, protein, or carbohydrates need to be extracted from the food. This can be done by choosing an appropriate solvent in which the fats/proteins/carbohydrates are soluble, and then using spectrophotometry, or one of a variety of other methods, to determine the content of each species in a solution.

Is a **calorie in food** the **same** as the unit of energy **used by scientists**?

The calories listed on food labels are actually kilocalories when converted to standard units for energy. So if the package for your snack says it contains 100 calories, that's really 100,000 calories in the units of energy used by scientists.

What is **smog**?

The word smog was originally derived from a combination of the words smoke and fog, and today it describes air pollution that, once in the atmosphere, reacts with sunlight to form additional pollutants. The main sources of this pollution are commonly motor vehicles, factories, power plants, and burning of fuels. The main chemical components of smog are ozone, nitric oxide, nitric dioxide, and volatile organic compounds. Smog is present in pretty much every large city around the world. It can be hazardous to human health and has been blamed for health problems and even deaths.

How does a **pregnancy test work**?

A pregnancy test works by detecting a hormone called human chorionic gonadotropin (hCG) that is produced in the placenta after fertilization. The hCG hormone will be present in the blood and urine about a week after conception, and it's detected by an indicator molecule on the pregnancy test that changes color upon binding to hCG. The tests you can buy in the store are actually pretty similar to those used at a doctor's office; both are based on hCG binding to a chemical indicator that changes color. The main difference is that the doctor's office has technicians who are more familiar with using the tests, so there's less chance of a mistake with using or interpreting the test.

BIOCHEMISTRY

MOLECULES OF LIFE

What is **biochemistry**?

Biochemistry is the field focused on elucidating and explaining the complex chemical processes that take place in biological systems. It's a diverse field, drawing on aspects from virtually every other subfield of chemistry to explain the molecular processes that drive living things. Often, biochemists study complex reaction sequences and catalytic processes involving molecules much larger than those routinely encountered in other subfields of chemistry. A few topics commonly encountered in biochemistry include studying how our cells obtain energy, understanding how our genetic material (DNA) dictates who we are, and explaining how our body regulates and stores nutrients from the foods we eat.

In what **professions** is knowledge of **biochemistry important**?

Biochemistry is important for people interested in careers in human medicine, veterinary medicine, dentistry, pharmacy, and food science, as well as research in almost any physical or biological science and many subfields of engineering as well. While this list isn't all-inclusive, you can see that biochemistry is used by people working in many different areas!

Where are **biomolecules found**?

Biomolecules are molecules that are specifically found inside of living things and have some function related to life. This includes molecules found in plants, animals, insects, bacteria, or even viruses (which are often considered not to be "alive" in a technical sense). Biomolecules span a wide range of sizes, some weighing only about 50 atomic mass units (amu), while others weigh millions of amu (see "Atoms and Molecules" to review the definition of amu). They are in our hair, skin, tissues, organs, and just about everywhere else in our bodies too.

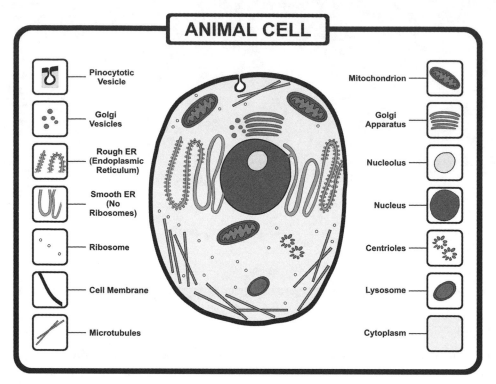

ANIMAL CELL

- Pinocytotic Vesicle
- Golgi Vesicles
- Rough ER (Endoplasmic Reticulum)
- Smooth ER (No Ribosomes)
- Ribosome
- Cell Membrane
- Microtubules
- Mitochondrion
- Golgi Apparatus
- Nucleolus
- Nucleus
- Centrioles
- Lysosome
- Cytoplasm

Cells contain a variety of organelles, which carry out specific functions for the cell to survive. Lipid bilayers separate the organelles from the rest of the cell.

What is a **cell**?

Cells are basic structures that make up all living things. The smallest organisms, like bacteria, can be made up of only a single cell; these are known as single-celled organisms. Other organisms, like animals or humans, are made up of many, many cells. The size of a typical cell is very small, usually on the scale of 10^{-4} or 10^{-5} meters. This is small enough that we can't see it with our eyes, but only under a microscope where cells are readily visible.

How **many cells** are there in **your body**?

An adult human is made up of approximately fifty trillion cells!

What is a **cellular organelle**?

An organelle is a compartmentalized subunit with a living cell that carries out a specific function. Organelles are separated from the rest of the cell by a lipid bilayer membrane that allows it to maintain different concentrations of solutes from the rest of the cell.

What are the main **classes of biomolecules**?

The four main classes of biomolecules are proteins, carbohydrates, lipids, and nucleic acids. Proteins are some of the most diverse biomolecules, and they can perform functions such as protecting the body from pathogens or acting as catalysts for chemical reactions. Carbohydrates can be a source of energy when your body needs it. Lipids include fat molecules and are used for energy storage as well as to form the membranes that surround each cell. Nucleic acids are what make up our genetic material and store the information that determines a lot about who we are. We'll go into more details about each of these types of biomolecules in the upcoming pages.

What is the basic **structure** of an **amino acid**?

The basic structure of an amino acid is shown below.

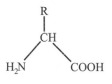

Each amino acid has an amine (NH_2) group, a central carbon atom known as the alpha carbon, a side chain (denoted R), and a carboxylic acid (COOH) group. The amine groups and carboxylic acid groups of amino acids can be joined in a chain to form polymers of amino acids, known as peptides. Two amino acids form a dipeptide, three form a tripeptide, and chains of four or more amino acids are commonly referred to as polypeptides. Long polypeptides fold into specific conformations, which is what makes up a protein.

How **many amino acids** are found **in humans**?

There are twenty amino acids commonly found in humans. They are classified into four groups based on the identity of their side chain. The four groups are polar, nonpolar, acidic, and basic. These twenty amino acids are what make up all of the proteins and enzymes in your body. The interactions between the individual amino acids determine the overall conformation of the protein, though the relationship between the sequence of amino acids and the structure adopted by the protein is very difficult to predict.

Is there a **reason all amino acids** in **biological systems** share the same **chirality**?

While the answer to this question is uncertain, there have been some interesting developments in this area. It has been shown that life (as it exists today anyway) cannot arise from a racemic mixture of amino acids because the presence of chiral centers in many biomolecules is crucial for biological function. The self-replication of DNA relies on the presence of chiral centers and without a shared chirality, the error rate in DNA replication would cause severe problems for many longer-lived plants and animals. One hypothesis for the origin of chirality is that molecules from outer space reached Earth with a net chirality already present. Another is that the net chirality in amino acids was established on Earth in a very short period of time. It's a pretty interesting thing to think about, and this topic is a subject of ongoing research and debate.

What is a **protein**?

Proteins consist of one or more long chains of amino acids that fold up into a specific arrangement, and each is presumably capable of some kind of biological function (though we don't know for sure that every protein has a biological function—the function of each protein is one thing many biochemists are still working hard to figure out). Some proteins act to protect an organism from invading pathogens, some help to carry messages between different areas of the body, some are responsible for movement of muscles, and some provide structural support for cells. Many proteins also fall into a class called enzymes, which is a name for proteins that catalyze specific chemical reactions in biological systems.

What is a **peptide bond**?

A peptide bond, or peptide linkage, is the bond that links the individual amino acids in a peptide. Have a look at the figure of the dipeptide below.

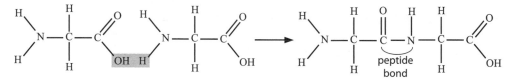

The formation of a peptide linkage is carried out by a structure called the ribosome, which is responsible for generating peptides and/or proteins based on the sequence of nucleotides present in an RNA molecule. We'll get to the details of this process a little later.

What is an **enzyme**?

An enzyme is a protein whose function is to serve as a catalyst for a chemical reaction within a biological system. Some functions that enzymes perform are the synthesis of proteins and other biomolecules, digestion of fats and other molecules, and sometimes they are even put to use in industrial applications outside of their natural biological environment.

What is an **active site**?

Each enzyme has an area called the active site in which catalytic activity is carried out. The active site is shaped in a way that reduces the energy barrier to carrying out the chemical reaction it is meant to perform. It's important to note that, while the reactant(s) must bind at the active site initially, they must also be released after the reaction has been carried out.

What is the **native state** of a **protein**?

Since there are so many atoms in a protein, there exist a very large number of possible conformations into which a protein can fold. The native state is the conformation in which the protein exists in its natural biological environment. This is most often the lowest energy conformation the protein can adopt. A protein in its native state is able to carry out its biological function, while a protein that is unable to reach its native state often will be unable to do so. If a protein is taken out of a cell, for example, the pH or other factors related to its environment may cause it to adopt a conformation other than its native state.

How many **different conformations** can a **protein adopt**?

A lot. For a typical protein consisting of 100 amino acids, there are approximately 3^{198} possible conformational states. This leads to something called Levinthal's paradox, which has to do with how a protein goes about sampling each of these possible states. If a protein just randomly went through each of 3^{198} possible states, it would, on average, take longer than the age of the universe to get through all of them!

Fortunately proteins don't sample each of the possible conformations randomly. Instead, the gradient (or slope) of the energy along the folding pathway, along with the formation of local interactions between the amino acids, guides the protein toward states of lower energy. This allows it to avoid sampling most of the unimportant conformations as it folds toward its native state.

What is a **sugar**?

Sugars are part of a class of biomolecules called carbohydrates that are made up of carbon, hydrogen, and oxygen atoms. The simplest sugars are called monosaccharides, but, like amino acids, they can polymerize to form disaccharides and oligosaccharides. Sugars can usually exist in either a ring form or an open chain form, as shown in the picture below. The sugar in this picture is called glucose, which is the sugar most commonly used for energy in the human body.

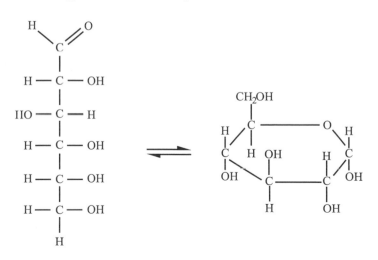

How are **sugars stored** in the **body**?

The body converts excess sugars into a branched polymer called glycogen for storage. In plants, sugars are instead stored as a polymer called starch.

How are **glucose levels regulated**?

In humans the amount of glucose in the blood at any given time is regulated very carefully. If your blood glucose level is too high, your body releases a chemical called insulin into the bloodstream, signaling that it's time to convert some of the extra glucose floating around into glycogen. Similarly, if your blood glucose level gets too low, your body releases a chemical called glucagon into the bloodstream, telling your body to start breaking down that glycogen to release more glucose into the blood.

What is a **glycosidic linkage**?

A glycosidic linkage is the chemical bond that bonds a carbohydrate to another carbohydrate molecule (or to another species). Glycosidic linkages are what hold together all of the individual glucose monomers that make up glycogen or starch. The figure below shows a glycosidic linkage.

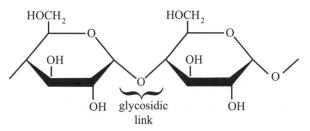

The enzymes that catalyze the breakage glycosidic linkages are called glycoside hydrolases. These are necessary to allow for glucose to be released from storage. The enzymes responsible for forming the linkages are called glycosyltransferases.

What is **saponification**?

Saponification is a type of reaction in which the ester functional groups of triglycerides are hydrolyzed under basic conditions. The term can also refer to the hydrolysis of any ester as well.

GENETICS

What is a **nucleotide**?

Nucleotides are one of the basic units for another class of biomolecules: the ones that join together to form your DNA and RNA (the long names for these are deoxyribonucleic acid and ribonucleic acid). DNA and RNA are the macromolecules that store the genetic information in your body. A nucleotide consists of three parts: a nitrogenous base, a sugar, and a phosphate group. The identity of the nitrogenous base determines which "letter" in the genetic code is associated with the nucleotide, while the type of sugar determines whether the nucleotide is a ribonucleotide or a deoxyribonucleotide (which essentially just tells us whether it's going to be a part of RNA or a part of DNA).

What is a **nucleic acid**?

Nucleic acids are the polymers formed by a series of linked nucleotides. DNA and RNA are the two most important types of nucleic acids for life as these are the molecules that carry and transmit our genetic information. In humans, DNA is found in the nucleus of our cells, and a complete copy of the genetic information is actually contained within each individual cell. That means we each have about fifty trillion copies of our genetic information in our body! RNA is made by enzymes called RNA polymerases, which are responsible for transcribing the information from DNA into a strand of RNA, where it can move about the cell to carry out its function.

How many **types of nucleotides** are found **in humans**?

Nucleotides are named for the nitrogenous (nitrogen-containing) base incorporated into the nucleotide, and there are only five types of bases used: adenine, guanine, cytosine (all found in both DNA and RNA), and thymine (DNA only) or uracil (RNA only). The corresponding nucleosides are called deoxyadenosine (DNA)/adenosine (RNA), deoxyguanosine (DNA)/guanosine (RNA), deoxycytidine (DNA)/cytidine (RNA), deoxythymidine (DNA), and uridine (RNA). All of the genetic information in our body is stored in our DNA using sequences of only these four nucleotides. Imagine trying to describe how to make and operate an entire machine using only four letters—that's just what our genetic information does.

What is the **structure of DNA**?

DNA is well known for its double helical structure. Two scientists named James Watson and Francis Crick first discovered this structure in 1953 using data from X-ray diffraction experiments.

The two strands of the helix are held together by hydrogen bonding as well as stacking interactions between the aromatic rings of the nitrogenous bases. These strong interactions make the double helical structure very stable. It's also true, however, that the double helical structure has to be temporarily pulled apart when it's time to transcribe the DNA into RNA. This is accomplished by an enzyme called helicase.

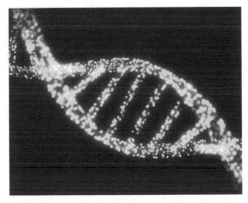

DNA looks like a ladder that has been twisted. The rungs of the ladder consist of pairs of nucleotides, and the way they are arranged creates a genetic code of instructions that tell our body how to grow and function.

119

What is a **phosphodiester bond**?

Phosphodiester bonds are what make up the backbone of DNA and RNA molecules. These bonds provide the linkage that holds together the individual nucleotides or nucleosides in DNA and RNA. Enzymes called polymerases catalyze their formation (DNA polymerases for DNA and RNA polymerases for RNA). Take a look at the picture below to see the chemical structure of a phosphodiester bond.

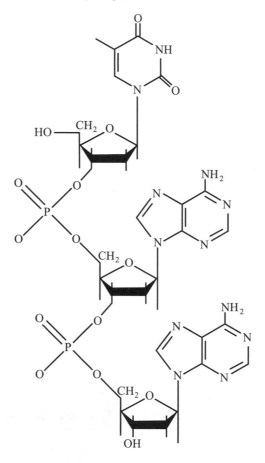

How is **information encoded** in **RNA** and **DNA**?

The sequence of nucleotides in DNA in can be read only by first separating the DNA double helix (performed by the enzyme helicase). The information is "read" by RNA polymerase, which then transcribes a corresponding piece of RNA. If the purpose of the RNA is to encode a protein, it can then be read by a structure called a ribosome, which creates proteins according to the specific ordering of the nucleotides in the RNA strand.

The ordering of the nucleotides in DNA (and thus the nucleosides in RNA) is extremely important in determining the function of a protein. Nucleotides are read in groups of three. These groups of three are called codons, and each codon tells the ribosome to incorporate a specific amino acid into the peptide/protein it creates. There are also specific codons that tell the ribosome to start or stop creating a peptide. Any mistake in the copying of DNA or RNA can potentially lead to major problems, so it's important that the cellular machinery that carries out these processes is extremely accurate.

What can go wrong if **errors** are present **in DNA**?

If errors do occur during DNA replication, they can have serious biological and physiological effects. Many diseases are believed to be associated with errors in genetic sequences, including cystic fibrosis, sickle cell anemia, hemophilia, Huntington's disease, Tay-Sachs disease, and a number of other genetic disorders. Other diseases or disorders may be linked to your genetics in a more complicated manner, such that you are more or less likely to suffer from them than the rest of the population. Some such diseases/disorders are cancers, mental/mood disorders, asthma, heart disease, and diabetes. Fortunately we have several DNA repair mechanisms that are constantly on the lookout for damaged DNA.

What are **genes**?

Genes are the basic unit of heredity. A gene is a sequence of nucleotides containing information on specific traits that shape the characteristics of an organism. Each person actually has two copies of each gene; one comes from each parent. There's not really a "typical size" for a gene, since the number of base pairs in a gene can range from just a couple hundred base pairs to millions of base pairs. Most of the genetic information in each of our genes is actually the same from person to person. In fact, differences in less than 1% of the DNA in our genes accounts for all of the physical differences between people.

What are **chromosomes**?

Chromosomes are a bundle of DNA and proteins packaged together. They are how your DNA is stored in your cells when it isn't being "read" by any enzymes. Each chromosome contains a large number of genes. Humans each have forty-six chromosomes, all of which are present in each cell in the body. Chromosomes need to be "unpacked" by enzymes before their DNA can be read by RNA polymerase or other enzymes.

What is **gene therapy**?

Gene therapy is a method of medical treatment that tries to correct defective or disease-causing genes. There are a few ways that this may be attempted, all of which are centered around getting a normal, functional copy of the gene inserted into the genome. Some methods involve repairing the mutated gene, while others involve just adding a working copy of the gene into a nonspecific location in the genome.

What is **genetic engineering**?

Genetic engineering is the process of changing the genetic makeup of cells or organisms to produce versions with specific traits or to produce new organisms altogether.

121

What is the **difference** between **eukaryotes** and **prokaryotes**?

Prokaryotes are organisms whose cells do not have a nucleus, while eukaryotes are organisms whose cells do have a nucleus.

Prokaryotes are most often, but not exclusively, unicellular (single-celled) organisms. In addition to lacking a nucleus, they also lack other separate membrane-bound organelles. All of the proteins, DNA, and other molecules in a prokaryotic cell float around within the cell membrane but are not separated into different compartments.

Eukaryotes do typically have separate membrane-bound organelles within their cells. They may be single-celled or multicellular organisms. Every large organism (animals, plants, fungi) is eukaryotic, and many small and single-celled organisms are in this category as well.

METABOLISM AND OTHER BIOCHEMICAL REACTIONS

What is a **fatty acid**, and what is the difference between **saturated and unsaturated fats**?

Fatty acids are long, organic molecules that contain a carboxylic acid functional group at one end (see the picture below) and a long, nonpolar tail at the other end. They are an important source of energy in the body because they can be metabolized to generate ATP, which the body uses as fuel. When you're looking at the nutrition information on your food, the concept of saturated versus unsaturated fats can be understood by looking at the structure of a fatty acid. Saturated fats are those that contain only single bonds connecting each carbon atom in the chain. Recall that double and triple bonds are called units of unsaturation. Unsaturated fats are any fats that do have units of unsaturation, or, in other words, that do have double bonds between some of the carbons in the chain. All fats are either saturated (top illustration) or unsaturated (bottom illustration).

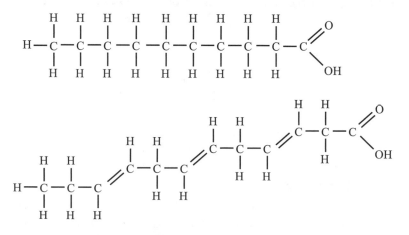

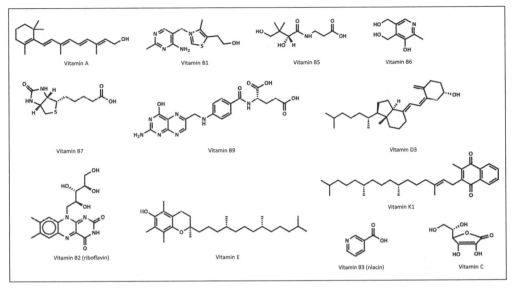

The structures of a sampling of vitamin molecules.

You might ask then, what is this "trans fat" everyone talks about in foods? First we should point out that most of the double bonds in unsaturated fats are in the cis conformation (where both carbon substituents are on the same side) in nature. Trans fats are fats that have had hydrogen artificially added, which can result in unsaturated fats with a trans conformation about the double bond. There is evidence that these trans fats can be more harmful to your health than other fats.

What is a **lipid**?

Lipids are a broad class of nonpolar or amphiphilic molecules including fatty acids, vitamins, sterols, and waxes, among others. An amphiphilic molecule is one that has both hydrophilic and hydrophobic groups, meaning that some parts of the molecule interact favorably with polar groups while other parts do not.

What do the **chemical structures** of **vitamin molecules** look like?

Above are the structures of some common vitamin molecules. They typically have a molecular weight in the neighborhood of 100–1500 g/mol.

What is a **lipid bilayer**?

Lipids are important in cells as they form bilayers that protect the cells and hold them together. In a lipid bilayer, the nonpolar tails gather on the inside of the bilayer, allowing the polar ends of each lipid molecule to interact favorably with the polar, aqueous environment of the cell and its surroundings. Within a bilayer, the lipid molecules can still slide around and can even flip from one side of the bilayer to the other. Lipid bilay-

In a lipid bilayer (the layers in the middle that look like rows of clothespins), the nonpolar tails gather on the inside of the bilayer, allowing the polar ends of each lipid molecule to interact favorably with the polar, aqueous environment of the cell and its surroundings.

ers generally don't allow molecules or ions to pass readily, allowing for the buildup of concentration gradients between the cell and its surroundings. Lipid bilayers are also found elsewhere; for example, they can also form separate compartments within cells.

What is a **biochemical pathway**?

A biochemical pathway is a cycle of chemical processes that mutually interact to effect some purpose important for biological function.

What is a **chemical-signaling molecule**?

Chemical signaling molecules are small molecules that act to carry a message in a biological system. A cell can secrete signaling molecules and allow them to diffuse through the bloodstream. Signaling molecules can also be stuck to the surfaces of cells. It would be impossible to give a comprehensive description, so we'll just mention a couple of examples of chemical signaling. Apoptosis, which is the intentional death of a cell, is a process that involves chemical signaling. An external signal reaches the cell, which sets off a series of reactions inside the cell, ultimately leading to its death. Calcium ions are another species often involved in cell signaling as their concentration affects the activity of many proteins and is also important for telling cells when to reproduce. Hormones are another type of chemical signals. They travel through our bodies, controlling growth of muscles and tissue, our reproductive systems, and our metabolism, among other things. These examples are just a small sample of the huge number of processes controlled by chemical signaling.

What is the **Krebs cycle**?

The Krebs cycle (also known as the citric acid cycle or the tricarboxylic acid cycle) is a biochemical process through which organisms can generate energy (in the form of ATP, or adenosine triphosphate) by oxidizing acetate that comes from other biomolecules (sugars, fats, and proteins). Since it uses oxygen, it is called an *aerobic* process. The Krebs cycle also generates other molecules, such as NADH (nicotinamide adenine dinucleotide), that are used in other biochemical processes. The names citric acid cycle and tricarboxylic acid cycle come from the fact that citric acid is used up and then regenerated in the reactions in the cycle. The name Krebs cycle is named after Hans Adolf Krebs, who was one of its discoverers.

What is **binding affinity**?

Binding affinity is used to characterize how strongly two molecules interact. Typically this will be a ligand and a receptor site, perhaps in a protein. Another common example would be a drug molecule binding to a receptor site somewhere in the brain. In simple cases where one drug molecule (D) binds to one receptor molecule (R) to form a complex (DR), the binding affinity can be described by the equilibrium constant:

$$K_{eq} = [DR]/([D][R])$$

The binding affinity can be looked at as a measure of how strongly a molecule binds to its receptor site. If a drug has higher affinity for its binding site, less of the drug will be required to achieve a response. Typically scientists designing pharmaceuticals would like a drug to have a high binding affinity for its receptor so that it can effect a strong response.

How is O_2 **transported** through the **body**?

Oxygen enters our body through our lungs in the air we breathe. O_2 makes up about 21% of the total volume of air on Earth. From the lungs, O_2 diffuses into the bloodstream, where it binds to a molecule called hemoglobin in our red blood cells. The binding affinity of O_2 to hemoglobin is pH-dependent such that oxygen can readily be picked up by red blood cells near the lungs and then released into tissues and other areas of the body that need it.

What is **cooperativity**?

Cooperativity describes how the binding affinity at one site of a protein affects the binding affinity at another site. In hemoglobin, for example, there are four sites at which oxygen molecules can bind. After the first oxygen molecule binds, conformational changes take place in the rest of the protein that increase the binding affinity at the other sites.

125

The second oxygen is then able to bind even more easily, and the third and fourth even more easily yet. This gives rise to a sigmoidal-curved shape in a plot of the fraction of hemoglobin bound by O_2 versus the partial pressure of O_2 (see illustration).

Similar binding curves result in other examples of cooperativity, though the example of hemoglobin is probably the most commonly discussed example of the phenomenon.

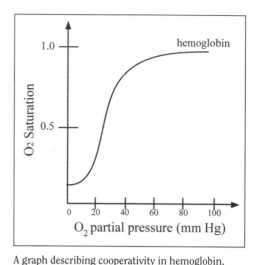

A graph describing cooperativity in hemoglobin.

What is **ATP**?

ATP, or adenosine triphosphate, is a molecule used as a source of energy in the body. The energy of an ATP molecule is stored in its chemical bonds, and it is released through hydrolysis of phosphate groups to form ADP (adenosine diphosphate). The ATP in our bodies is continuously recycled, and on average, each ATP molecule in our body will be used and regenerated over one thousand times each day.

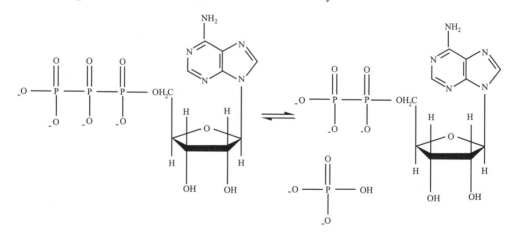

How does **blood clotting** take place?

Platelets in the blood send a message to constrict blood vessels in the area surrounding a wound. Platelets gather to block the flow of blood coming out. At the same time, a chemical messenger called prothrombin activates an enzyme called thrombin, which then produces a species called fibrin. Fibrin forms threads to block the wound even better.

What is a **kinase**?

Kinases are a class of enzyme that is responsible for transferring phosphate groups (see image below) from donor molecules, such as ATP, to substrates during a reaction called phosphorylation. These are typically named for their substrate: for example, a tyrosine kinase catalyzes the transfer of a phosphate group to a tyrosine residue of a protein. Kinases are part of a larger group of enzymes called phosphotransferases, all of which carry out chemistry involving phosphate groups, shown below.

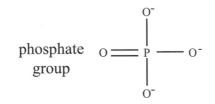

How do your **muscles work**?

Muscles allow you to exercise and move heavy objects and are necessary for basic activities like breathing, pumping blood, and pretty much anything else you do. On a molecular level, muscles work based on the binding, movement, and rebinding of molecules called actin and myosin. This process involves the hydrolysis of ATP to generate energy.

For your muscles to move, myosin first attaches to actin, forming a bridge. At this point, ADP and a phosphate group are attached to the myosin. The myosin bends (this is what actually controls movement), releasing the ADP and phosphate. Then a new ATP molecule binds again, and then the myosin releases the actin. The ATP is then hydrolyzed, putting the myosin back into its original position, at which point the cycle can begin again.

What is **rigor mortis**?

Rigor mortis is the stiffening of the muscles that occurs shortly after death. Since very little or no ATP is present, muscles are left contracted, stiff, and unable to relax.

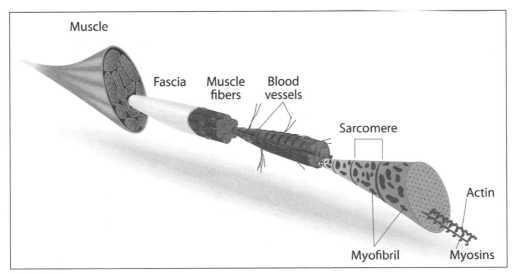

The molecules actin and myosin bind, move, and rebind within your muscle fibers in order to provide movement.

> **What determines whether a cell is a skin cell, blood cell, etc.?**
>
> The process that new cells undergo to become a specific type of cell is called cellular differentiation. Interestingly, different types of cells don't contain different genetic information, but rather they just express different parts of their genetic information. Signal molecules tell the cell which parts of its DNA to express, which in turn controls what types of proteins and other molecules are present in the cell, and ultimately these factors determine its function.

How does **photosynthesis work**?

Photosynthesis is the process carried out by plants to harvest energy from sunlight. The most important molecule in photosynthesis is called chlorophyll; it's what collects sunlight and gives plants their green color. Carbon dioxide (CO_2) is taken in through cells called stomata, and the plants also draw water up through the roots and into the leaves. The reaction catalyzes when chlorophyll absorbs light and makes ATP along with another molecular energy source called NADPH (nicotine adenine dinucleotide phosphate). In the process, CO_2 is used up and water molecules are split, releasing O_2 gas, which other organisms (like humans and animals) can then breathe.

How do living things **store fat**?

Fat is stored in a type of tissue called adipose tissue. This is made up of cells called adipocytes that store lipid molecules to be used for longer-term energy storage.

What **causes addiction** to a substance?

Drugs that cause addiction change the ability of receptors in your brain to cause you to feel pleasure. There are a few ways this can happen. Depressants often work by increasing the affinity of a receptor for a small molecule called GABA (gamma-aminobutryic acid). Stimulants make you feel happy or better than you would otherwise. They can do this in a few different ways, but two common ones are to either cause more dopamine to be released or to prevent the reabsorption of dopamine so that it stays around and keeps you happy for longer. Narcotics act in a similar manner to stimulants in that they mimic the molecules that make you happy normally.

What gives some people such a **high alcohol tolerance**?

From the biochemical point of view, the amount of the enzyme alcohol dehydrogenase in a person's body determines how rapidly their body can process alcohol. People with more alcohol dehydrogenase can convert ethanol (which is what gets you drunk) to acetaldehyde faster. A person's body size also is important, though. Bigger people have more mass to spread the alcohol around so their blood concentration of ethanol doesn't rise as fast as that of smaller people.

BRAINS!

What are **neurons**?

Neurons are cells responsible for transmitting information between different locations in the body. Their functions are diverse and include telling muscles to move, passing information from your sensory organs to your brain, making you experience pleasure, and processing information in your brain.

What is **neuroscience**?

Neuroscience refers to science that studies the way the brain and nervous system works.

What are the **different parts** of the **brain**?

The brain is divided into several areas that specialize in different functions. Some of the major sections of the brain are the frontal lobe, the parietal lobe, the temporal lobe, the pons, the medulla oblongata, the occipital lobe, and the cerebellum. The spinal cord connects the brain to the body via millions of nerves that transmit information between the brain and the rest of the body.

What is the **cerebrum**?

The cerebrum is the top part of the brain, and it contains your memories, knowledge, and languages, and it manages your senses as well. The cerebrum is the part of your brain that controls your movements and emotions and it's where your thoughts take place.

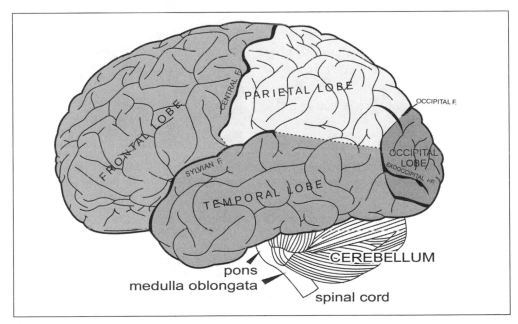

The major parts of the human brain.

What happens in the **frontal lobe**?

The frontal lobe is responsible for decision-making processes and problem-solving as well as managing your active memory.

What is the **parietal lobe** for?

Your parietal lobe controls your speech, visual perception, pain and touch perception, spatial orientation (like knowing what direction is up), and other cognitive processes. As you can probably tell by now, there is some degree of overlap between the broad types of functions carried out by different parts of the brain.

And the **temporal lobe**?

The temporal lobe is involved in hearing, memory, and language skills.

What about the **occipital lobe**?

The occipital lobe is in charge of visual perception; it's responsible for processing the information received by our eyes.

What happens in the **cerebellum**?

The cerebellum is important for controlling your sense of balance, motion, and your motor skills in general.

A part of my brain is called the **pons**?

Yes. The pons is located in the brainstem, and its function is to move information between the cerebellum and the cerebrum.

Medulla oblongata? Now you're just making words up.

Nope. The medulla oblongata is a bunch of neurons packed together in the back of the brain. These neurons control bodily functions such as your heartbeat, breathing patterns, the constriction or dilation of blood vessels, sneezing, and swallowing.

What is a **prion disease**?

Prions are misfolded forms of proteins that behave in an infectious manner. They do so by causing other proteins they en-

Many fruits contain ascorbic acid (vitamin C), which is important to our health. It aids in preventing inflammations, boosts our immune systems, and helps us in the digestion of foods, among other benefits.

counter to misfold as well. A prion disease is a disease that results from this misfolding. Some examples of prion diseases include Mad Cow Disease, Scrapie, and Creutzfeldt-Jakob disease. In mammals, all known prion diseases are actually caused by the same prion protein, known as PrP; the abbreviation actually stands for "prion protein."

What are some **examples** of **naturally occurring molecules** that affect our **biochemistry**?

Ascorbic acid, also known as vitamin C, plays an important role in keeping us healthy. It was isolated for the first time in 1932 from citrus fruits (oranges, lemons, limes, grapefruits), and can be synthesized naturally from D-glucose via two distinct biological pathways. The hydroxyl groups allow it to be soluble in water (i.e., in biologically relevant environments), and it is used as a coenzyme in the synthesis of collagen.

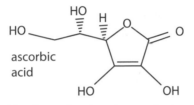

Benzaldehyde is found in almonds, cherries, apricots, and peach pits. It is often used as an artificial oil of almond (perhaps not surprisingly) in making perfumes, dyes, and food flavorings. Researchers are also continuing to look into its utility as a pesticide and anticancer agent. Benzaldehyde can be readily synthesized in the laboratory using toluene as a precursor.

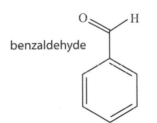

We all know about caffeine. For coffee drinkers, this is probably one of our favorite chemicals! In its pure form, caffeine is just another white crystalline powder, though most people have probably never seen it that way. It's more commonly found in the cocoa plant, coffee beans, and tea leaves, where people have been consuming it for thousands of years. Historically people have consumed caffeine to increase their heart rate, body temperature, mental alertness, and attention span. Today, people still use caffeine for most of the same reasons. Caffeine can be extracted using chemical solvents from sources like tea leaves and coffee beans to be used in other caffeinated products like soft drinks. Not to scare you, but it's worth mentioning that caffeine is a potentially addictive substance that stimulates the central nervous system, and if you consume too much at a given time, you may suffer from headaches, irritation, and insomnia.

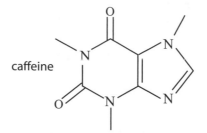

caffeine

Calanolide A comes from a tree found in the rainforests of Malaysia. It was originally tested as an anticancer agent, but was found to be unsuccessful in this regard. However, Calanolide A has been found to be very effective in fighting the HIV–1 virus (which causes AIDS). Due to the rarity of this chemical, a synthesis was designed soon after its utility was realized. This drug acts by preventing the transcription of the viral RNA into DNA in a cell, which serves to prevent the HIV virus from replicating. Fortunately, it does this with relatively mild, and temporary, side effects.

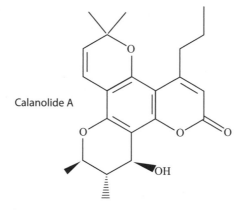

Calanolide A

Dopamine is synthesized in our bodies from an amino acid precursor. It is an important neurotransmitter that balances our feelings of happiness. Imbalances or deficiencies in dopamine production or its regulation have been linked to a number of diseases, including Parkinson's disease, schizophrenia, and Tourette's syndrome. While dopamine was recognized for its role as a neurotransmitter in the 1950s, it took decades of research before its role was more completely understood. The work that led to its connection to the disorders we mentioned, and to an understanding of its exact function, was awarded a Nobel Prize in Physiology or Medicine in the year 2000. An understanding of the role of dopamine in physiology has been extremely important in understanding several neurological diseases.

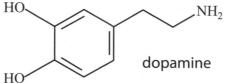

dopamine

Ethanol is a molecule you are probably familiar with: it's the alcohol found in alcoholic beverages that makes you feel intoxicated, and humans have consumed it for hundreds of years. Additionally, it is useful as a solvent, an antiseptic, a sedative, and a component in perfumes, lacquers, cosmetics, aerosols, antifreeze, and mouthwash. Ethanol can be produced from a raw feedstock, like corn or grain, by fermenting it in the presence of microbes, which can readily digest sugars and produce ethanol as a byproduct.

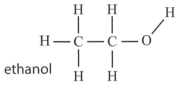
ethanol

Oxytocin is a hormone produced naturally in females' posterior pituitary glands, which are located at the back of the brain. It is responsible for causing lactation and uterine contractions in pregnant women. Oxytocin is also used to induce labor when a pregnancy does not begin on time naturally.

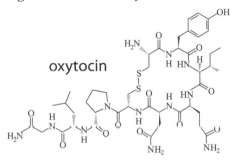
oxytocin

Pyridoxal phosphate is more commonly known as vitamin B6. It helps your nerves and brain to function properly and to maintain the right chemical balances in your body. It is also necessary for the enzymatic reaction that frees up glucose (the sugar monomer) from glycogen (the sugar storage polymer). Vitamin B6 can be obtained from many types of foods including meats, grains, nuts, vegetables, and bananas.

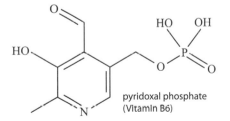
pyridoxal phosphate
(Vitamin B6)

Quinine is a molecule that has been used to treat malaria and nighttime leg cramps as well. It was first discovered in South America, in the bark of a type of tree called the Cinchona tree, by Spanish explorers, who used it as a medicine. The high demand for quinine eventually led to Cinchona trees becoming very hard to find, but thankfully a synthetic method for its production was eventually developed.

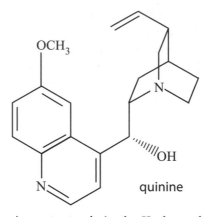

quinine

Succinic acid plays an important role in the Krebs cycle and can deliver electrons to the electron transport chain. Succinic acid has been found in a very wide range of plant and animal tissues, though its purification challenged chemists for a long time before it was finally successfully purified from tissues. Today succinic acid can be produced readily in the laboratory, and one method even can use corn as a feedstock. In addition to its role in the Krebs cycle, succinic acid has been used as an intermediate in dyes, perfumes, paints, inks, and fibers.

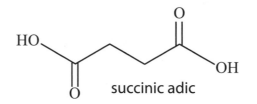

succinic adic

PHYSICAL AND THEORETICAL CHEMISTRY

ENERGY IS EVERYTHING

What is **physical chemistry**?

Physical chemistry is a branch of chemistry primarily concerned with developing a better understanding of the fundamental principles that govern chemical processes. It is an empirical science, meaning that it is based on experimental observations, though it is probably the most closely linked experimental branch of chemistry to developing new theories in chemistry. As the name implies, physical chemistry is intrinsically concerned with topics in physics that are also relevant to the study of chemistry.

What is **energy**?

In chemistry, energy serves as the "currency" for making or breaking chemical bonds and moving molecules (or matter) from one place to another.

What is **potential energy**?

Potential energy describes all of the nonkinetic energy associated with an object. This energy can be the energy stored in chemical bonds, in a compressed spring, or in a variety of other ways. Another example is gravitational potential energy, like that associated with a ball sitting at the top of a hill. Since there are many types of potential energy, there isn't a single equation that describes them all. Since the value we assign to potential energy is always inherently described relative to some choice of a reference value, we can only actually measure *changes* in potential energy in a meaningful way. A closed system can exchange potential energy for kinetic and vice versa, but the total energy must always remain constant. This is stated in the First Law of Thermodynamics, which we'll get to soon.

What is **kinetic energy**?

Kinetic energy is the type of energy associated with the movement of an object. Faster-moving objects have more kinetic energy, and the kinetic energy of an object is related to its mass, m, and velocity, v, by the equation:

$$E = \tfrac{1}{2}mv^2$$

This tells us that, for example, if we have two objects of equal mass and one is moving twice as fast as the other, the faster-moving object will have four times the energy of the slower object.

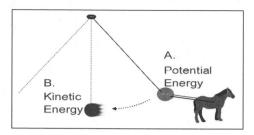

In this illustration, a horse pulls a pendulum into a position where it is about to be released to swing freely. Before it is released, the weight at the end of the pendulum has potential energy (A), and when the pendulum is in full swing, it has kinetic energy (B).

Can **molecules** have any **arbitrary energy**?

No, molecules actually have a discrete number of possible energy levels. Another way to say this is that their energies are quantized. To illustrate why this is so different from situations we're used to in everyday life, consider what happens when you're throwing a baseball. You could throw it at any speed between 0 meters per second (m/s) and however fast you are capable of throwing it. In molecules, though, only a discrete set of energies are possible. It's as if you could throw the baseball either 2 m/s or 40 m/s, but not 20 m/s or any other speed in between. There aren't many situations we encounter in everyday life in which the possible energies associated with objects come in a discrete set of values.

What **types of energy levels** exist in **molecules**?

There are three main types of energy levels that physical chemists are concerned with. These are electronic, vibrational, and rotational energies. Changes in electronic energy levels occur when an electron undergoes a transition from one molecular orbital to another. Vibrational energy levels are associated with vibrations of chemical bonds in the molecule, and rotational energy levels involve the molecule rotating in space. As you could probably guess, atoms don't have vibrational energy levels since there aren't chemical bonds present in single atoms. Physical chemists can often learn about the structure and reactivity of molecules by studying the transitions between these energy levels.

What is **quantum mechanics**?

Quantum mechanics is a branch of physics that is needed to provide an accurate description of objects with very small mass, such as electrons. It does so using an approach that describes matter as being both similar to a particle and similar to a wave. The description of a particle in quantum mechanics is contained in something called a wave function, which can be related to the probability of an object being in any of its possible states.

One interesting and counterintuitive thing we learn from quantum mechanics is that, for particles with very small mass, the position, velocity, and other quantities defining the state of the particle cannot all be precisely specified at the same time. The wave-like description offered by quantum mechanics is needed to explain why molecules have discrete energy levels, along with many other experimental observations from physical chemistry that are inconsistent with classical mechanics.

What is **work**?

Work is a name used in physics for processes that transfer energy between objects by the application of a force over a distance. Take throwing a baseball, for example. As your arm moves, your hand applies a force in the direction the baseball is moving. During the time the ball is in your hand and moving forward, you are doing work on the ball. The total amount of work done can be calculated as the product of the force you apply times the distance over which you applied force. Once it leaves your hand, you're no longer applying a force, so you aren't doing work on it anymore.

What is **heat**?

Heat is responsible for all types of energy transfer other than those that fall under the definition of work. One easy example to think about is ice cream melting on a hot day. Because the ice cream is at a lower temperature than its surroundings, heat flows from the surroundings to the ice cream, causing its temperature to increase, and eventually it melts. There are lots of other examples of heat flow too; it's a pretty big category since it covers all types of energy transfer that aren't defined as work.

What is the **zeroth law of thermodynamics**?

The zeroth law of thermodynamics states that any two systems, call them A and B, that are each in thermal equilibrium with a third system, call it C, must be in thermal equilibrium with each other. Thermal equilibrium implies that the systems must have the same temperature, and therefore systems A and B must have the same temperature. This might seem totally obvious, but it is what puts our use of thermometers to compare the temperatures of different objects on a sound footing. If object C is our thermometer, we can use it to compare the temperatures of other objects.

What is the **first law of thermodynamics**?

The first law of thermodynamics is a statement of the conservation of energy, which tells us that energy can be transferred from one form to another but never created or destroyed. It tells us how energy is related to work and heat, and it is typically stated through the equation:

$$\delta E = \delta Q - \delta W$$

This equation tells us that the differential change in energy (δE) is equal to the heat (δQ) that flows into the system minus the work (δW) the system does on its surroundings. **137**

What is **entropy**?

Entropy is a measure of the total number of microstates in a system. There have been two widely used definitions of entropy, which were suggested by Ludwig Boltzmann and J. Willard Gibbs. We'll just look at the one specified by Boltzmann, since it's a little more straightforward to understand. The equation for Boltzmann's definition of entropy is:

$$S = k_b \ln(\ )$$

In this equation, k_b is Boltzmann's constant, and   is the number of microstates accessible to a system.

To get an idea of how entropy works, consider the example of rolling one or more six-sided die. On the first roll, there are six possible outcomes, so the entropy associated with rolling one die is $k_b \ln(6)$. If we roll two dice, there are $6^2 = 36$ possible outcomes, and the associated entropy is $k_b \ln(36)$. For three, it's $6^3 = 216$ possible outcomes, and the associated entropy is $k_b \ln(216)$. As you can see, the number of outcomes for statistically independent events grows very rapidly (exponentially) with the size of our system, which is also true for molecules. By taking the logarithm of the number of outcomes, we make the entropy scale linearly with system size. While the number of possible outcomes/configurations grows exponentially with system size, the entropy grows linearly, which means that if we double the system size we double the entropy. This property makes entropy fall into a category of variables known as extensive variables, which just means that they scale with the size of a system in this simple way.

What is the **second law of thermodynamics**?

There are several different statements of the second law of thermodynamics, but they are all centered on the idea of identifying what things can happen spontaneously in nature.

One formulation of the second law states that, for a closed system, the entropy of the system can only increase or remain the same. In plain language, this says that nature favors having more accessible configurations or arrangements. It's why, for example, a drop of ink in water tends to spread out to fill its accessible volume but won't spontaneously reform a drop of ink.

Another statement of the second law says that heat cannot spontaneously flow from a colder body to a warmer body. Work would have to be done for this to happen, which would imply the process was not spontaneous.

What is the **third law of thermodynamics**?

The most common statement of the third law of thermodynamics is that the entropy of a perfectly crystalline system approaches zero as the temperature of the system approaches zero. (Recall from "Macroscopic Properties: The World We See" that a perfect crystal is a regularly ordered lattice of atoms that exist in a repeating pattern in three dimensions with no defects or irregularities in the lattice.) This is equivalent to saying that a perfectly crystalline system has only one accessible state as the temperature approaches zero. In truth, this isn't always strictly true since there can be multiple low-energy states that all have the similar energy, but let's ignore this for now.

What **effects cause deviations** from the **ideal gas law**?

Deviations from the ideal gas law occur due to intermolecular forces between the gas particles as well as the fact that gas particles do actually occupy volume. There is a modified version of the ideal gas law (see "Atoms and Molecules"), called the van der Waals equation of state, that uses constants specific to each molecule or atom to adjust for these factors. Deviations from ideal gas law behavior become more important at relatively high pressures and/or low temperatures.

What is the **average kinetic energy** of a molecule?

The average kinetic energy of a molecule is closely related to the temperature of its environment, and this determines how fast the molecule is moving. On average, the speed of any molecule is close to 300 m/s, which is equivalent to covering the distance of a few football fields every second! The other important thing to realize, though, is that collisions with other molecules are constantly causing changes in direction, which slows the overall distance in a given direction that a molecule travels.

What is an **ideal solution**?

An ideal solution describes a solution of dilute solute particles that do not interact with one another. It is very similar to an ideal gas, except that instead of empty space occupying the space between the gas particles, a weakly interacting solvent occupies the space between solute particles.

What is **osmosis**?

Osmosis is the movement of solvent molecules in a solution to establish an equal concentration of solute throughout the solution. Solvent molecules move from areas of low-solute concentration to areas of high-solute concentration, which tends to remove any gradient in solute concentration.

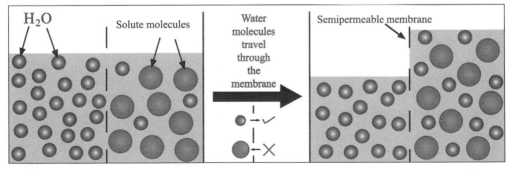

In osmosis, solvent molecules move from areas of low solute concentration to areas of high solute concentration through a permeable membrane to equalize the solute concentration on either side.

What is an **isothermal process**?

An isothermal process is a process in which the temperature remains constant throughout the process.

What is an **isobaric process**?

An isobaric process is a process that is carried out at a constant pressure.

What is an **adiabatic process**?

An adiabatic process is a process in which no heat is exchanged with the surroundings.

What is an **isochoric process**?

An isochoric process is a process that takes place at a constant volume.

KINETICS

What is the **transition state** for a reaction?

The transition state of a chemical reaction is the highest energy structure through which the reactant molecule(s) must pass to complete the reaction. Since this is the highest energy point along the path of the reaction, this configuration is the most "difficult" to reach along the reaction path and thus the energy barrier to reach the transition state limits how quickly the reaction can proceed.

What is the **rate constant** for a reaction?

The rate constant for a chemical reaction is a quantity that describes how rapidly the reaction proceeds. Rate constants can have different units, depending on how many molecules are involved in the reaction. Consider a simple reaction where a single molecule of a species A becomes a molecule of species B. The rate of the reaction will depend on the concentration of species A (denoted [A]) present, and the rate constant (k) for this reaction. The rate equation for this reaction would be:

$$\text{Reaction rate} = k[A]$$

This tells us that the reaction rate depends only on the concentration of A, and that the reaction rate will increase as the concentration of A is increased. In truth, the reaction rate also depends on the temperature, pressure, and perhaps other factors as well, but these are all bundled into the rate constant, k.

How is the **rate of a reaction affected** by **temperature**?

The rate of a chemical reaction will generally increase with increasing temperature. This is because a higher temperature translates into a higher average energy per molecule, which makes it easier for molecules to surmount the energetic barrier to the reaction. In terms of how this fits into the rate equation, the rate constant k depends on temperature, and k almost always (but there are exceptions) increases with increasing temperature.

FASTER THAN A SPEEDING WAVE

How fast does light travel?

In a vacuum, the speed of light is about 3×10^8 meters per second, which is very, very fast. That's so fast that a beam of light could travel around the whole world in only about 0.13 seconds!

It's interesting to consider just how far away from the Earth stars really are. After the Sun, the next closest star to the Earth is over four light years away (a light year is the distance light travels in one year). This means the next closest star is over 20,000,000,000,000 miles away. Because light must reach our eyes for us to see anything, if that star exploded, we wouldn't see it until over four years after it actually happened!

Can anything travel faster than the speed of light in a vacuum?

No, or at least it's presently thought that this would be impossible. It's interesting, though, that there have been a couple of experiments in recent years which have observed particles called neutrinos moving faster than the speed of light. Even the scientists who carried out these experiments questioned the results, however, and they encouraged others to try to confirm their results or to find where a mistake might have been made in their measurements. In the end, they did find out that these results were caused by a mistake; there was a loose electrical cable that caused enough error to invalidate the results.

Is the speed of light always the same?

The speed of light actually depends on what material the light is moving through! Each material has a property called an index of refraction. From the index of refraction, we can calculate the speed of light passing through a material from the following equation:

$$v = c/n$$

In this equation, c is the speed of light in a vacuum (about 2.998×10^8 m/s), n is the index of refraction of the material in question, and v tells us the velocity of light in the material.

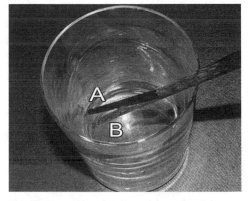

This photo combines two images: one of a stick partly submersed in a glass of water (A), and the same picture without water in the glass (B). The stick seems bent in A because light is refracted as it leaves the water and is perceived by your eyes.

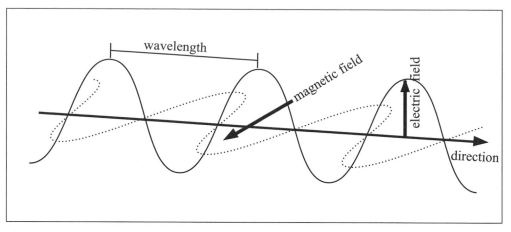

Electromagnetic radiation consists of electric and magnetic fields that run perpendicular to one another. The wavelength of light is the distance measured between wave crests.

What are the **wavelength** and **frequency of light**?

The light we see is a form of energy called electromagnetic radiation. It may sound like something complicated, but it's nothing too exotic or unfamiliar. Everything you see is because of electromagnetic radiation. Electromagnetic radiation consists of perpendicular electric and magnetic fields that oscillate in amplitude. The number of times per second that the fields oscillate is the frequency of the radiation. This is measured in Hertz (Hz), or inverse seconds. The wavelength is the distance that the light travels through space during one oscillation of the electric or magnetic field.

What is the **electromagnetic spectrum**?

The electromagnetic spectrum describes the entire range of frequencies (or wavelengths) possible for electromagnetic radiation to have. In principle, the spectrum is practically infinite, though there are limitations on how high or low of frequencies we can practically achieve and work with. On the high end of the frequency spectrum are usually gamma rays, with frequencies of around 10^{20} Hz, while on the low end are "extremely low frequencies" of only a few Hz.

Electromagnetic Spectrum

Type	Frequency (Hz)	Wavelength (cm)
Radio	$< 3 \times 10^{11}$	> 10
Microwave	$3 \times 10^{11} - 10^{13}$	$10 - 0.01$
Infrared	$10^{13} - 4 \times 10^{14}$	$0.01 - 7 \times 10^{-5}$
Visible	$4 - 7.5 \times 10^{14}$	$7 \times 10^{-5} - 4 \times 10^{-5}$
Ultraviolet	$10^{15} - 10^{17}$	$4 \times 10^{-5} - 10^{-7}$
X–rays	$10^{17} - 10^{20}$	$10^{-7} - 10^{-9}$
Gamma Rays	$10^{20} - 10^{24}$	$< 10^{-9}$

How is the **frequency of electromagnetic radiation** related to its **energy**?

The frequency of a photon is related to its energy, E, by a pretty straightforward equation:

$$E = h$$

In this equation, h is Planck's constant, which has a value of 6.626×10^{-34} J•s. The frequency term, , is the frequency of the radiation in Hz. As we can see from this equation, electromagnetic radiation with higher frequency has higher energy.

BIG FREAKING LASERS

What is **spectroscopy**?

Spectroscopy is a branch of science associated with using light to study transitions between energy levels. Not all scientists who use spectroscopy (spectroscopists) are physical chemists, though physical chemists (and physicists) are typically the people who develop new spectroscopic methods and experimentally investigate the details of how light interacts with matter. Data collected in spectroscopic experiments is typically presented as some response of an atomic or molecular system as a function of frequency/wavelength or time. When the response is plotted as a function of frequency/wavelength, it is called a spectrum.

What are **Fraunhofer lines**?

When scientists first began observing the spectrum of light reaching Earth from the Sun, the spectrum contained many dark lines, which indicated that light of certain wavelengths wasn't present in the sunlight. These are now called Fraunhofer lines (named after their discoverer), and they are caused by the elements in the outer atmosphere of the Sun absorbing certain wavelengths of light, which prevents those wavelengths from reaching the Earth. Understanding that Fraunhofer lines were caused by atomic absorptions was one of the earliest examples of atomic spectroscopy.

What are **ground** and **excited electronic states**?

The ground electronic state of an atom or molecule is the lowest energy electronic state. Excited states are any electronic states that have a higher energy than the lowest energy electronic state.

How can **light cause transitions** between **energy levels**?

Light comes in discrete units called photons, and each photon has a particular en-

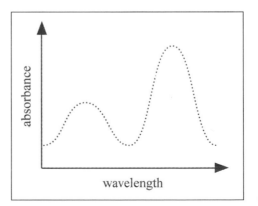

An example of a typical spectrum plotting absorbance versus wavelength.

A laser beam is simply a beam of light that has been intensified by stimulating the emission of photons. They can be used for many purposes, ranging from cutting metal to delicate surgery to aiding scientists with complex measurements.

ergy associated with it. When a photon's energy matches the energy spacing between two energy levels in an atom or molecule, it can cause a transition between these energy levels. This results in absorption of the photon, which transfers its energy to the atom or molecule. For example, the energy spacing between the ground and first excited electronic states of a hydrogen atom is 1.64×10^{-18} J which corresponds to a photon frequency of 2.47×10^{15} Hz. So photons with this frequency can excite the electron in a hydrogen atom from the ground to first excited electronic state.

What is a **laser**?

A laser is a light source that emits light amplified through the stimulated emission of photons. The acronym, LASER, actually stands for Light Amplification by Stimulated Emission of Radiation. Lasers come in many shapes and sizes. Some can fit in your pocket, and some are huge and take up entire rooms. Some emit pulses of light, while others emit continuous beams of light. Since there are so many types of lasers, it's not surprising that they find wide-ranging applications from simply pointing at a screen during a presentation to carrying out complex measurements in physics and chemistry experiments.

Why are **lasers** useful for **physical chemists**?

Chemists use lasers to study how molecules interact with light. In some cases, a chemist may want to know how a molecule reacts when a pulse of light is used to excite it. In

Can lasers be dangerous?

Definitely. Many lasers used in modern chemistry laboratories are powerful enough to cause a person to go blind after only a fraction of a second of direct exposure to the eye. Some are even so powerful that they can burn or ignite objects placed in their path. Laser pointers you can buy in the store are not this powerful, however, so you don't have to worry about a laser you personally own being quite this dangerous. You should still definitely avoid shining them in your eye, though, because they can be damaging.

other cases, lasers can be used to gain information about the structure of the molecule. One of the reasons lasers are good for these purposes is that they can provide pulsed light to gain information about how molecules are changing over time. Another is that many "tricks" exist for controlling and manipulating the wavelength of the light produced by a laser, making them versatile light sources.

What is the **biggest laser** in the world?

The largest laser in the world is located at the Lawrence Livermore National Laboratory in Livermore, California. This laser is so large that it covers the size of three football fields! The scientists who use this giant laser for their research are hoping to show that a nuclear fusion reaction (see "Nuclear Chemistry") can be controlled and used as a source of energy. If that's possible, it could revolutionize the way power plants make energy.

OTHER SPECTROSCOPY

What is **microwave spectroscopy**?

Microwave spectroscopy, as the name implies, is spectroscopy carried out using electromagnetic radiation in the microwave region of the spectrum (0.3 to 300 GHz). The energies associated with microwaves are relatively low, and these energies are typically a good match for energy-level spacing between the different rotational levels of molecules. Thus, microwave spectroscopy is typically used to study the rotational energy levels of molecules. The rotational energy levels of molecules are typically studied in the gas phase.

What is **infrared spectroscopy**?

Infrared spectroscopy is carried out using somewhat higher energy electromagnetic radiation (300 GHz to 400 THz) than microwave spectroscopy. The infrared region of the spectrum is usually a good match for the vibrational energy-level spacing in molecules, so infrared spectroscopy is typically used to study the vibrational energy levels of mol-

How does RADAR work?

RADAR, which stands for RAdio Detection And Ranging, works by sending out electromagnetic radiation, allowing it to bounce off of objects, and receiving it again after it has been reflected. The RADAR system measures things like how long it took for the signal to make it back, how the frequency of the signal has changed, and how the strength of the signal has changed. From this information, the RADAR system can "see" where the objects that reflected the light are located. This can also be used to determine the speed of an object, such as when the police use a RADAR gun to track the speed of a vehicle.

ecules. Vibrational spectroscopy can be used to study molecules in the gas, liquid, and solid phases, as well as molecules on surfaces.

What is **UV/Vis spectroscopy**?

Electromagnetic radiation in the ultraviolet and visible region of the spectrum (40 to 1000 THz) is higher in frequency (and thus energy) than that in either the microwave or infrared. This makes it a good match for the larger energy-level spacing associated with transitions between electronic energy levels. UV/Vis spectroscopy can be used to study molecules in any phase; however, it is most commonly used for liquid samples.

What is **Beer's law**?

Beer's law tells us how the amount of electromagnetic radiation absorbed by a sample is related to the concentration of the absorbing species. Beer's law tells us that the absorbance is equal to the length of the sample, l, times the concentration, c, of the absorbing species in the sample, times the molar absorptivity coefficient, ε, of the species.

$$A = \varepsilon l c$$

In this equation A is absorbance, which is defined as the negative of the logarithm of the ratio of the intensity of light passing through a sample to that incident on it. Basically, this gives us a measure of how much light a sample is absorbing and how much light is passing through it.

What is **fluorescence**?

Fluorescence is a process by which molecules that have absorbed light can re-emit light to release some of the energy they absorbed. For fluorescence to take place, a molecule must first absorb a photon of light, which causes an electron to be excited to a higher energy level. At the same time, this process will typically also cause some vibrational excitations to take place. Some of the energy associated with this absorption will be given off through relaxation of the excited vibrational energy levels. For fluorescence to occur,

> ## Why do "black" lights make white materials appear to glow?
>
> **"B**lack" lights are lights that emit ultraviolet light at frequencies on the upper-edge, or higher, relative to what our eyes can see. Many objects can absorb these frequencies and then undergo fluorescence, giving off light at lower frequencies that our eyes can see. This is the reason "black" lights cause these materials to appear to glow.

the electronic excitation relaxes by emitting a photon of light. Some of the energy was dissipated as the vibrational energy levels relaxed, so the photon that is emitted has less energy than the photon that was initially absorbed. Remember that less energy means a lower frequency, so the photons that are emitted have a lower frequency than those that were absorbed.

What is **mass spectrometry**?

Mass spectrometry is a method of chemical analysis that involves determining the molecular mass of charged particles by measuring the mass-to-charge ratio of an ionized molecule or molecular fragment. There are several ways of performing mass spectrometry, but the general procedure involves making the sample into a vapor, ionizing the sample, and then detecting the ions that form in a way that separates them according to their mass-to-charge ratio. After being ionized, the molecules in the sample will often fragment into smaller ions, and these too are detected according to their mass-to-charge ratio.

This technique can be useful for carrying out an accurate determination of the mass of a molecule as well as for obtaining structural information about molecules via their fragmentation patterns. It also allows the elemental composition of a sample to be determined.

How does a **microscope work**?

A microscope is all about lenses. The lens near the sample you're looking at is called the objective lens, and this lens is responsible for collecting the light from the sample and focusing it. Typically there will be a light under or behind the sample that provides the light used to view the sample. At the other end is another lens called the ocular lens, and the total magnification of the microscope is determined by multiplying the magnification of the objective lens by that provided by the ocular lens. The apparatus we typically think of when we think of a microscope is essentially just a big framework used to hold the lenses, the sample, and perhaps other optical devices used to improve the image of the sample.

What is **electron microscopy**?

An electron microscope is a microscope that uses a beam of electrons to produce an image of a sample (rather than using light like in a standard microscope). There are

147

several ways of obtaining an image, but the original was the transmission electron microscope (TEM), which produces an image by passing an electron beam directly through the sample. Electron microscopes offer a significant advantage in resolution over traditional light microscopes. This is due to the fact that the wavelengths associated with electrons are much shorter than those associated with visible light. It possible to achieve resolutions of up to roughly 10,000,000 times magnification using electron microscopes, as compared to about 2,000 times magnification in the best light microscopes.

Simple optical microscopes work by magnifying images using lenses.

What is **electrical resistance**?

Electrical resistance describes how a material opposes an electric current through the material. The resistance is related to the applied voltage through the relationship:

$$R = V/I$$

where V is the applied voltage and I is the current through the material. Typically the resistance is a constant, and thus the current will increase linearly with an increase in the applied voltage. This relationship is Ohm's law. As you can see, a material with a higher resistance (R) will have a lower current (I) for a given applied voltage (V).

What is a **voltage**?

A voltage, or potential difference, is the difference in electric potential energy between two points. The voltage describes the amount of work that would need to be done, per unit charge, to move a charged object between the two positions. A voltage may be present due to a static electric field, to electric current flowing through a magnetic field, or due to magnetic fields that change over time.

IF YOU CAN IMAGINE IT

What is the **focus of theoretical chemistry**?

Theoretical chemistry is a branch of chemistry that, as the name suggests, develops and applies theories to explain chemical observations and also to make predictions about things chemists cannot directly study by experiment. Theoretical chemists work on a wide variety of problems that cover pretty much all of the other branches of chemistry. Two of the major subfields within theoretical chemistry are electronic structure theory and molecular dynamics.

What is **electronic structure theory**?

Electronic structure theory is an area of theoretical chemistry that is focused on calculating the arrangement and energies associated with configurations of electrons in molecules. This can include predicting the structure of a molecule, the most probable arrangement of its electrons, its reactivity, and different excited states of the molecule. For reasons we won't go into, this is not an easy task, and the electronic structure of molecules cannot be solved exactly, even with very powerful computers. Most theoretical chemists working in this area develop approximations for calculating the true electronic structure of molecules and on testing these approximations against available experimental data to continue to improve existing methods. The electronic properties of molecules play a vital role in determining their stability and reactivity, so despite being challenging, it's a problem well worth trying to solve.

What **molecular properties** do **theoretical chemists** try to **calculate**?

Theoretical chemists try to calculate pretty much every molecular property there is! If there's a property we've discussed somewhere in this book, odds are that a theoretical chemist has worked on ways to calculate its value for atoms or molecules.

How much **error** is associated with **electronic structure theory calculations**?

There can be a pretty large amount of error associated with these calculations, and the main goal is to keep the errors as consistent as possible and to obtain properties via differences in calculations. For example, the energy of the metal–carbon bond in $Cr(CO)_6$ would be calculated by comparing the energy calculated for $Cr(CO)_6$ to that calculated for $Cr(CO)_5$ and CO separated from one another at infinite distance.

What is a **molecular dynamics simulation**?

Molecular dynamics simulations are a computational model of a collection of molecules that interact with one another under a specified set of conditions (temperature, pressure, etc.). While electronic structure theory calculations typically only involve one or a few molecules, molecular dynamics simulations can involve hundreds or thousands of molecules all at once. The purpose of molecular dynamics simulations is typically to investigate how molecules interact and react while surrounded by a collection of other molecules. While the energies of individual molecules can be investigated by electronic structure theory, a molecular dynamics simulation allows theoretical chemists to also study how molecules are influenced by their surroundings. These effects can be especially important in liquids, where solvent molecules can have a significant impact on reactivity.

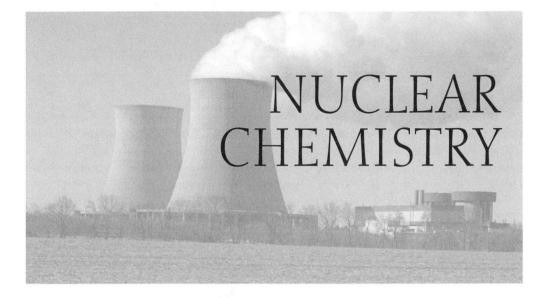

NUCLEAR CHEMISTRY

CHEMISTRY INSIDE THE ATOM

How is **nuclear chemistry different** than **other types** of **chemistry**?

As the name implies, nuclear chemistry deals specifically with chemical events involving the nucleus itself, while most other areas of chemistry involve rearrangements of the electrons. Nuclear chemistry is focused on radioactivity and the properties of nuclei, and it finds some of its most important applications in energy production, weapons, and medicine.

What is an **isotope**?

Isotopes are atoms with the same number of protons and electrons, but with different numbers of neutrons. The most important thing to keep in mind is that the number of protons determines the element we're dealing with. In most areas of chemistry, this is enough to determine the reactivity of the atom, though in nuclear chemistry the number of neutrons is also quite important in determining the nuclear processes an isotope can undergo. (Please refer back to the "Atoms and Molecules" chapter for additional information on isotopes.)

Do **electrons, protons,** and **neutrons** all have the **same mass**?

Protons, neutrons, and electrons each have different masses. Electrons are, by far, the lightest of the three, with a mass of only about 1/2000th that of a proton or neutron. Protons and neutrons have similar masses, with that of the neutron being just slightly higher than that of the proton. The masses of the three particles in kilograms are:

Electron mass: 9.1094×10^{-31} kg
Proton mass: 1.6726×10^{-27} kg
Neutron mass: 1.6749×10^{-27} kg

Nuclear decay processes can involve the release of electrons, protons, neutrons, or combinations of these basic particles.

Are all **isotopes stable**?

Not all isotopes of a given element are stable. For example, tin has twenty-two different known isotopes, ten of which are stable and twelve of which are unstable (though there is some debate about just how stable those ten are). Stable is, of course, a relative term. Usually when one says an isotope of an element is stable, it means that it has a decay half-life that is too long to be measured by current methods. There are some elements, such as technetium, radon, and plutonium that do not have any stable known isotopes. In fact, no elements with an atomic number of over 83 (i.e., more than eighty-three protons) have any known isotopes that are considered to be stable!

What is an **antiparticle**, and what is **antimatter**?

For most kinds of particles there is postulated to exist a corresponding antiparticle, which is of the same mass but an opposite charge. These antiparticles have only recently been observed in laboratory settings for the first time, and they are very difficult to isolate and study experimentally. This is because particle and antiparticle pairs collide to generate photons of light in a process that annihilates the particle and antiparticle pair. Antiparticles are not well understood and are an active area of research related to nuclear chemistry. Antimatter is just matter made up of antiparticles, in the same way that normal matter is made up of particles. There has been postulated to be an equal amount of matter and antimatter in the universe, though the observations made to date do not suggest this to be the case. This represents an unresolved dilemma that scientists hope to someday better understand. These types of fundamental, unresolved problems are a big part of the reason science is so interesting!

What is a **positron**?

A positron is the antimatter counterpart of the electron. It has the same mass and spin as an electron, but with a charge opposite in sign and equal in magnitude to that of the electron. If an electron and positron collide, they can annihilate each other and release their energy in the form of a photon.

What are **particle accelerators** used for?

Particle accelerators are used to generate beams of particles moving at very high speeds, which are typically then collided with matter or other particles to learn about fundamental interactions. Most of the time the particles in question are subatomic particles, though atoms can also be used. Such experiments are used to address fundamental questions in physics surrounding the structure of matter and space. Typical modern particle accelerators are several kilometers long, with some operating in a linear fashion and others in a large ring.

What is a **quark**?

Quarks are the fundamental particles that make up protons and neutrons, as well as several other types of particles. There are six types of quarks, which are referred to as different "flavors." These are named up, down, top, bottom, charm, and strange. Protons and neutrons are each made up of three quarks. Two up and one down quark make up a proton. Two down and one up quark make up a neutron.

How do **nuclei spontaneously decay**?

Nuclei can undergo several types of decay through spontaneous means without colliding or interacting with nuclei of other atoms. The most common types of nuclear decay are called alpha radiation, beta radiation, and gamma radiation. These differ by the type of fragmentation the nucleus undergoes during the decay process.

What is **alpha radiation**?

Alpha radiation involves the fragmentation of the nucleus into two particles, one consisting of two protons and two neutrons (an alpha particle, or in other words, a helium nucleus), and the other consisting of the remaining protons, neutrons, and electrons initially present in the parent nucleus. Alpha decay decreases the number of protons in the nucleus by two and decreases the atomic mass of the nucleus by four amu.

What is a **beta particle**?

Beta particles are another type of particle that can be emitted during a nuclear decay process. A beta particle can be either an electron or a positron, which is the antiparticle of an electron. If it is an electron being emitted, one of the neutrons in the nucleus must become a proton to conserve charge in the process. Beta decay increases the number of protons in the nucleus by one and leaves the atomic mass essentially unchanged.

How is nuclear chemistry related to the alchemists' goal of transmutation?

Alchemists sought a way to turn common metals into gold, which we now know is not possible to do in any simple way. The reason is that transmutation would involve converting one element into another, which can't be done by simple chemical processes. It would require a nuclear reaction to take place; either a heavy nucleus would have to divide into a gold nucleus and another byproduct, or two lighter nuclei would have to combine to form one of gold. Neither of these things happen readily. If early alchemists had recognized the distinction between more ordinary chemical reactions and the nuclear reaction they were looking for, it would likely have saved a lot of time and effort.

What is **gamma radiation**?

While alpha and beta radiation is the loss of some particle from an atom, gamma radiation is the release of electromagnetic radiation (called gamma rays). This energy is typically of a high frequency ($>10^{19}$ Hz), which means it's high energy (>100 keV) and can cause significant damage. Gamma radiation can easily penetrate deep into your body, unlike alpha and beta particles, causing damage to your cells and the DNA inside them. Sometimes this damage is useful, though, and some radiation therapies for cancer treatment make use of gamma radiation to kill the malignant cells.

What **holds nuclei** together?

The nucleus of an atom consists of neutrons, which are uncharged, and protons, which are positively charged. While the uncharged neutrons don't feel an electrostatic attraction or repulsion to other particles, the positively charged protons should repel each other. In fact, this repulsive force between the protons is quite strong because protons in the same nucleus are very close together. Thus the force that holds them together must be a very strong force. Indeed it is, and it's even named the *strong force*. This strong force acts only over distances on the order of 10^{-15} m—a very very short distance. If the protons were to become separated by a more substantial distance, the strong force would decrease in magnitude faster than the repulsive force, and the protons would be pushed apart. It's also often said that neutrons act as a sort of "glue" to help bind all of the neutrons and protons together, since there seem to be favored relationships between the number of neutrons and protons present in stable nuclei.

Do all **isotopes** of an **element decay** at the **same rate**?

No, actually each isotope decays at a unique rate. The most radioactive isotopes are those isotopes which decay most quickly. There are some elements (especially the heaviest ones) that don't have any truly stable isotopes, and these can only be synthesized for fleeting amounts of time in laboratory settings.

What is the **half-life** of a **radioactive species**?

The half-life of a radioactive species is the amount of time it takes the quantity of the species to decrease by half. After one half-life, ½ of the initial quantity of material will remain, after two ¼ will remain, after three ⅛ will remain, and so on. Half-lives of radioactive nuclei vary widely, and we'll list just a few values below to give an idea of the range of timescales covered.

Radioactive Nucleus	Half-Life
Carbon-14	5,730 years
Lead-210	22.3 years
Mercury-203	46.6 days
Lead-214	26.8 minutes
Nitrogen-16	7.13 seconds
Polonium-213	0.000305 seconds

What defines how long one second lasts?

One second is defined as 9,192,631,770 times the period of the electromagnetic radiation (see the "Physical and Theoretical Chemistry" chapter for more on electromagnetic radiation) corresponding to the difference in hyperfine energy levels in the ground state of a Cesium–133 atom.

What does this mean? To begin, the difference in two closely spaced energy levels of a Cesium–133 atom defines a specific gap in energy. Using the relationship between the energy and frequency of a photon of light, this energy gap can be converted to a frequency of light. Recall that light is electromagnetic radiation. Also recall that the reciprocal of the frequency of light tells us the period of the oscillation of the electromagnetic fields that make up the light. The period tells us how long it takes the electric and magnetic fields to oscillate a single time, and one second is defined as 9,192,631,770 times this (extremely brief) time interval. As those 9,192,631,770 oscillations of the electric field of the light take place, the second hand of each clock on Earth moves 1/60th of a rotation forward.

What is **electron capture**?

Electron capture is a process that involves an electron combining with a proton to form a neutron. This decreases the atomic number of the element by 1 and leaves the atomic mass unchanged.

Who was **Marie Curie**?

Marie Curie was a famous French-Polish scientist, and she was the first person ever to be awarded two Nobel prizes, one in chemistry and the other in physics. She was also the first woman to ever win the Nobel Prize, and remains the only woman to have ever won two Nobel prizes in different fields. Curie was responsible for much of the pioneering work in nuclear chemistry during the late nineteenth and early twentieth centuries. Much of her work focused on studying radioactive elements, and she discovered radium and polonium. Tragically, it was Curie's work that also led to her death. During her career, the dangerous effects of radiation were not yet known, so she worked without the same safety precautions that would be taken today. Her death was the result of a condition known as aplastic anemia, brought on by her prolonged exposure to radiation in the laboratory.

How is **radiation exposure quantified**?

The scientific unit for radiation exposure is the sievert (Sv), though several other types of units do exist. The maximum radiation exposure that is allowable for occupational exposure in the U.S. is 50 millisieverts (mSv). For comparison, the average natural background level of exposure is roughly 3 mSv.

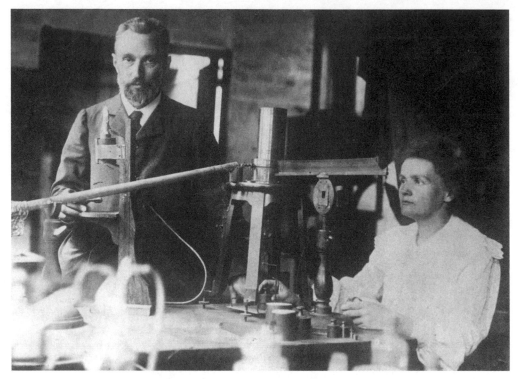

Pierre and Marie Curie working in their laboratory. Marie Curie was the first woman to ever win a Nobel Prize. In fact, she won two of the prestigious prizes. She studied, among other things, radioactive elements and discovered polonium and radium.

NUCLEAR CHEMISTRY AT WORK

What is **nuclear fusion**?

Nuclear fusion is the process by which two nuclei combine to form a single, heavier nucleus. Energy is usually released when two lightweight nuclei fuse, though for heavier nuclei, fusion generally requires an input of energy. Nuclear fusion can be used in bombs to cause a massive and rapid release of energy. Fusion is also responsible for the fact that stars burn bright and give off light and heat.

What is **cold fusion**?

A cold fusion reaction is one that takes place under ambient conditions using simple equipment. Such a fusion reaction would be extremely desirable, since it could allow for a simple and efficient means of energy production.

What is **nuclear fission**?

Nuclear fission is essentially the opposite of nuclear fusion. Here, a single nucleus divides into two smaller nuclei. In the case of heavy atoms, this is often accompanied by

Is cold fusion really possible?

In the late 1980s, reports surfaced of experimentally realized cold fusion, exciting the scientific community. It turned out, however, that these reports were false, and nobody was able to reproduce the results of what were initially reported as relatively simple experiments. Since these experiments were disproved, other credible reports of cold fusion experiments have indeed surfaced, and thus it does appear that cold fusion is possible in principle. Unfortunately, the energy released from the few successful experiments has been much smaller than the amount of energy needed to actually run the experiments, making the feasibility of cold fusion as a source of energy production unlikely. Compared to the initial burst of interest, mainstream scientists have generally lost interest in the topic, though there remains a group of fringe experimentalists who still seek to make cold fusion for energy production a reality. If such experiments could work, they would certainly be of great interest to the scientific community, but today most believe that it just isn't possible to generate enough energy from cold fusion sources to make it a viable source of energy production.

the release of heat. For example, the radioactive decay of uranium-235 can be used to generate the heat used to drive turbines to generate electricity in nuclear power plants. The use of nuclear fission to harness energy for use by humans is typically considered the much more viable choice (as opposed to nuclear fusion).

Is **mass conserved** during a **fission** process?

Almost, but not quite. A small amount of mass is given off in the form of energy. Specifically, the relationship between the amount of energy, E, released and the amount of mass, m, that becomes energy is given by the famous relationship $E = mc^2$, where c is the speed of light.

How can **radioactivity** be **measured**?

Radioactivity is measured by detecting the products of radioactive decay processes. The most well-known instrument used for this purpose is the Geiger counter. A Geiger counter is sensitive to the products of nuclear decay, including alpha and beta particles and gamma rays. The units used to quantify radiation are the Curie or the Becquerel, which describe the number of nuclear decays a substance undergoes per unit of time.

In many cases, it may not be necessary to directly detect the radiation being given off at this instant, but rather to just determine the isotopic ratio of an element present in a sample. This can be done using techniques borrowed from analytical chemistry, such as mass spectrometry. Information on the isotopic ratio present, along with knowl-

A Geiger counter is a useful tool for measuring radioactivity in almost anything. It can detect alpha and beta particles, as well as gamma rays.

edge of the half-life of the isotope in question, can be related to the age of the sample being studied.

How does **radioactive dating** work?

Radioactive dating (also called radiometric dating) is a technique used to determine the age of a sample based on the ratio of isotopes of an element present in the sample. Using the known half-life of the isotope being studied, along with knowledge of the natural abundance of the isotopes present at the time the sample was formed, the age of the sample can be determined. To obtain an accurate age for a sample, it is required that none of the isotopes being measured have been able to escape or re-enter the sample over the course of its lifetime. Otherwise this could serve to establish a ratio of isotopic abundances that is not representative of that based purely on the half-life of the isotope whose decay is being measured.

What is a **nuclear chain reaction**?

A nuclear chain reaction is a string of reactions that occurs when a given nuclear reaction causes, on average, at least one more nuclear reaction to take place. Such chain reactions are important for the generation of nuclear power and also for nuclear weapons. Uranium-235 is responsible for the chain reaction that generates power in nuclear reactors and in some bombs as well. Uranium-238 is the more common isotope, so it is typically necessary to first enrich the uranium to be used in the 235 isotope. When a neutron collides with uranium-235 it generates uranium-236, which then undergoes fission to release energy and further neutrons that can collide with other uranium-235 atoms, causing the chain reaction to continue.

How does an **atomic bomb work**?

Atomic bombs (A-bombs) are based on nuclear chain reactions that occur very rapidly, causing a huge release of energy in a very short amount of time. In early designs, two pieces of uranium would be fired at one another in the core of the bomb, initiating the fission chain reaction responsible for the explosion of the bomb. As the bomb starts to detonate the core of the bomb expands, and it is necessary that pressure be applied against the expanding core while the fission process takes place. Within a fraction of a second after detonation, the explosion takes place. These are the type of bombs that were used at Hiroshima and Nagasaki in World War II and are the only nuclear weapons that have been used in war to this day.

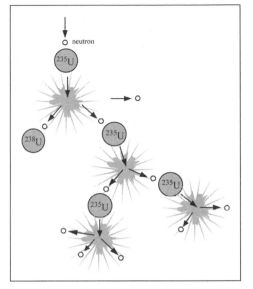

Uranium-235 chain reactions start because the radioactive substance naturally emits neutrons that then collide with other atoms. In a nuclear bomb, the goal is to let the reaction reach a critical point where there is an explosion, but with a nuclear reactor the tricky part is controlling the reaction.

What's the **difference** between an **H-bomb** and an **A-bomb**?

The hydrogen bomb (H-bomb) is actually significantly more destructive than even an A-bomb. While A-bombs release energy via chain fission reactions (breaking apart heavy nuclei), H-bombs release energy through fusion of light nuclei. This energy comes from an overall increase in stability due to the strong force that holds nuclei together as the light nuclei fuse to create heavier ones. To give an idea of the relative powers of these two weapons of mass destruction, consider that the A-bomb dropped on Hiroshima had a force on the order of 10 kilotons (explosive force equivalent to 10,000 tons of TNT), while a common H-bomb has a force on the order of 10 megatons, or 1,000 times the explosive force of the A-bomb used at Hiroshima.

How is **radiation used** in **medicine**?

We should begin by pointing out the distinction between radiation used in nuclear medicine/radiopharmaceuticals (more akin to the other topics of this chapter) and electromagnetic radiation (light of different wavelengths).

Nuclear medicine is the branch of medicine most closely tied to the concepts of nuclear chemistry discussed in this chapter. Diagnosis via nuclear medicine typically involves the injection of a radiopharmaceutical into the body, and the radiation released by this drug can

An atomic bomb blast (illustration shown here) releases huge amounts of energy by creating a fision chain reaction within the bomb.

then be monitored to gain information about organ function, blood flow, the location of a tumor, or to locate a fractured bone. In some cases, the use of nuclear medicine can allow for earlier diagnosis than with other imaging techniques.

In terms of using electromagnetic radiation for medical applications, perhaps one of the first treatments that come to mind is radiation therapy, which is used to fight against a broad range of cancers. This involves using focused electromagnetic radiation to damage the DNA in the tissue of a tumor while hopefully not causing too much damage to the surrounding healthy tissue. The goal is to damage the DNA of cancerous cells so that they are unable to reproduce, hopefully killing the tumor with time. Beams of radiation are focused onto the tumor from different angles to minimize the effect on any one area of healthy tissue.

X-rays and CT scans are two commonly used, noninvasive medical techniques that make use of electromagnetic radiation to take pictures of what's going on inside the human body. It should be noted that prolonged exposure to the X-rays used in these procedures can be harmful and are capable of causing cancer themselves over long periods of time.

How are **isotopes made**?

Specific isotopes of an element can be obtained in one of two ways: either by separation of the desired isotope from a naturally occurring sample or by synthesis of the desired isotope.

Since the different isotopes of an element all have the same chemical properties, they can be quite difficult to separate. The separation techniques used to separate different isotopes are thus based on their differences in mass, rather than on differences in chemical properties. Some of the methods used include separation by diffusion in the gas or liquid phases, centrifugation, ionization and mass spectrometry, or chemical methods based on differences in reaction rates due to different atomic masses.

Different isotopes of an element can also be generated synthetically. One way to do this is to fire high-energy particles at the nucleus of an atom. Depending on the situation, this can either cause a particle to be emitted from the parent nucleus (generating a lighter nucleus) or the fired particle can be absorbed (generating a heavier nucleus). It is also possible to synthesize isotopes of some elements by making use of another naturally occurring nuclear reaction, such as when the particles released by one nuclear fission reaction are absorbed by another nucleus.

Which **elements** are **man-made**?

Actually, we can make a lot of elements synthetically. Here's a list of all of the man-made elements:

technetium (Tc)—43 (the first man-made element)
promethium (Pm)—61
neptunium (Np)—93
plutonium (Pu)—94
americium (Am)—95
curium (Cm)—96
berkelium (Bk)—97
californium (Cf)—98
einsteinium (Es)—99
fermium (Fm)—100
mendelevium (Md)—101
nobelium (No)—102
lawrencium (Lr)—103
rutherfordium (Rf)—104
dubnium (Db)—105
seaborgium (Sg)—106
bohrium (Bh)—107
hassium (Hs)—108
meitnerium (Mt)—109
darmstadtium (Ds)—110
roentgenium (Rg)—111
copernicium (Cn)—112
ununtrium (Uut)—113
ununquadium (Uuq)—114
ununpentium (Uup)—115
ununhexium (Uuh)—116
ununseptium (Uus)—117
ununoctium (Uuo)—118

How do **nuclear power reactors** work?

A nuclear power reactor works by generating heat from a controlled fission reaction, which then generates steam used to drive turbines to generate electricity. The fuel for the nuclear reactor is typically uranium-235 or plutonium–239.

What is a **thorium reactor**?

A thorium fuel cycle is also possible for use in nuclear power reactors. This involves using thorium–232 to generate uranium-233, which is capable of undergoing fission processes to generate energy in the form of heat.

161

Nuclear reactors work by generating heat from controlled fission reactions. Breeder reactors actually create more fissionable material than they use and are self-sustaining.

What is a **breeder reactor**?

A breeder reactor is a type of nuclear reactor that is capable of generating fissile material (material that can sustain a chain fission reaction) faster than it uses it up. This is accomplished by using the neutrons given off in the fission reaction to generate additional isotopes capable of fusion. Typically this involves the use of either thorium to generate fissile uranium or uranium to generate fissile plutonium.

What is **radon**?

Radon is an element that is widely known for its potential to cause cancer. It is the heaviest gas known to man, with a density roughly nine times greater than that of air. It is usually found in soil and rocks, though it can also be found in water. Fortunately, radon detectors are commonly available that allow you to test your home for elevated radon levels.

What are some of the **worst nuclear disasters** in history?

A few of the worst nuclear disasters in history are those which took place at Three Mile Island in the USA in 1979, at Chernobyl in the Ukraine in 1986, and more recently following an earthquake in Fukushima, Japan, in 2011. Nuclear disasters are very dangerous if they do occur, and the possibility of a nuclear disaster represents a primary reason that some people oppose the construction of new nuclear power plants.

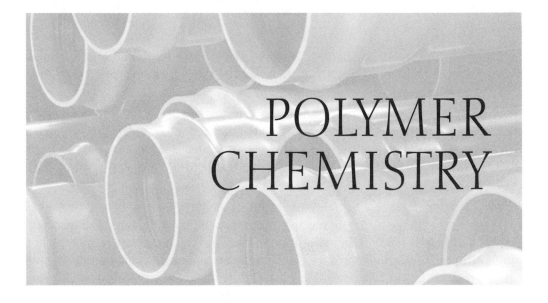

POLYMER CHEMISTRY

POLYMERS ARE MOLECULES TOO!

What is a **polymer**?

Polymers are large molecules, usually made up of smaller repeating units. The word itself, polymer, means "many parts" in Greek. You probably started thinking about plastics (like milk jugs and plastic cups) when you read the title of this chapter. Plastics are common examples, but polymers also play important roles in all plants and animals, including you.

What is a **monomer**?

If polymer means "many parts," a monomer is "one part" of that whole. A monomer is a molecule that is attached to many copies of itself to make a polymer molecule. Usually these bonds are covalent, but not always.

How are **polymers different** than **small molecules**?

So many ways! Polymer chemistry and polymer physics are big areas of research both in the recent past and today because connecting a bunch of small molecules into one big one results in lots of interesting changes.

To give you a metaphor, let's talk about pasta. Start with uncooked macaroni and uncooked spaghetti: If you try to move your hand through a bowl of uncooked macaroni you won't have much trouble, but if you had spaghetti noodles all lined up and you tried to move your hand through them (in either direction!), you'll run into problems. You either need to break the noodles or you need to carefully thread the noodles through your fingers. Both of these actions require energy (enthalpy in the first case and entropy in the second).

163

Now let's cook those noodles. Stick a fork in each of the bowls and spin it around. With macaroni, nothing happens, but the spaghetti starts to wind around your fork, gets tangled up, and so on.

Macaroni, a collection of small molecules…I mean, noodles, is totally different than polymers (spaghetti) which are also made up of flour and water, but are much longer. The raw and cooked spaghetti aren't just easy to imagine, they're great ways to think about polymers in different states (solids and liquids, glassy states and polymer melts).

Are all **polymer chains** the **same size**?

No. Let's stick with the macaroni metaphor to understand this. Imagine you're stringing noodles together to make a macaroni necklace. You can put as many noodles on a single string as you want. If you have two strings, you can put an equal number of noodles on each string or make one longer than another. Again, the macaroni noodles are monomers, which form polymers when we string them together.

So if all polymers are not the same size, what is the **weight of a polymer**?

Good question. If we know the number of monomers that make up a polymer chain (technical term: degree of polymerization), then the molecular weight of the polymer is the molecular weight of the monomer multiplied by that number of monomers.

How do you **measure molecular weight** of a **polymer**?

The most common way is based on size. The technique is known as size exclusion chromatography, or gel permeation chromatography. The sample is passed through a column that has a porous solid material. The smaller polymers can work their way into those pores, while larger molecules don't interact with the solid material. The biggest molecules, because they don't interact with the solid phase, come out of the column first followed by smaller and smaller molecules. The time it takes for a polymer to get through the porous column is related to its molecular weight (okay, technically it's based on the hydrodynamic volume, but let's let this approximation slide). In practice, these instruments are calibrated using standard polymer samples of a known molecular weight.

What is **molecular weight distribution**?

We just talked about the fact that polymers can have different molecular weights. Oftentimes in reactions that make polymers a range of molecular weights are produced. The molecules may be composed of the same repeating unit (monomer), but for a number of reasons the chains are different in length. It turns out that this distribution of lengths is important to a number of polymer properties. The details of how this number is calculated are not worth going into; it's sufficient to know that a higher molecular weight distribution means that there is a larger spread of polymer chain lengths. A distribution of 1.0 would mean that every single polymer chain has the exact same molecular weight.

Does polymer stereochemistry matter?

In lots of examples, it does. Let's stick with polypropylene for now. Isotactic polypropylene is a crystalline material with a melting point around 160 °C. The crystallinity is due to the perfect arrangement of the methyl groups along the polymer backbone. This crystallinity makes the material very tough, so it's used in all sorts of applications—from pipes to plastic chairs and carpeting. But if there are errors along the polymer chain (methyl groups pointing in the wrong direction) the melting point decreases, and the plastic loses its strength.

Do **polymers** have **stereochemistry** like small molecules?

Yes! The most common example is polypropylene. This polymer has a methyl group attached to the backbone of the polymer. If the methyl groups are all on the same side of the chain, the stereochemistry is known as isotactic (top structure below). If the arrangement of the methyl groups alternates which side of the chain it's on, the polymer is called syndiotactic (bottom structure below). If there's no order at all to the substituents, we call the polymer atactic.

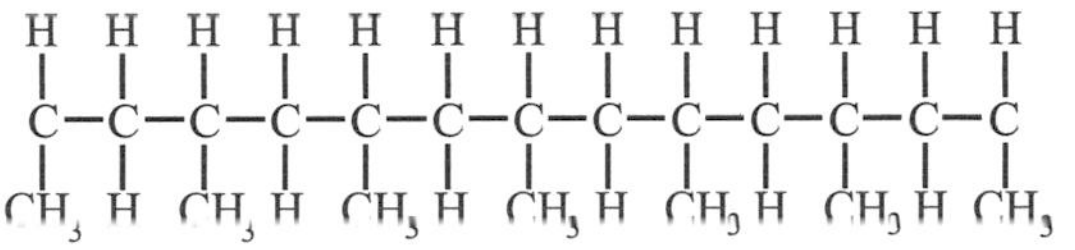

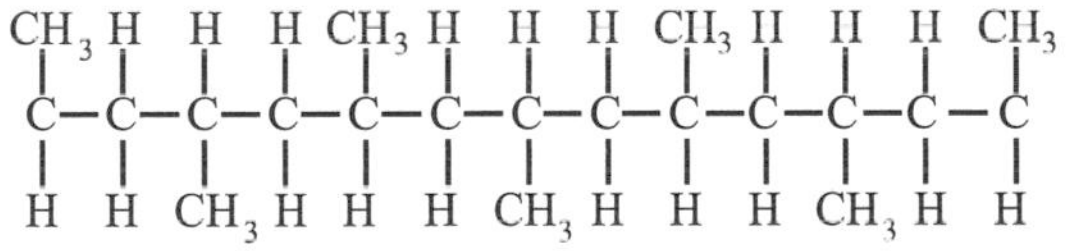

Are all **polymers linear chains**?

No, and this is another way that chemists classify these really big molecules. The major types of polymer shapes (technical term: topology) are linear, branched, and crosslinked networks. Linear polymers are chains of monomers joined together, like a noodle or a rope. If there is a point along a polymer chain where a second chain starts, like a fork in the road, this arrangement is referred to as branched.

What is a **crosslinked polymer**?

When a bond is formed between two polymer chains (and technically not at the chain ends), the product is called a crosslinked polymer. Creating linkages between chains usually increases their viscosity (so more like molasses than olive oil) and creates elastic properties like those found in rubber bands. At higher levels of crosslinking, polymers can even become stiff or glassy.

165

POLYMERS IN AND AROUND YOU

What **polymers** are found in **nature**?

There are a ton! Proteins, enzymes, cellulose, starch, and silk are all polymers.

Is **DNA** a **polymer**?

It is. DNA contains two long polymers of sugars (called nucleotides). Attached to each sugar molecule are phosphate groups and a nitrogen base (technically a nucleobase). The sequence of these nucleobases encodes the information in DNA. (For more on DNA, se the "Biochemistry" chapter.)

What is **cellulose**?

Cellulose (see diagram below) is linear polysaccharide—"poly" for many and "saccharide" means sugar, so cellulose is a chain of sugar molecules. It's an amazing molecule: it's the most abundant organic compound on the planet because it is the main component in plant cell walls. Cellulose is highly crystalline because of the way the sugar molecules are connected and because of the fact that it's made up of a single enantiomer of glucose. Like polypropylene, highly crystalline polymers like cellulose are very strong—strong enough to make trees stand up straight.

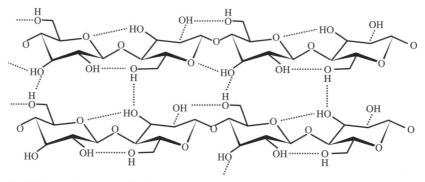

Is **starch different** than **cellulose**?

Starch is also a polysaccharide, like cellulose, but it is much less crystalline. The major component of starch is amylopectin, which is a highly branched polymer, while cellulose is strictly linear. These branches prevent amylopectin from crystallizing as well as cellulose. Starch is an excellent source of energy (or stored sugar) for plants and animals for these reasons: Because it is less crystalline than cellulose it is more soluble than cellulose, and the branched structure also means there are more end groups at which enzymes can start "chewing" the polymer apart.

What is **rayon**?

You probably know what rayon fabric looks or feels like—the best Hawaiian shirts (and much of 1980s fashion) were made of it. What's fascinating about rayon is that it is not really a synthetic *or* a natural fiber. Rayon is a chemically modified cellulose polymer, first prepared in the 1850s. While there have been a number of ways of preparing this "artificial silk," the Viscose method led to the first commercial production of rayon. This method treated cellulose with a combination of sodium hydroxide and carbon disulfide as indicated below.

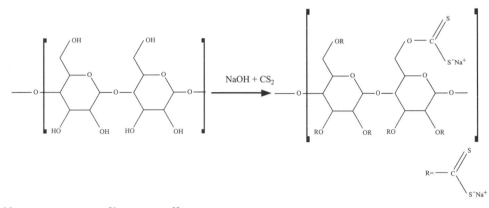

How was **rayon discovered**?

The first artificial silk was probably prepared by a Swiss chemist, Georges Audemars, in 1855. Audemars mixed the pulp of mulberry bark (chosen likely because silkworms eat mulberry leaves) and a rubber gum and used a needle to pull out long fibers of material. This was a rather labor-intensive and difficult process and could not be done in any economic way. Some accounts also claim that Audemars drew fibers of nitrocellulose (the product of mixing nitric acid with cellulose); in addition to being a delicate process, the resulting fibers of nitrocellulose were highly flammable.

Hilaire de Chardonnet, a French engineer, was another key player in the history of artificial silk. Working with Louis Pasteur in the 1870s, the legend claims that he spilled a bottle of nitrocellulose while working in a photography darkroom. The spilled solution was left to evaporate, and Chardonnet returned later to clean up his mess. Wiping up the residue, he noticed long, thin fibers had formed. Chardonnet received a patent on this material, but again the flammability kept it from achieving large market adoption.

The Viscose method mentioned earlier was finally worked out in 1894 by English chemists Charles Frederick Cross, Edward John Bevan, and Clayton Beadle. This method was a commercial success, and the fabric was manufactured first by Courtaulds Fibers in the United Kingdom and then Avtex Fibers in the United States.

Where does **rubber come from**?

Rubber trees! No kidding. Natural rubber is collected from rubber trees like maple syrup comes from maple trees, except the syrup is latex sap. It's a polymer of isoprene where each carbon–carbon double bond along the chain has *cis*-stereochemistry. Although there are man-made alternatives produced artificially, even today about half of the rubber produced each year on our planet does come from rubber trees.

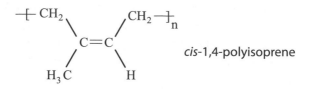

cis-1,4-polyisoprene

What is **vulcanization**?

Natural rubber that comes directly from rubber trees looks nothing like your car or bike tires. It's sticky, doesn't hold its shape when it gets warm, and if you live where it snows, it can get brittle. That sounds like an awful material to make a tire out of! Vulcanization creates crosslinks in the rubber with the addition of sulfur to the natural rubber chains and improves all of these properties.

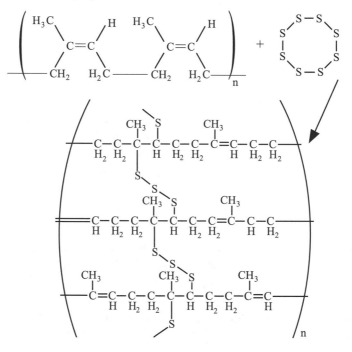

What is an **addition polymerization reaction**?

The easiest way to describe an addition polymerization reaction is that monomers are bonded together without the loss of any atoms of the monomer. There are more complicated ways to classify this type of reaction based on kinetics, but it essentially boils down to this fact.

How is a **condensation polymerization reaction** different?

Addition polymerization reactions do not involve the loss of any atoms from the monomer, but condensation polymerization reactions do. The molecule that is lost is almost always water.

What do the recycle numbers mean on plastic bottles?

These are technically called Resin Identification Codes (RSI) and were introduced in the 1980s to make it easy to separate plastics for recycling. The numbers correspond to what kind of polymer the item is made of and have no other meaning. They are not in any sort of order based on how easy or hard it is to recycle the resins, despite rumors you might have heard about this.

RSI	Plastic
1	Polyethylene terephthalate
2	High-density polyethylene
3	Polyvinylchloride
4	Low-density polyethylene
5	Polypropylene
6	Polystyrene
7	All others

What is a **thermoplastic**?

If a polymer becomes soft when it is heated (and then hardens again when cooled), it's known as a thermoplastic. The temperature at which the polymer softens depends on what it's made of and on the size of the polymer chains. Thermoplastics are easy to recycle because they can be remolded when they're hot.

What is a **thermoset**?

Unlike a thermoplastic, thermoset materials are cured such that they don't soften when exposed to heat (at least up to a certain point). This curing step usually introduces a lot of crosslinks between polymer chains, which creates a rigid network. These materials are much more difficult to recycle, but are used where high-temperature stability is needed.

What is **PET** (polyethylene terephthalate)?

Polyethylene terephthalate is a thermoplastic that is used in both textiles, where it's called polyester, and in bottling, where

According to the EPA, in 2010 Americans only recycled eight percent of their plastic waste. We can do better. About seventy-five percent of the packaging we use is recyclable.

the same polymer is called PET. PET is an alternating polymer of ethylene and tereph-thalate monomers. This material is really good at preventing gases from diffusing through it, so it's great for keeping carbonated drinks fizzy.

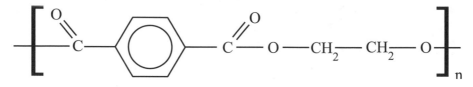

What is **HDPE** (high-density polyethylene)?

High-density polyethylene is technically any polyethylene with a density of 0.93 to 0.97 g/cm^3. The density in polyethylene is controlled mostly by the number of branch points along the polymer chain. HDPE has very few branches, so the chains can stack together very closely. This tight packing makes it a very strong polymer, so it's used to make things like bottlecaps, milk jugs, and Hula Hoops.

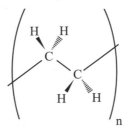

What is **LDPE** (low-density polyethylene)?

If the density of a polyethylene is between 0.91 to 0.94 g/cm^3 (yes, there's a little over-lap in these ranges) it's called low-density polyethylene. To get to this density, the poly-ethylene chains have more branching than in HDPE but still only a few percent of the atoms along an entire chain. These branches prevent the chains from stacking together quite as well, which makes the material softer and more flexible. With those properties, LDPE finds use as trash, grocery, and sandwich bags, and that "clingy" food wrap (al-though the original Saran® Wrap was not LDPE—see below).

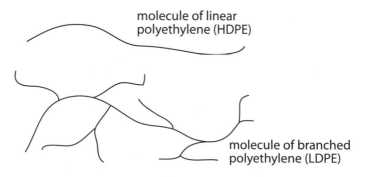

molecule of linear
polyethylene (HDPE)

molecule of branched
polyethylene (LDPE)

What is **PVC** (polyvinyl chloride)?

If one of the hydrogen atoms on every ethylene monomer in polyethylene is replaced by a chlorine atom (note that this is not how this material is actually made!), you get PVC, or polyvinylchloride. It's the third-largest-volume polymer produced each year behind polyethylene and polypropylene. It is a very tough polymer, so it is used to make pipes and flooring among many other things. PVC can also be softened (technical term: plasticized) by introducing small organic molecules, like phthalates (a benzene ring with two esters). Among other applications, plasticized PVC is used to insulate electric wires and to make your garden hose.

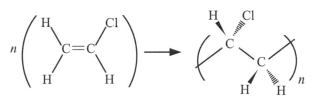

What is my **credit card made of**?

Also PVC—but for the material in credit cards, no plasticizer is added. To manufacture a credit card, usually a few thin sheets of PVC are glued together.

What is **PP** (polypropylene)?

If instead of substituting a chlorine atom we add a methyl group to each ethylene monomer, we get polypropylene. Recall from earlier that this introduces stereochemistry along the polymer. We've already mentioned that the arrangement of the methyl groups along the polymer chain can have large effects on the melting point and other physical properties. You can likely find polypropylene all over your house from dishwasher-safe food containers to synthetic carpets (especially outdoor carpeting) and an increasing weight fraction of your car, including the bumper and the casing for the battery. It can also be made into ropes, which are quite strong and resistant to weather, so they are frequently used in fishing and farming. Polypropylene is also used for many medical applications because it is capable of withstanding the high temperatures required to sterilize.

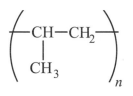

What is **PS** (polystyrene)?

If one of the hydrogen atoms on every ethylene monomer in polyethylene is replaced by a benzene ring (note again that this is not how this material is actually made!), you get PS, or polystyrene. Polystyrene is usually the fourth-largest-volume polymer produced globally, with billions of pounds made annually. Polystyrene can be manufactured into parts (like CD cases, furniture, and eating utensils), or air can be mixed with the polymer to make a foam used in insulation both for your house and your coffee cup. Styrofoam® is a trademarked name held by the Dow Chemical Company for foamed polystyrene.

What is **Saran® Wrap**?

Saran® Wrap is a trade name (another held by the Dow Chemical Company) for polyvinylidene chloride. If two of the hydrogen atoms on every other carbon in poly-ethylene are replaced by chlorine atoms (note *yet* again that this is not how this material is actually made!), you get PVDC, or polyvinylidene chloride. It was discovered by accident in 1933 by Ralph Wiley, who was having trouble washing this strange material out of the bottom of a piece of glassware. The actual polymer they were trying to make was poly(perchloroethylene)—where every hydrogen is replaced by a chlorine atom. It was just before WWII that a breakthrough was made that allowed the scientists to make film from this new material. It was quickly adopted by the Army to wrap equipment being transported by sea in order to prevent corrosion from saltwater and other applications to keep soldiers dry in jungle environments. After WWII, Dow found a new use for the material and introduced a PVDC film product for wrapping food called Saran® Wrap. The clingy food wrap you buy today is not PVDC, however. This material was phased out due to environmental and health concerns of those chlorine atoms, and low-density polyethylene took its place.

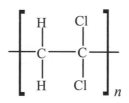

Okay, but **why** was **PVDC** ever **called Saran® Wrap**?

Many industrial trade names have no interesting story behind the creation. Saran® Wrap is an exception. You might think that Ralph Wiley was responsible for naming this material, having discovered it. Nope. Ralph Wiley's boss, John Reilly, named the material after his wife and daughter—Sarah and Ann.

What is **nylon**?

Nylon is a synthetic polymer formed by the condensation of a dicarboxylic acid with a diamine. This reaction forms an amide bond and releases a molecule of water. The term

What makes my cooking pans "nonstick"?

The coating that is placed on cookware is usually polytetrafluoroethylene (PTFE), marketed as Teflon® by DuPont®. The strength of the C–F bond, and its reluctance to interact with just about anything else, makes Teflon® very heat-resistant and slippery stuff. Aside from coating cookware, it's used to make gears and bearings, and it's a key component of Gore-Tex (the material your waterproof jacket is made of).

"nylon" is a generic name for these types of polymers, but one of the common nylons is called "Nylon 66." The numbers signify the number of carbon atoms in the amine (6) and the acid (6) reactants.

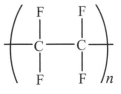

When was **nylon discovered**?

Nylon 66 was first made by Wallace Carothers, a scientist working at DuPont®, on February 28, 1935. Dr. Carothers also contributed to the discovery of neoprene, which is used to make suits used for scuba diving.

What was **nylon first used** for?

The first commercial application was probably toothbrush bristles. For centuries toothbrushes were made of coarse animal hairs (usually boar) until Dupont® introduced Doctor West's Miracle Toothbrush in 1938.

What is **silicone**? Is that the same as **silicon**?

Silicon is an element, while silicone is a polymer with a backbone of silicon and oxygen atoms. These polymers are very resistant to heat and have a rubbery feel. The latest squishy cookware and bakeware is made out of silicone.

What is **glue**?

There are many different types of adhesives, but you're probably thinking of that white glue you had as a kid. This type of glue is known as a "drying adhesive" because it hardens by evaporation of a solvent. In the case of white glue, the solvent is water and the sticky stuff that gets left behind is polyvinyl acetate.

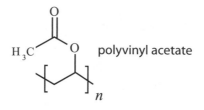

polyvinyl acetate

Does **hairspray** contain **polymers**?

Yep—and almost the same ones that make white glue and acrylic paints. Many hairsprays contain vinyl acetate (or something similar), polyvinyl pyrrolidone, and/or lots of other variants. Like glue, you also need a liquid to disperse or dissolve the polymers, but in hairspray a mixture of alcohol and water is normally used.

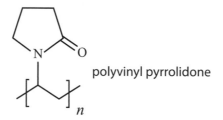

polyvinyl pyrrolidone

What's in **paint**?

There are three main ingredients in paint: binder, solvent, and pigment. Binder is the stuff in paint that forms an adhesive so that it sticks to the wall. Unlike glue, it's not a polymer in paint (at least not a fully formed one), but instead monomers or short polymer chains that react (crosslink) to form larger polymer networks as the solvent evaporates. The solvent in paint is there to make the paint solution the right thickness so that you can easily put it on the wall without it dripping all over the floor. Then the solvent evaporates, driving the formation of the crosslinked polymer networks. Pigments are, of course, used in paint so that not everything is painted white (though white paint still needs pigments to be white!).

How is **recycled plastic** used to **make fleece**?

Fleece can be made from polyethylene terephthalate (PET) bottles. The first step is to wash and then mechanically crush the bottles, shaping the plastic into small chips. The chips can then be heated and forced through tiny holes in a metal plate (called a spinneret), which forms fibers that harden as they cool to room temperature. These fibers are wound onto a spool as they are formed, and they can subsequently be stretched to improve their strength. Machines can then be used to texturize and cut the fibers to their desired length and be used to make fleece cloth for clothing, blankets, etc.

There are **polymers** in my **shampoo** and **conditioner**, too?

There are! Many ingredients in shampoos and conditioners are similar to most soaps (surfactants, etc.), but cationic polymers play important roles in these products. One family of these polymers is named "polyquaterniums," which is closer to a trade name than a technical chemistry name (the structure of one member of this family, "polyquaternium 1," is shown below). All of these polymers contain positive charges, which allows them to form ionic bonds with hair strands. This prevents the polymers from being rinsed away with water. Once coated, your hair strands are now less likely to stick to adjacent strands and appear shinier.

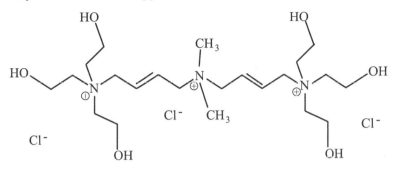

What is **fiberglass**?

Fiberglass is a polymer made from a plastic matrix that is reinforced by fibers of glass. It is a popular material because it is inexpensive to make, and its strength and weight properties compare favorably against those of metals for many applications. Fiberglass has a wide range of applications and is commonly used in glider aircraft, boats, cars, showers and bathtubs, roofs, pipes, and surfboards.

What **absorbs liquid** in **diapers**?

The general term for the materials that absorb water in diapers is "super-absorbent polymers"; these same compounds are used to chill drinks and to make fire retardants and fake snow. Modern absorbing materials are typically sodium salts of polyacrylic acids, and these can become almost entirely water by weight and as much as 30–60% by volume.

Wait—**absorbent polymers** are also used to **chill drinks**? How does that work?

If you stuff a cup holder with absorbent polymer and then soak it in water, the polymer will obviously swell up. The water will slowly evaporate out of the polymer, which reduces the temperature of the polymer gel and ultimately your drink.

What is **Styrofoam®**?

Styrofoam® is a brand name (owned by the Dow Chemical Company) for expanded polystyrene foam. It's 98% air, which is why those styrofoam® coffee cups are so light (and actually buoyant). In addition to disposable food containers, polystyrene foam is used in building and pipe insulation, packing peanuts, and that green stuff they use for holding together fake flower arrangements.

And **spandex**? What's that?

Spandex (as it's called in North America; Europe knows the same material as "elastane" and the Brits refer to it as "lycra," but that's confusing because Lycra® is actually a trade name) is a rigid copolymer of polyurethane and polyurea and is a rubbery material, like polypropylene oxide. These two polymers do not mix well, so tiny domains of each polymer form. It's this separation (of the hard bits from the soft bits) that gives Spandex its stretchy, yet strong, behavior.

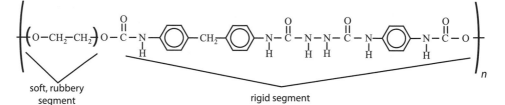

ENERGY

ENERGY SOURCES

What is the breakdown for **fuel source usage globally**?

Roughly 32.4% of the world's energy is used in the form of oil. About another 27% comes from coal sources, and another 21% from natural gas. Together, these three sources account for a total of over 80% of the energy sources used globally. The remaining 20% comes primarily from combustible renewables and waste (10%), nuclear energy (6%), and hydroelectric sources (2%). As you can see, some of the commonly discussed green energy sources, like solar energy or energy collected by wind turbines, do not yet make significant contributions to the breakdown of energy sources commonly used on a global scale.

Energy Sources Worldwide*

Energy Source	% of Total Used
oil	32.4
coal/peat	27.3
natural gas	21.4
biofuels and waste	10.0
nuclear	5.7
hydro	2.3
other	0.9

*According to a 2010 survey published by the International Energy Agency (http://www.iea.org/publications/freepublications/publication/kwes.pdf).

What is **crude oil**?

Crude oil is the oil that is found underground in natural reserves. It is formed from the natural decay of living things that were in the sea millions of years ago, which is why it's commonly called a "fossil fuel." The key ingredient that makes oil a valuable energy source is hydrocarbons. As we discussed briefly in "Chemical Reactions," hydrocarbons release substantial amounts of energy when they undergo combustion reactions.

How is **oil refined**?

There are several processes that can be used to refine oil. The oldest way is through distillation, which involves heating up the crude oil slowly, allowing the different hydrocarbons to boil off one at a time, and then collecting them as vapors. There are also newer chemical refinement methods that can use a chemical reaction to convert one type of hydrocarbon into another.

Oil refineries like this one can either refine oil through distillation processes or through chemical methods.

What is **cracking**?

Fluid catalytic cracking is the name for the process that converts high-boiling hydrocarbons found in crude oil into lighter hydrocarbons that are more useful as gasoline and other products. These units operate continuously for years at a time, and there are hundreds of them around the world. In addition to the high temperature and heat that these reactors use to break down long hydrocarbons, there are catalysts used to help speed up the process. These catalysts are usually strong acids in the form of zeolites, specifically faujasite, which is a mixture of silica and alumina.

What is **fracking**?

Hydraulic fracturing, or "fracking," is a technique used to break up rock deep underground in order to release the natural gas trapped inside it. The fluid injected into the rock is generally water-based, but the specific mixtures of chemicals dissolved in that water are generally a closely guarded secret. Using fluids to crack rocks is an old technique dating back over a hundred years, but modern techniques trace back to experiments in the 1940s by Floyd Farris and J.B. Clark, who worked for Stanolind Oil and Gas.

Where does **coal come from**?

Coal is formed from trees and plants that died hundreds of millions of years ago. The dead plant material eventually became buried deep underground, which placed a large amount of pressure on it. Over time the plant material became compressed and hardened, eventually forming coal. While this resource can be replaced over very long periods of time, we use it much, much faster than it can ever be regenerated.

If fracking is an old technique, why does it seem to be taking off now?

The second key innovation that has allowed fracking to grow so fast in recent years is the ability to drill wells horizontally. Technically any well that isn't vertical is called directional or slant drilling. Hydraulic fracturing can release trapped natural gas from a much wider area this way, which has led to the recent surge in this technique.

How does a **nuclear reactor generate energy**?

Nuclear power comes from the energy released in fission processes, like the ones we discussed in "Nuclear Chemistry." The nuclear reactors used for commercial energy production make use of controlled chain reactions of uranium-235. Upon colliding with a neutron, uranium-235 decays spontaneously into two lighter atoms, releasing energy for each atom of uranium that is split. The amount of energy released in the fission process is very large compared to that released from the combustion reactions that we use to obtain energy from fossil fuels. The result is that we can get a lot of energy from a relatively small amount of uranium. In fact, a kilogram of enriched uranium can generate about the same amount of energy as eight million liters of gasoline!

Where do we get **uranium**?

Uranium is mined largely in Kazakhstan, Australia, and Canada, although the U.S., South Africa, Namibia, Niger, Brazil, and Russia are also significant producers. Uranium is mined using a leaching process that allows it to be dissolved from ore deep in the mine and then pumped to the surface, where it is extracted and concentrated for use.

POLLUTION

What are the **potential dangers** of **nuclear energy**?

The main risks associated with nuclear energy have to do with health hazards of radiation poisoning. This isn't the same type of radiation we're talking about with light and microwaves (that type is electromagnetic radiation). The radiation associated with nuclear energy involves subatomic particles, like neutrons, given off from nuclear decay processes. Exposure to this type of radiation can cause cancer or genetic defects, among other problems. Since the fission processes used to generate energy are normally well controlled, though, nuclear energy is generally one of the cleanest energy sources available. Nuclear reactors only become dangerous in the event of a meltdown or other disaster, but the chances of these happening are extremely small. Modern nuclear power plants are built with a series of fail-safes such that a series of multiple, extremely unlikely events would have to happen, one after another, for a meltdown to occur.

What **pollutants** do **nuclear power plants** generate?

Not many, unless something goes wrong. No greenhouse gases (like carbon dioxide) are generated in nuclear power plants, and each only produces roughly one cubic meter of waste per year. As long as this is properly taken care of, nuclear power plants are one of the cleanest forms of energy.

What **pollutants** do **coal power plants** generate?

Burning coal introduces pollutants such as nitrogen oxides, sulfur dioxide, and mercury into the atmosphere. A large amount of carbon dioxide, the principal pollutant implicated in global warming, is also produced by coal power plants. Aside from global warming, many of the pollutants from coal power plants are bad for human health and also cause smog, soot, and acid rain. The solid waste produced by burning coal can also be harmful to the environment. The ash from coal power plants is typically composed of about 5% pollutants or dangerous substances such as arsenic, cadmium, chromium, lead, and mercury. This can also cause pollution in water if it is not disposed of properly, and there have been many cases of water contamination caused by improper disposal of this ash.

Which **energy sources produce** the **most CO_2**?

By far, it's the fossil fuels: oil, gas, and coal. It's the CO_2 from all those combustion reactions that makes up about 96.5% of all the carbon dioxide emissions.

Which **countries produce** the **most CO_2**?

Based on data for the year 2009, the top twenty-five, in order, are:

1. China
2. United States
3. India
4. Russia
5. Japan
6. Germany
7. Canada
8. South Korea
9. Iran
10. United Kingdom
11. Saudi Arabia
12. South Africa
13. Mexico
14. Brazil
15. Australia
16. Indonesia
17. Italy
18. France
19. Spain
20. Taiwan
21. Poland
22. Ukraine
23. Thailand
24. Turkey
25. Netherlands

How much CO_2 production comes from automobiles?

According to the Environmental Protection Agency (EPA), the transportation industry as a whole accounts for 33% of all CO_2 emissions in the U.S. Automobiles used for personal transportation account for 60% of this, or 20% of total U.S. CO_2 emissions. The remainder comes from other sources of transportation, such as large diesel vehicles and airplanes burning jet fuel.

What is **carbon sequestration**?

Carbon sequestration and carbon capture are processes to remove carbon dioxide (CO_2) from the atmosphere and store it. This is exactly what plants do when they convert CO_2 into other molecules like sugars and proteins. Carbon dioxide can also react with water and limestone ($CaCO_3$) to form calcium bicarbonate ($Ca(HCO_3)_2$) in an inorganic example of natural carbon sequestration. Generally, though, this term is used today to refer to man-made processes for removing CO_2 from the atmosphere or for capturing it before it gets released (like from power plants). A number of approaches to long-term carbon dioxide storage are being, or have been, tried, including pumping the gas deep underground into natural rock formations, scrubbing the CO_2 out from flue gas by reacting it with bases, or transforming the CO_2 into other useful molecules for making polymers, just to name a few.

What **causes smog**?

Smog is caused by chemical reactions involving volatile organic chemicals and nitrogen oxides that take place in the presence of sunlight. These pollutants can come from a variety of sources, but in urban areas, a large quantity usually comes from motor vehicles. These are the reasons that smog often becomes a bigger problem when there is heavy traffic and lots of sunlight.

What causes **acid rain**?

Similar to smog, acid rain is caused by chemical reactions of pollutants like sulfur dioxide and nitrogen oxides with oxygen and water in the air. The acidic pollutants formed in these reactions are the cause of acid rain. In addition to environmental pollution due to human activities, volcanic activity can also cause acid rain. Acid rain can be harmful to natural

Trees and other plants are seen here, damaged or killed by acid rain. Acid rain results when water in the atmosphere combines with chemicals like sulfur dioxide, acidifying rainwater.

181

plant life, farm crops, animals, and marine life. It can also be damaging to buildings, depending on what materials were used in their construction.

What is the **air quality index** (AQI)?

The air quality index, or AQI, is a measure describing the amount of particulate matter found in the air. Since this value can vary significantly over short periods of time, the AQI is typically reported for a given city at least once per day. The higher the AQI value, the greater the associated health concerns. Since a few different measures exist out there for characterizing the air quality, we will spare you the details of exactly how the number is calculated. It is worth paying attention to the AQI when you are considering traveling or relocating to new places. Some large cities have developed major ongoing problems with their air quality, and looking at a city's recent AQI history is a useful indicator of the air quality you can expect to find.

The table below summarizes relevant ranges of AQI values used in the United States and their health implications.

Air Quality Index	Level of Health Concern	Color (AQI)
0-50	Good	Green
51-100	Moderate	Yellow
101-150	Unhealthy for Sensitive Groups	Orange
151-200	Unhealthy	Red
201-300	Very Unhealthy	Purple
301-500	Hazardous	Maroon

What is the **ozone layer** and why is it so **important**?

Ozone has the chemical formula O_3, and it is a gas that is naturally present in the Earth's atmosphere. Most of the ozone in the atmosphere exists as part of a layer that sits a few miles above the ground. This is the ozone layer, and it serves to protect the Earth from a significant fraction of the potentially harmful UV light from the Sun. Depletion of the ozone layer means that more harmful UV light reaches the Earth's surface. Ozone is also a greenhouse gas, which means that it also plays an important role in controlling the Earth's climate.

What **pollutants** are **dangerous** to the **ozone layer**?

The pollutants principally responsible for depletion of the ozone are volatile halogenated organic compounds. The most well known among these are chlorofluorocarbons (CFCs) and hydrochlorofluorocarbons (HCFCs), which used to be used in virtually all air conditioning, refrigeration, and cooling systems. Another is methyl bromide, which is used commercially as an agricultural fumigant.

EMERGING SOURCES OF ENERGY

How do **solar cells work**?

The Sun gives off 1,000 watts of energy per square meter of the Earth's surface, which, if we could harness all of it, is more than enough to fulfill all of the world's current electricity needs. It's not easy to do this, however, which is why we still rely primarily on other energy sources. A solar cell, or photovoltaic cell, harnesses energy from photons of sunlight by having them excite electrons in a material. An electric field is established within the solar cell using a process called doping, and this makes it so that the excited electrons can only flow in one direction. Thus, when sunlight strikes the solar cell, a current flows, and this is the basic principle on which solar cells operate to capture energy. The harnessed energy can either be used immediately or it can be stored for later use. Designing more efficient solar cells is a very active area of research, and you can expect their efficiency to continue to increase in coming years.

How do **wind turbines** generate **energy**?

Wind can be converted into energy by a wind turbine, which typically consists of a wheel with large blades connected to a series of gears. As wind causes the blades of the turbine to spin the turbine collects mechanical/kinetic energy, which can either perform mechanical processes or be converted into electrical/potential energy. Some mechanical processes that make use of wind turbines are pumping water and grinding grain. To store electricity the turbine must be connected to a generator, which converts the mechanical energy from the blades into stored electrical energy. Typically, wind speeds of 7–10 mph are needed to generate energy using a wind turbine. The orientation of a wind turbine can be controlled by a computer to optimize the amount of energy it collects.

What are **carbon offsets**?

A carbon offset involves a party committed to reducing emissions of greenhouse gases in order to compensate for emissions made elsewhere. The gases involved include carbon dioxide (CO_2), methane (CH_4), nitrous oxide (N_2O), sulfur hexafluoride (SF_6), perfluorocarbons, and hydrofluorocarbons. Conversion factors are used to attempt to equate the negative effects of the different gases on the atmosphere. Carbon offsets are typically purchased by either companies or governments under regulations with regard to greenhouse gas

Photovoltaic cells, or solar cells, work by containing materials such as mono- or polycrystaline silicon, cadmium telluride, or copper indium selenide whose electrons are easily excited by photons from the sun, creating electricity.

183

emissions or by personal consumers who wish to offset their own contributions to greenhouse gas emissions.

What is **biodiesel**?

Biodiesel is a type of fuel made from vegetable oil and/or animal fat which consists of long alkyl esters (see "Organic Chemistry" to review functional groups). Biodiesel fuel is commonly distributed for sale as a mixture with petrodiesel and is labeled with a "B factor" describing the fraction of biodiesel present. B100 is pure biodiesel, while B20 would represent a mixture of 20% biodiesel with 80% petrodiesel.

How is **biodiesel produced**?

The production of biodiesel fuel involves a chemical reaction called transesterification of lipids (those from the vegetable oil or fat) with alcohols to produce alkyl esters. Methanol is most commonly used as the alcohol in these reactions, which leads to methyl esters, though other alcohols have also been used.

Glycerol is a byproduct of the transesterification reaction, and this compound is actually formed in fairly substantial quantities (ca. 10% by mass). This has given rise to research directed toward finding ways to carry out chemical reactions beginning with, or involving, glycerol, such that the cost efficiency of the overall process of biodiesel production might be improved.

How is **ethanol produced** for use as a **fuel**?

Ethanol is produced by fermenting sugars from plants like corn, soybeans, and sugarcane. The sugars in the plant material are first broken down and then "fed" to yeast, which ferments them to produce ethanol as a byproduct.

How does **ethanol work** as a **fuel**?

Ethanol works as a fuel in much the same way as gasoline; it is burned in a combustion reaction to release energy. In automobiles, ethanol is usually mixed with gasoline. Most

---

How could algae potentially be used as a source of energy?

Algae are potentially a very useful source for biofuels, but currently the cost of using them as a source of biofuels is too high to make them practically useful. One current area of research involves using the algae biofuel production process to produce a by-product that is rich in protein and that could be used to feed farm animals. This would help to offset the cost associated with algae biofuel production. Since corn is currently used to both feed animals and to produce ethanol for fuel, the protein-rich by-product could also help by reducing the amount of corn that needs to be grown.

cars can run on an ethanol-gasoline mixture with 10% ethanol (E10)—but to use an 85% ethanol mixture (E85) requires a specially designed system. Ethanol burns much cleaner than gasoline, making it less hazardous to the environment. It is also a renewable resource that can be produced by growing crops, so it has the potential to reduce dependence on foreign oil sources.

An experimental fuel cell car is shown here. Fuel cell technology has been deployed commercially in recent years, but is more often seen in public vehicles such as buses than in private cars.

How is **hydrogen** produced for use as a **fuel**?

Most of the current hydrogen production in the U.S. is carried out by steam reforming natural gas (methane), which involves reacting steam with methane to generate H_2. To make hydrogen a viable fuel source, more efficient ways of producing hydrogen on the large scale will be needed. Many scientists are currently investigating chemical and biological catalysts capable of carrying out a process called water splitting, which is the production of H_2 and O_2 from water (H_2O). Water splitting may have the potential to make hydrogen into a viable fuel source for vehicles.

How do **hydrogen-powered cars** get their **energy**?

Hydrogen-powered cars are based on fuel cells that store hydrogen, or H_2 gas, inside a material called a polymer exchange membrane. The fuel cell contains two electrodes: an anode (negative side) and a cathode (positive side). At the anode, the H_2 molecules are split into protons and electrons. The protons pass through a polymer exchange membrane, while the electrons are unable to pass through this membrane and thus have to flow in a different direction. This creates a current of electricity by which the car is powered.

What **wavelengths of light** from the Sun **reach Earth's surface**?

The light from the Sun reaching the Earth's surface spans a range of wavelengths between roughly 300 to 2,500 nanometers with a few gaps in between where atmospheric water and carbon dioxide absorb radiation. The range of wavelengths spanned by light in the Earth's upper atmosphere is slightly broader, demonstrating that gases present in the atmosphere absorb a notable amount of light before they reach the Earth's surface.

How much energy in the **United States** comes from **hydroelectric power** sources?

About ten percent of the electricity in the United States comes from hydroelectric power sources. The state of Washington leads the United States in hydroelectric en-

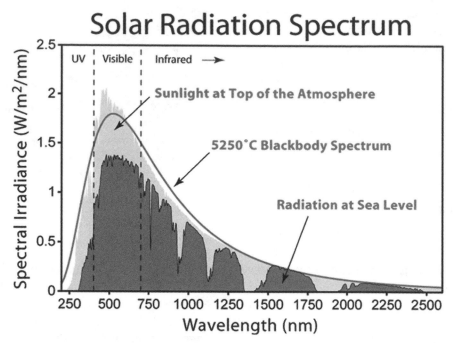

Solar Radiation Spectrum

As sunlight reaches the earth, it must penetrate the atmosphere, which absorbs or reflects much of the radiation before it reaches the surface of the planet.

ergy production, with roughly 87% of this state's electricity coming from hydroelectric power!

What is natural gas, and where does it come from?

Natural gas is primarily composed of methane (CH_4), and like other hydrocarbons, methane provides energy via combustion reactions. Natural gas is typically found underground, near sources of petroleum, and can be pumped up via pipelines. Actually, natural gas does not have a smell, so it is mixed with a small amount of a strong-smelling thiol compound so that you can tell if you have a gas leak in your home.

What is fusion and how might it be used as an energy source?

We looked at fusion in "Nuclear Chemistry." To recap, fusion involves the combination of two nuclei to form a single nucleus. For lightweight nuclei, this process typically involves the release of energy. Unfortunately, it's currently very difficult to make fusion happen under practical conditions; nuclei typically have to be accelerated to very, very high speeds to make fusion take place.

What is a renewable energy source?

A renewable energy source is one that is provided by resources that are continuously replenished or which will always be available during the foreseeable future of the planet.

So if the radiation coming from the sun more or less covers all wavelengths of visible light, why aren't plants black?

Let's start with explaining the question a bit more. You can see from the graph on the preceding page that the sun provides light in almost every wavelength. The vast majority of plants, however, appear green, which means they are absorbing blue and red light and reflecting the green light back into your eyeballs. If plants were taking in all the energy the sun is providing, they would appear black, as no light would be reflected. So, to restate the question, why do plants reflect any light at all?

The easiest answer here is that absorbing red and blue energy obviously works well enough, so evolution stuck with it. Evolution, after all, doesn't provide the best solution, but rather a solution. Nonetheless, let's make some guesses as to why plants are green. Chlorophyll isn't just after energy in general, but needs very specific wavelengths to pass along energy to the reactive centers of Photosystems I and II. The second reasonable guess is that too much energy is not a good thing. Absorption of light creates high-energy species, and if they aren't used in some productive way the energy will find something else to do. These other reactions could be destructive to cells. In sum, chlorophyll only takes what it needs or what it can use.

Which **energy sources** are **renewable**?

Renewable energy sources include wind, hydroelectric, solar, biomass/biodiesel, and geothermally derived power.

What **fraction** of power **globally** comes from **renewable energy sources**?

It is estimated that about 16% of global energy consumption currently comes from renewable resources. Hopefully this number will increase significantly in the near future.

QUANTIFYING POWER

What is a **watt**?

A watt is the standard SI unit of power, representing 1 Joule of energy per second. It is named after a Scottish engineer by the name of James Watt. To give an idea of how much energy this is, lightbulbs in your house typically use about 25 to 100 watts (new types of lightbulbs are doing a bit better than this, actually). For comparison, if you are doing manual labor over the course of a long day, your body will average somewhere in the range of 75 watts of power output.

What is a **kilowatt hour**?

A kilowatt hour (KWH) is a unit of energy representing the amount of work that is done by 1,000 watts of power operating for one hour. Notice that a watt represents a quantity of energy per unit time, also referred to as power, while a kilowatt hour simply represents a quantity of energy.

How **many nuclear power plants** are there in the **United States**?

There are currently sixty-five nuclear power plants in the U.S. and a total of 104 nuclear reactors. This is because thirty-six plants have more than one reactor.

How **many nuclear power reactors** are there in the **world**, and how many are currently **under construction**?

There are currently 436 nuclear power reactors around the world. The breakdown of the distribution by country is shown in the table below. Aside from the United States, a few other countries with large numbers of nuclear reactors are France, Russia, China, Japan, and Korea—each of these nations has at least twenty nuclear reactors currently in operation.

Number of Nuclear Power Reactors in Countries*

Country	Number of Active Reactors	Number under Construction
Argentina	2	1
Armenia	1	
Belgium	7	
Brazil	2	1
Bulgaria	2	
Canada	19	
China	19	29
Finland	4	1
France	58	1
Germany	9	
Hungary	4	
India	20	7
Iran	1	
Japan	50	3
Mexico	2	
Netherlands	1	
Pakistan	3	2
Romania	2	
Russian Federation	33	11
Slovakia	4	2
Slovenia	1	
South Africa	2	
South Korea	23	3
Spain	8	

Number of Nuclear Power Reactors in Countries* (continued)

Country	Number of Active Reactors	Number under Construction
Sweden	10	
Switzerland	5	
Taiwan	6	2
Ukraine	15	2
United Arab Emirates		1
United Kingdom	16	
United States of America	104	1

*As of January 2013.

How much energy is produced from a given amount of coal, oil, or gas?

One ton of coal can produce 6,182 kWh of energy

One barrel of oil can produce 1,699 kWh of energy

One cubic foot of gas can produce 0.3 kWh of energy

How much does it cost to produce energy from gas, coal, solar energy, etc.?

If one ton of coal costs $36, which gives an energy cost of $0.006 per kWh

And one barrel of oil costs $70, which gives an energy cost of $0.05 per kWh

And one cubic foot of gas costs $0.008, which gives an energy cost of $0.03 per kWh

Then a 4 kW solar panel system providing energy for an average ranch-style home and costing $25,000 and lasting about 20 years would provide about 120,000 kWh for the life of the system (depending on climate) and cost about $0.21 per kWh.

You can see that solar power is still more currently more expensive than other fuel sources.

THE MODERN
CHEMISTRY LAB

PURIFICATION IS ESSENTIAL

How do **chemists purify compounds**?

A few of the most commonly used purification methods are chromatography (see following question), recrystallization, or extraction methods. These methods typically exploit a difference in how a property (like polarity, for example) of one chemical species causes it to interact differently with the surrounding material than the others from which it is being separated.

What is **chromatography**?

Chromatography is a method for separating chemical compounds based on chemical properties as they travel over a distance. The sample to be separated may be in a solution or in the gas phase.

How is **gas chromatography different** than **liquid chromatography**?

Gas chromatography involves first vaporizing the sample, while liquid chromatography typically involves making a solution or suspension of the mixture to be separated. In gas chromatography all of the molecules in the vaporized sample will have the same average kinetic energy, but the different chemical species will each have different average velocities which are determined by their molecular weight. Heavier molecules will move with slower average velocities than lighter ones. This is how separation of different chemical components is achieved in a gas sample. In liquid samples, the mixture is dissolved in a solution that flows over a stationary phase, which interacts differentially with different chemical species in the sample. These different interactions with the stationary phase cause some compounds to move faster than others through the chromatography column, and this is the basis for separation via liquid chromatography.

What is the **"mobile phase"** in chromatography?

The mobile phase is the vaporized sample in gas chromatography or the solvent used to elute the sample over the stationary phase in liquid chromatography.

How is **liquid–liquid extraction** used in the chemistry lab?

Liquid–liquid extraction is a technique that exploits the differential solubilities of compounds between two liquid phases to extract the compound we want into a single phase. The liquids being used must be immiscible, meaning that they form two separate layers when placed in the same container. The goal is to have our compound of interest dissolved exclusively in a single liquid phase and also to have it be the only chemical compound dissolved in that phase. If this can be achieved, the liquid layer containing the desired compound is then separated and the solvent is removed, yielding the pure substance.

How does **crystallization purify compounds**?

Crystallization as a purification technique relies on the fact that it is usually much easier to form a crystal from a single chemical species than it is to form a crystal from a mixture of chemical compounds. Recrystallization is the process of dissolving a crude, impure product in a hot solvent (or mixture of solvents) and allowing it to crystallize out of the solution form as the solvent cools. Once a small crystal, even one so small we can't see it, begins to form, it is relatively easy for other molecules of the same compound to add to the crystal. Other compounds will not be able to readily add to this crystal, which results in the formation of a crystal containing a single, pure chemical compound.

What is **polymorphism**?

Polymorphism is the capacity for a compound to exist in different crystalline arrangements. This can arise due to different packing of the individual units of the crystal or due to different conformations of the molecules making up the crystal.

Why do **polymorphs matter**?

Different crystalline polymorphs will have different material properties and physical properties. In medicine, for example, the body may absorb some polymorphs of a drug more readily, making them more potent or more effective. Polymorphism can also change the fundamental physical properties of a material, affecting things like its conductivity and thermal stability.

How are **compounds separated** by **distillation**?

Distillation is a purification technique that involves using the different boiling points of compounds to effect their separation. As a solution containing a mixture of chemicals is heated the compound with the lowest boiling point will evaporate first, and this can be collected on a cool surface after it has diffused away from the solution. In this way, a pure liquid component can be isolated from a mixture of liquids.

Does every **compound** have a **unique boiling point**?

No, and this is one limitation of distillation as a separation technique. If two chemicals coincidentally have very similar boiling points, then it will be difficult to separate them using distillation.

Why would a chemist **measure** the **melting point** of a **chemical** sample?

The melting point of a compound can provide information on whether it is pure and about whether the correct compound has been made. This, of course, assumes that the melting point of the desired compound is already known. If a compound is being made for the first time, knowledge of the melting point can be useful to the next chemist who tries to synthesize it.

What **affects melting** and **boiling points** of chemical compounds?

Intermolecular (between molecules) forces govern the melting point of a chemical substance. These include Van der Waal's interactions, dipole–dipole interactions,

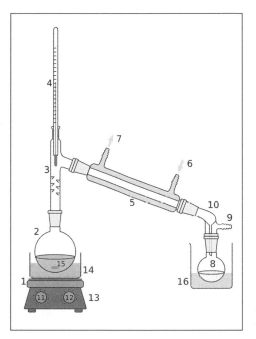

A simple distillation device (or "still") takes advantage of the fact that compounds usually have different boiling points and can be separated by applying heat to a solution. In this diagram, the parts include: 1) a heat source, 2) still pot, 3) still head, 4) thermometer/ boiling point temperature, 5) condenser, 6) cooling water in, 7) cooling water out, 8) distillate/receiving flask, 9) vacuum/gas inlet, 10) still receiver, 11) heat control, 12) stirrer speed control, 13) stirrer/heat plate, 14) heating (oil/sand) bath, 15) stirrer bar/ antibumping granules, and 16) cooling bath.

hydrogen bonding, and, in the case of ionic compounds or ionic solutions, ionic bonds or Coulombic interactions. The stronger the intermolecular forces between the molecules in a solid, the harder they will be to melt, so stronger intermolecular forces lead to higher melting points. The same is true with boiling points: stronger intermolecular forces make the molecules harder to separate, leading to higher boiling points.

For solids, the shape of a molecule can also affect its ability to pack into an ordered lattice. Having a shape that allows a well-ordered lattice to form will tend to stabilize the solid phase of a compound, leading to a higher melting point. The shape of a molecule can also affect the boiling point of a compound. In liquids that are able to form hydrogen bonds, the location of the hydrogen bond donor or acceptor can affect its spatial availability to serve as a donor or acceptor. In organic liquids, where Van der Waal's interactions are important, molecules with larger surface areas will have stronger Van der Waal's interactions, leading to a higher boiling point.

SPECTROSCOPY
AND SPECTROMETRY

What does **infrared (IR) spectroscopy measure**, and what does it tell you about a compound?

Infrared spectroscopy measures the absorption of light by molecules in the infrared region of the spectrum, which provides information on the types of functional groups present in a molecule. For example, looking at the infrared spectrum can provide evidence as to whether certain pairs of atoms are bonded by one, two, or three pairs of electrons. The shapes of the peaks in an infrared spectrum can also be used to gain information about intermolecular interactions between the species we are looking at and the surrounding medium.

How does **mass spectrometry** (MS) work?

The purpose of mass spectrometry is to ionize a chemical sample, causing it to fragment, and to characterize the mass of the fragment ions that form to gain chemical information about the sample. The first thing that happens is that the sample molecules need to be ionized, and this is done by removing an electron from the sample to yield a positively charged species. Once ionized, a sample of molecules will typically fragment in a characteristic way. The ions that form, whether they are fragments or the original ionized species, are accelerated and then deflected in a magnetic field. Ions of different masses are deflected by different amounts, and this is how the masses of the different ions are distinguished.

Why do **mass spectrometers need a vacuum** to operate?

As the ions travel, they must not collide with any other atoms or molecules before they reach the detector. If they collide with anything else this will change their direction and kinetic energy, which will make detection either unreliable or impossible.

What is **nuclear magnetic resonance (NMR) spectroscopy**?

Nuclear magnetic resonance spectroscopy is a type of spectroscopy that involves nuclei in a magnetic field absorbing and re-emitting radiation.

How does **NMR help chemists** determine the **structure** of a **compound**?

The peaks in an NMR spectrum can be related to certain properties of a molecule's structure. The number of peaks present tells how many types of chemically distinct atoms of a given element are present. A single NMR spectrum typically only contains information about a single element (for organic molecules, hydrogen or carbon NMR spectra are the most common), so it can be useful to record multiple NMR spectra for the same com-

When was NMR invented?

The first time NMR spectra were recorded was in 1945, and two different research groups accomplished this independently (one at Stanford and the other at Harvard). These groups were led by Felix Bloch (Stanford) and Edward Purcell (Harvard), who shared the 1952 Nobel Prize in Physics in recognition of their great discovery. See the list of Nobel Prize winners in Chemistry in the back of this book.

pound. The location of the peaks along the x-axis is known as the chemical shift, and this value can be related to the electron density surrounding each nucleus. In some NMR spectra, the intensities of each peak (their height on the y-axis) can be related to the number of nuclei of a given element that are present. There are more and more details that one learns after spending lots of time interpreting NMR spectra, but these are the basic principles that allow a chemist to relate an NMR spectrum to a chemical structure.

Why do **NMR instruments** use **big magnets**?

The purpose of using big magnets is, not surprisingly, to establish a strong, uniform magnetic field in the spectrometer. This creates an energy difference for the nuclear spins that are aligned either with or against the magnetic field. It is this energy difference that is being measured in an NMR spectrum, which can be related to molecular properties like the electron density surrounding a given nucleus.

What **compound** was the **first to be analyzed** by **proton NMR**?

The first recorded example where the shifts of the different nuclei in a molecule were separated by chemical shift was ethanol (CH_3CH_2OH).

What determines whether a particular type of nucleus can be monitored by NMR spectroscopy?

Similar to electrons, nuclei can have a net "spin," or spin angular momentum. NMR spectroscopy can be used to study any nucleus that has a nonzero spin angular momentum. Some common nuclei that chemists study by NMR are ^1H, ^2H (deuterium), ^{13}C, ^{11}B, ^{15}N, ^{19}F, and ^{31}P.

Are only **small main group elements detectable** by NMR?

The list of nuclei in the previous question are just commonly used ones in NMR experiments. Other elements are observable, but sometimes these require special hardware to obtain useful signal levels. Some of these other nuclei that can be studied by NMR are ^{17}O, ^{29}Si, ^{33}S, ^{77}Se, ^{89}Y, ^{103}Rh, ^{117}Sn, ^{119}Sn, ^{125}Te, ^{195}Pt, ^{111}Cd, ^{113}Cd, ^{129}Xe, ^{199}Hg, ^{203}Tl, ^{205}Tl, and ^{207}Pb.

Is **NMR** the **same** as **MRI**?

MRI, or magnetic resonance imaging, is based on the same principles as NMR spectroscopy. While NMR is typically used to investigate problems related to chemistry and physics, the goal of MRI is to image nuclei in living things. This is accomplished in a way similar to how an NMR spectrum is collected except that the applied magnetic field has a gradient, meaning the field strength is different in different parts of the sample being imaged (usually human or animal tissue). This allows the use of microwave radiation to excite nuclei in individual slices of tissue, one at a time. The gradient in the magnetic field can be varied to collect the MRI of different slices of tissue, which can then be analyzed to get a picture of what is going on inside the body.

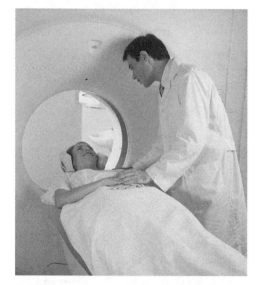

An MRI (magnetic resonance imaging) machine can be used to noninvasively examine the inside of a patient's body using a magnetic field. MRIs are very useful in diagnosing illnesses such as cancer, the effects of strokes, and torn ligaments.

Who built the first **MRI machine**?

Dr. Raymond Damadian, who was formally trained as a medical doctor, led construction of the first MRI. The first MRI capable of imaging a human was completed in 1977.

When I get an **MRI**, what is the **machine actually measuring**?

An MRI is basically just performing an NMR measurement on the hydrogen atoms in your body. Similar to an NMR instrument, an MRI instrument uses a large magnet and a radio frequency pulse to generate a rotating net magnetization in the hydrogen atoms in your body. The resulting magnetization generates an electric current in a receiver in the MRI instrument that can be processed to generate an image of what's going on inside your body.

OTHER MEASUREMENTS

How do those new **scanners at airports work**?

There are two main types of these scanners: one uses radio waves to generate a three-dimensional image of what's underneath a person's clothes, while the other uses low-intensity X-rays to generate a two-dimensional image. Both of these techniques rely on measuring the radiation that is scattered back off of one's body to generate an image. The purpose of these scanners is to look for basically the same things a security officer would look for in a pat-down: weapons, explosive devices, or anything else someone is trying to hide.

How is **X-ray diffraction used** by chemists?

Chemists use X-ray crystallography to determine the exact structure of chemical compounds. This involves taking a solid crystal of a pure compound and diffracting X-rays off of it, which produces a complex diffraction pattern. Using computer software, the diffraction pattern can be processed to yield a structure that describes the structure of an individual molecule of the compound making up the crystal. This is a powerful technique, but it can only be used on compounds that can be crystallized. It's also worth pointing out that the solid phase structure of a molecule is not always the same as that in a solution, so caution should be used when relating crystal structures to chemical reactivity in the solution phase.

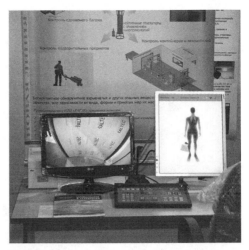

Airport scanners work by using radiation to see under clothing, whether it is X-ray or radio-wave radiation. The levels of radiation are kept low enough to be safe.

How do you **measure conductivity** of a solution?

Conductivity describes the ability of a solution to conduct an electric current. There are a few methods for measuring the conductivity of a solution, and the most straightforward to understand is probably the amperometric method. This method simply applies a voltage between two electrodes and measures the current. While this method is simple to describe, it can have complications in practice and it isn't always the most accurate method. Another way of measuring conductivity is with a potentiometric method, which makes use of two pairs of rings. There are two outer rings that apply an alternating voltage, and this results in a loop of current being generated in the solution. The other pair of rings sits inside the first and measures the change in voltage between the pairs of rings; this change in voltage is directly related to the magnitude of the current loop induced by the outer rings, which is in turn directly related to the conductivity of the solution. Other methods exist to measure conductivity, but these are probably the easiest to both describe and understand.

Why would a chemist want to **measure conductivity**?

Most practical applications of solution-conductivity measurements involve determining the quality of water samples. Conductivity measurements can provide a measure of the total amount of dissolved solids in a water sample. This information can be put to use in different ways, depending on the context, but in general it provides a measure-

A simple way to test the pH of a solution is with pH paper; the color of the paper is then compared to samples that show how acidic or alkaline a solution is.

ment of the purity of the water. Chemists sometimes use a method called ion chromatography, which is a type of liquid chromatography (LC). Ion chromatography often uses a detector that measures conductivity to detect when different analytes pass through the detector.

How do you **measure the pH** of a solution?

One of the first ways science students usually learn to test the pH of a solution is by using pH paper. This is a pretty simple test that only requires you to place a drop of the solution onto the paper and to look at its color. The color change accompanying changes in pH is due to a chemical indicator whose absorption spectrum changes with changes in pH. Another way to measure pH is with a pH-sensitive electrode. This can provide a more accurate measure of pH, as it digitally outputs the pH as a number value and does not rely on a person visually inspecting a color change to interpret the result.

What is a **centrifuge**?

A centrifuge is a machine that spins its contents very quickly and is used to separate the components of a mixture. The rapid rotation applies a force that causes the more-dense components of a mixture to collect on the bottom of the centrifuge tube and the less-dense components to rise toward the top.

What is a **centrifuge used for** in a chemistry lab?

In chemistry labs, centrifuges are typically used for separating suspensions. A suspension is a heterogeneous mixture containing small solid particles suspended in a liquid. The number of applications in biochemistry and biology laboratories probably far outweighs those in pure synthetic chemistry; centrifuges are often used to separate the contents of homogenized cellular material to isolate the proteins or cellular organelles. Centrifuges have also found applications in controlling the rates of reactions by simple partitioning of reactants; a centrifuge can be used to separate enzymes from their substrates in solution, which can serve to stop or significantly slow a reaction that is already taking place. In other cases, centrifuges have been used to attempt to accelerate reactions by forcing reactants together at the bottom of the centrifuge tube.

How are **centrifuges used** for **nuclear power**?

Centrifuges are used to enrich the uranium that is used in nuclear power plants. The two main isotopes of uranium are U–235 and U–238, with U–235 being the isotope used to generate nuclear power via fission processes. Unfortunately, over 99% of naturally oc-

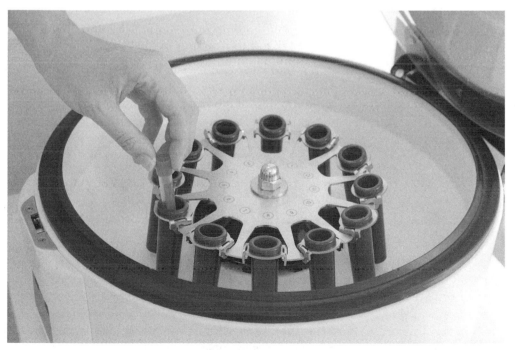

Centrifuges like this one are used to separate components in solution by spinning them rapidly, causing denser components to collect at the bottom of tubes.

curring uranium is U–238, so a lot of effort has to go into enriching the fraction of U–235 present in a sample. Centrifuges are often used to isotopically enrich a sample of uranium in the U–235 isotope. As described in an earlier question, this is accomplished by spinning a centrifuge tube, and in this case the heavier U–238 isotope is weighed down more, allowing a greater fraction of U–235 to be collected (in the gas phase) from the top of the centrifuge. This process can also be facilitated by heating the bottom of the centrifuge tube, which also helps the U–235 to move toward the top of the tube where it is collected. The process is typically repeated many times before the desired fraction of U–235 is reached. Highly enriched uranium often contains >85% Uranium-235 though, so clearly people have gotten pretty good at carrying out the isotopic enrichment process.

SAFETY FIRST!

What kinds of **safety precautions** are typically taken when working in a **chemistry laboratory**?

Chemists working with chemicals in a laboratory typically wear safety goggles to protect their eyes, a lab coat to protect their skin and clothes, and a pair of gloves to protect their hands. Of course, there are many situations when additional specialized protective gear is necessary.

What makes **strong acids and bases dangerous** to work with in the laboratory?

Strong acids, like nitric acid, are strong oxidizing agents and can cause severe burns on your skin. Strong bases, like sodium hydroxide, can also cause burns as well as nerve damage. Strong acids and bases destroy your cells by reacting with the membranes, proteins, and other components that make up your cells. They are also dangerous to your eyes and can permanently damage your vision.

Why should a chemist add acid to water and not the other way around?

This is an issue of safety. When an acid mixes with water, it will react very quickly. This reaction can release a large amount of heat and can cause the solution to bubble or splash. It's important to pour an acid into water, since acidic solutions are almost always more dense than pure water, meaning that the acidic solution will sink down into the water as it reacts. If you were to add water to a more dense acidic solution, the water would react at the surface of the solution, and there is a much greater chance it could bubble or splash up at you.

What is a **glovebox** and what is its purpose?

A glovebox is a method of carrying out chemical reactions under an inert atmosphere. It basically consists of a large box that has had the air removed and replaced with an inert gas (like pure nitrogen or argon). One side of the box is typically made of a clear, hard, transparent plastic, which has openings covered by large rubber gloves, into which a researcher can insert their arms. This allows a researcher to manipulate items inside the inert atmosphere of the glove box without letting any air inside. To move items in or out of the box, a purgeable antechamber is attached to one side of the box. While it is certainly not as easy to manipulate items inside a glovebox as it is to manipulate items outside the box, a glovebox provides one of the most straightforward approaches to carrying out chemistry in an air-free environment.

THE WORLD AROUND US

Many of the questions in this chapter were submitted by undergraduate students from around the country. We've listed all of their names in the Acknowledgement section at the beginning of the book. We want to thank them again for submitting such interesting questions, and if you have questions about chemistry that you'd like to see answered, please send them to us!

MEDICINE AND DRUGS

How does **Viagra® work**?

Viagra® (Sildenafil) interferes with an enzyme located in the erectile tissue that breaks down a molecule called cGMP (cyclic gunaosine monophosphate). Viagra® slows down this enzyme, allowing the amount of cGMP to build up. cGMP is responsible for a process known as vasodilation, or widening of the blood vessels in smooth muscle cells. More blood in these vessels means a stronger erection.

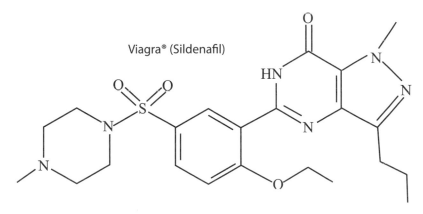

Viagra® (Sildenafil)

How does my **ADHD medication work?**

Ritalin® (Methylphenidate) is probably the most common drug for the treatment of ADHD (attention-deficit hyperactivity disorder). Ritalin® is a stimulant, increasing the levels of dopamine in your brain. Your brain normally releases dopamine to communicate feelings of pleasure and to increase the rate at which your neurons fire, which is supposed to help you focus.

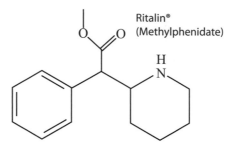

Ritalin®
(Methylphenidate)

How do **painkillers know** what part of the body to **target?**

Or phrased another way—why doesn't your whole body get a little numb when you take a painkiller? Let's limit this discussion to what are called nonsteroidal drugs, like ibuprofen. Ibuprofen, and other drugs like it, work by interrupting the series of signals that your body uses to communicate pain to your brain. The COX (cyclooxygenase) family of enzymes play an important, but intermediate, role in this pathway. This means that the COX enzymes take one type of signal and convert it into another that your body recognizes as pain. By working on an intermediate messenger, ibuprofen only takes effect where the pain is occurring—that first signal has to already be there—so your whole body doesn't get numb when you take ibuprofen.

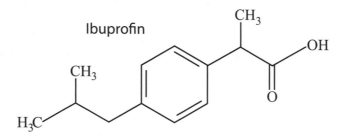

Ibuprofin

What's the **difference** between acetaminophen and ibuprofen?

While the structure of acetaminophen and ibuprofen are very different, the mode of action is similar. Both of these painkillers interfere with the COX family of enzymes we just talked about. Ibuprofen interacts more broadly with members of this family of enzymes, which ends up making it a good painkiller as well as a good antiinflammatory medication. Acetaminophen binds primarily to one of the members of this family (COX–2), so it's basically just a painkiller.

Why do some prescription pills come in such elaborate colors/capsules?

There's no chemical reason for this. Sometimes pills are colored for branding purposes—Pepto-Bismol® is pink, Viagra® is blue, Cymbalta® is green—and you recognize those colors. It can also help people keep track of which pill is which, and also to remember which ones to take.

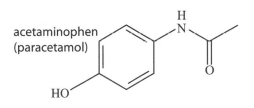

acetaminophen (paracetamol)

How does **cold medication work**?

Unfortunately, doctors and scientists still have not found a cure for the common cold! Cold medicines work by trying to minimize the symptoms of a cold while your body is busy fighting the virus that caused it in the first place. Some medicines, like those you buy at your local drugstore, work by using antihistamines, pain medicines, and decongestants to help relieve symptoms. These include all of the things you typically associate with a cold: a runny nose, a scratchy throat, sneezing, and dry or itchy eyes. Other cold remedies try to help strengthen or support your body's immune system while it fights the cold. These include vitamins (often vitamin C), zinc supplements, and echinacea. You shouldn't expect these to bring immediate relief to your symptoms as you would with over-the-counter medications, though.

What is **echinacea**?

Echinacea is a group of flowers, related to daisies, sometimes also called coneflowers. Different members of this group are used in herbal medicines to stimulate the body's immune system. Scientific studies on the effectiveness of echinacea are contradictory, however, as some have shown clear effects in the prevention of or shortening of the length of colds, while others have concluded that it is mostly ineffective.

What is typically **analyzed** when a doctor takes your **blood sample**?

There are actually several types of blood tests that your doctor may perform, depending on your symptoms and what they are looking for. These include tests to look at your blood chemistry, the enzymes present in your blood, tests of how well your blood is able to form a clot, tests to assess risk for heart disease, or a complete blood count. A complete blood count can detect the presence of many diseases and immune system disorders by measuring the numbers of red and white blood cells, platelets (blood cell fragments that promote clotting), hemoglobin (the protein that carries oxygen), hematocrit (the amount of space taken up by red blood cells), and mean corpuscular volume (a measure of the size of red blood cells).

Blood chemistry tests provide information about the health of your muscles, bones, and organs. This test reports blood glucose, calcium, and electrolyte levels. It will also test the function of your kidneys. This is often a test that requires you to not eat any food for some length of time before the test so that the doctor can get an accurate measure of your blood chemistry that is not influenced too heavily by what you have eaten recently.

Blood tests to assess heart disease focus on measuring your cholesterol levels. This includes low-density lipoproteins (LDLs, often called "bad" cholesterol), high-density lipoproteins (HDLs, often called "good" cholesterol), and triglycerides (a type of fat). This test also typically requires you to not eat for about half a day prior to having your blood taken, as your cholesterol levels can be easily influenced by what you have recently had to eat.

While blood tests typically cannot diagnose diseases themselves, they can provide a strong indication of what disease(s) you may have and direct your physician to provide other tests that can confirm a diagnosis.

What chemicals are typically analyzed in a urine analysis?

Just like a blood analysis, there are several analytes that can be looked at by urine analysis: pH, density, proteins, glucose, ketones, leukocytes, blood, or human chorionic gonadotropin (the presence of which can indicate pregnancy).

The results can indicate several things about a person's health. If the person is well-hydrated, for example, a lower density would be expected. The presence of proteins in the urine is uncommon, and this may indicate that the patient's kidneys are not functioning properly. Similarly, glucose and ketones should not be present in the urine, but if they are, their presence could be a symptom of diabetes.

Can cold weather really cause a cold?

This actually is not true! Colds are caused by a virus that must enter your body to cause cold symptoms. Contracting the cold virus really has nothing (directly) to do with cold weather. The only reason colds seem to be more prevalent during colder seasons is that people spend more time indoors, placing them in closer proximity to one another, which makes it easier for the virus to spread between people.

Why does drinking alcohol make people loopy?

Alcohol, specifically ethanol (CH_3CH_2OH), is a mild depressant that affects your central nervous system. The specific biological effects are pretty complicated—some systems are enhanced by ethanol, others are inhibited. This combination of effects is why drinking can relax your muscles but also make you more animated. It may seem that alcohol lowers your inhibitions, but some experiments suggest this is a psychological effect and not a chemical one.

> ## Can cracking your knuckles really lead to arthritis?
>
> No. In your joints there is a liquid called synovial fluid, which serves as a lubricant. When you cause your knuckles to crack the synovial fluid has to fill more space, and this is what causes your knuckles to make a cracking noise. Arthritis comes about when your immune system starts to cause harm to your joints. Of course, cracking your knuckles too much can still cause other problems for your joints, just not arthritis.

How does your **liver process alcohol**?

The liver contains an enzyme called alcohol dehydrogenase, which is responsible for metabolizing ethanol. Alcohol dehydrogenase converts the ethanol to another molecule called acetaldehyde, which is then excreted from the body. A healthy liver can process about half an ounce of pure ethanol each hour, which equates to roughly one beer, one glass of wine, or an ounce of liquor each hour.

What happens chemically that **causes** a **hangover**?

Hangovers are believed to be caused by the buildup of acetaldehyde. This aldehyde is formed from the oxidation of ethanol, which is of course the "alcohol" you drank in the first place. Acetaldehyde is an intermediate in your body's process for dealing with all that booze you drank last night. An enzyme, acetaldehyde dehydrogenase, further oxidizes acetaldehyde to acetic acid.

Of course there are a lot of other factors that might make you feel terrible the morning after a big night out. Your liver gets pretty taxed dealing with all of the toxins, which is why some people think that distilled alcohol (which removes heavier molecules in the alcoholic beverage) leads to less severe hangovers.

Is it true that **coffee** can help you to **sober up** faster?

Unfortunately, no. Intoxication is related to the amount of alcohol (ethanol) in your body, and the quantity of alcohol will only decrease as your liver works to metabolize the liquor. Drinking caffeine won't really do much to accelerate your liver function, so coffee cannot, in fact, sober you up. It may help you to stay more awake or alert, but you'll still be intoxicated.

Does putting **urine** on a **jellyfish sting** really help?

You may have heard of this before, but it seems that it may just be an old wives' tale. Washing the area of the sting with saltwater is recommended so as to deactivate any of the stinging cells from the jellyfish that may be present on your own skin. Fresh water may actually reactivate the cells that caused the sting, causing you further pain.

Does **eating chocolate** or **fried food cause acne**?

No. Acne is caused by oil glands in a person's skin over-producing an oil called sebum, which is what the body naturally uses to keep skin lubricated. If this excess oil, along with dead skin cells, blocks your pores, then your skin can become irritated and a pimple may form. While the causes of excess sebum production are not definitively established, there is no reason to think that chocolate or fried foods are at fault.

What **causes dandruff**?

Despite what you may have heard, dandruff is actually not caused by dry skin but rather by a specific type of fungus or yeast. While dandruff does cause flaky scales of skin to shed off of the top of a person's head, dandruff is actually due to the interplay between yeast organisms and natural oil glands. Unfortunately there is no genuine cure for dandruff, but it can be controlled by using special shampoos containing zinc pyrithione, selenium sulfide, coal tar, or ketoconazole. To be effective, these products typically need to remain on the skin for several minutes before being washed away. In fact, the scalp is not the only place that dandruff can occur! Dandruff can also manifest on a person's eyebrows, mustache, and beard area, as well as the ears and the nose.

A common myth is that eating some foods, such as chocolate, causes acne. Pimples are actually caused by a build up of sebum, but why this happens is not clearly understood.

Can wearing a **copper bracelet** help with the symptoms of **arthritis**?

Arthritis is caused by deterioration of the cartilage in a person's joints, where deterioration happens faster than the body is able to repair it. Copper bracelets are sold based on the idea that a person may have a copper deficiency that leads to the pain in their joints, thus implying that copper from the bracelet could be absorbed through the skin to help correct the deficiency. In fact, copper deficiencies are extremely rare, and most people eat plenty of copper in their regular diet. Rarely are additional copper supplements necessary. It has not been proven that copper can be absorbed through the skin, nor has it been proven that the bracelets can help with any symptoms of arthritis or joint pain.

Moreover, excess intake of copper can result in poisoning, so if copper can be absorbed through the skin from a bracelet, one would want to monitor their dosage carefully. It is also worth pointing out that we know of no cases of poisoning resulting from wearing copper bracelets.

Can **eating carrots** help your **eyesight**?

It does seem likely that carrots help your eyesight, so it's not a myth! Carrots, and other colorful vegetables, tend to have a lot of vitamin A, and vitamin A helps the retina to stay healthy and working well. This is because it helps to generate rhodopsin, which is a light-sensing molecule located in the retina. Developing a deficiency of vitamin A can actually lead to night blindness!

Does **spinach provide** a good source of dietary **iron**?

Spinach has the same dietary content as many other green vegetables. However, spinach additionally contains oxalic acid, which actually prevents iron from being

Unlike the myth about chocolate causing acne, carrots actually do help your eyesight because they contain a lot of vitamin A, which is important for the health of your retinas.

absorbed by the body! Thus it may actually be a worse source of iron than other foods. An interesting anecdote: The iron content of spinach was initially reported to be about ten times higher than it actually is, all due only to a misplaced decimal point. Spinach does contain plenty of good antioxidants and vitamins, though, so we don't mean to suggest you should stop eating it. It's just that if you heard it was an exceptionally great source of iron, you may have been misled by the original mixup over how much iron is actually present in this vegetable.

Will you **contract poison ivy** if you come into contact with a person who has it?

Poison ivy is caused be a an oil called urushiol, found on the leaves of poison ivy plants. The only way to contract poison ivy is to come into contact with this oil. Thus, if a person who has poison ivy still has some of the oil from the plant present on their skin, they could spread that oil to another person. In most cases, though, by the time a person has a rash, the oil will have been washed away. To be clear, the blisters on a person's skin who is suffering from a poison ivy rash do not themselves have the potential to spread poison ivy to other people.

CHEMICALS IN FOOD

Why do **diet Coke®** and **Mentos® fizz** like that?

In diet Coke®, and all soda, there are a lot of CO_2 molecules trapped in solution. The slow release of CO_2 from solution is what normally makes the calm bubbling in soda. But what if you were to put a catalyst for gas release into a bottle of diet Coke®? That's exactly what a Mentos® does. Dissolved gases need a surface to start forming a bubble (let's just assume that's true, which it is), and the Mentos® candy provides a huge amount of surface area because it is a very porous material. So add a Mentos® and many, many more CO_2 bubbles can form at the same time, which leads to a sugary eruption.

Is it good to **drink chocolate milk** after a **workout**?

Yes! Milk contains a significant amount of protein: roughly eight to eleven grams per cup. It is often recommended to consume fifteen to twenty-five grams of protein after a workout session, so drinking two to three cups of chocolate milk would meet this suggested quantity. In comparison to regular milk, chocolate milk has about twice as many carbohydrates, which is good for soothing sore and tired muscles. Of course, it also serves to rehydrate your body from the water lost during the workout. Moreover, milk provides you with nutrients like vitamin D and calcium.

Why **don't oil** and **water mix**, anyway?

The simplest way to answer this is with the phrase you probably learned in high school—"like dissolves like." Water is a polar molecule, so it prefers to interact with other polar molecules. Oil, a hydrocarbon of some type, lacks polar groups, and forms weak Van der Waals interactions with other nonpolar molecules.

This is only partially correct, and the actual situation is quite complicated. The major force at work here is the stability of the water phase due to the hydrogen bond interactions. When a molecule of a hydrocarbon is dissolved in water, some number of hydrogen bonds must be broken. This bond breaking costs energy. When oil and water don't mix, these hydrogen bonds outweigh the entropy gained by mixing the phases so the water molecules stick together, and the oil remains separated.

What will happen to the **gum** I just **accidentally swallowed**?

When you were a kid, your parents or teachers probably told you (or at least they told us!) not to swallow your chewing gum because it would take years to digest. Actually your body can do just fine with digesting the flavor components, sugars and sweeteners, and softening ingredients in chewing gum. It's only the gum base that you cannot digest. However, just because you cannot digest it does not mean it will stay around in your body for several years. The gum base will just typically pass through your digestive system in just a few days' time. But please don't try to swallow a bunch of gum at the same time, though—it can get stuck and cause all sorts of problems.

Is it true **fresh eggs** will **sink in water**, while **bad ones** will **float**? Why?

Yes. As an egg goes bad, proteins and other chemicals decompose and release volatile molecules like carbon dioxide (CO_2). The shell of an egg is somewhat porous, so these gases can escape, decreasing the mass of the egg. Since the size of the egg doesn't change (assuming you don't break it), the density decreases as the egg ages, and it eventually becomes less dense than water. Whether it becomes buoyant exactly when it is spoiled or weeks after, we can't say. That's why your nose is so good at smelling that awful smell—so that you don't eat rotten eggs!

Why does my mom put **salt in the water** to make it **boil faster**?

Probably because her mom told her to. There are actually two things that are affected when you add salt to water. First, the boiling point increases (remember boiling-point elevation from the chapter "Macroscopic Properties: The World We See"?). Second, the heat capacity, or how much energy it takes to raise the temperature, decreases. While you might think this would change how fast your pot of water boils, the amount of salt you would have to add to see a significant change is pretty large. The real reason you add salt to water for cooking? Flavor. Salt tastes good.

Why does **asparagus** make **pee smell** weird?

The compounds you smell in your pee after eating asparagus are most likely thioethers (although the literature over the last one hundred years on this subject includes some debate—no joke). That's a sulfur atom with two carbon substituents. Asparagus, for some reason, has high levels of sulfur-containing amino acids. Your body breaks these down into chemicals that smell like rotten eggs or other foul odors.

Why does my **chocolate turn white** when I wait too long to eat it?

Chocolatiers refer to this as a bloomed chocolate, which is a wonderfully obscure and appealing way of describing the white stuff that forms on old Halloween candy. There

are two processes that could be happening here, either sugar bloom or fat bloom. Sugar bloom is the candy world's way of saying sugar crystallization. If your candy is exposed to moisture, sugar molecules dissolve out of the fat in the chocolate and once that moisture evaporates, the separated sugars have a chance to crystallize. If your candy has stayed dry but underwent a quick temperature change or was stored warm, it's probably fat from the cocoa butter that has separated from the chocolate. In either case, it is usually still fine to eat.

What does a **preservative do**?

Preservatives keep food fresh by slowing the growth of mold and bacteria (antimicrobial preservatives), preventing oxidation (antioxidants), or slowing enzymes from continuing the ripening process after a fruit or vegetable was picked or cut.

Asparagus has a lot of sulfur-containing amino acids, and when you urinate after eating this vegetable, the sulfur can be smelled in your waste.

What is the difference between **artificial** and **natural flavoring**?

The difference between calling a molecule artificial or natural is a legal definition, not a chemical or biological one. If a molecule of vanillin is isolated in a lab by extraction of a particular seed pod, the chemical is called "natural vanilla." If that same molecule is made from lignin, which is a polymer found in naturally in wood, that substance is called "artificial vanilla." The difference is that transforming lignin to vanillin requires chemical steps not covered by the legal definition of the word "natural." To be clear, the "natural" and "artificial" versions of a molecule are exactly the same chemical species (same atoms, same bonds, same stereochemistry, etc.), but there may be other chemicals present that may differ in the natural and artificial versions of a product.

vanillin

So is **natural vanilla** worth the **extra money**?

Well, that's a different question…. When the seed pods of the Mexican orchid *Vanilla planifola* are collected, they are put through an elaborate series of curing and aging steps. These steps allow additional flavor molecules to develop beyond just vanillin. While vanillin is a major flavor component of natural vanilla, there are hundreds of other tasty chemicals in "natural vanilla." So you're not buying exactly the same thing when you buy artificial and natural vanilla. Which one tastes better in your cookies is up to you, though.

Why does **milk go sour**?

Milk, even milk that has been pasteurized, contains a bacteria called lactobacillus. The first part of the word, "lacto," refers to the sugar that these particular bacteria eat—lactose. When these bacteria process lactose, they secrete lactic acid. It's this acid that makes milk taste sour and leads to curdling.

It actually doesn't have to be lactic acid that curdles milk; any acid will do. Try taking a small glass of milk and adding lemon juice or vinegar to it. The milk will start to curdle just the same.

Is **lactobacillus harmful**?

Lactobacillus is used in the production of lots of foods, actually—cheese and yogurt, sourdough bread, pickled vegetables, and wine and beer. There are helpful strains of lactobacillus living in your digestive tract too. So, to answer the question, lactobacillus is not necessarily bad for you in appropriate quantities.

Why can't **lactose intolerant** people eat dairy?

Most every dairy product contains lactose, which is a type of sugar. Specifically, lactose is a disaccharide made up of one galactose molecule and one glucose molecule. Lactase is an enzyme that breaks the bond that joins the galactose sugar to the glucose sugar to form this "double sugar." Lactose-intolerant people lack the enzyme lactase, so they can't metabolize this particular molecule.

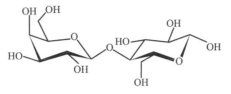

All mammals can digest lactose when they are first born, but it is very rare for adult mammals to continue to be able to absorb the sugars found in dairy products. Recent studies have suggested that humans only obtained this ability around the same time they began domesticating animals. This makes it possibly the most recent example of evolution in our species.

Does eating **turkey** make you **sleepy**?

You've probably heard that turkey contains a lot of tryptophan, which makes you sleepy. It's true there is tryptophan in turkey, but not more than is contained in most meats. It's also true that tryptophan is used to make serotonin, a neurotransmitter that makes humans sleepy. The problem with connecting these dots ("I'm sleepy after Thanksgiving" and "Tryptophan makes serotonin, which makes me sleepy") is that there are also a bunch of other amino acids in turkey. For amino acids to get into your brain, they need to use transporters to cross the barrier. Tryptophan is competing for a ride on the transporter molecules with all of those other amino acids you just stuffed in your face.

So why are you sleepy after Thanksgiving dinner? It's not the turkey, but the extra servings of carbohydrates. Carbohydrates, or sugars, cause your pancreas to release insulin. Insulin helps the body deal with the huge amount of sugar and amino acids you just consumed, but interestingly it has no effect on tryptophan. So other amino acids are taken out of the bloodstream, and tryptophan is free to use the transporters to get into your brain, making you sleepy.

A chemical called tetrodotoxin in the fugu fish can be released into the fish if it is not prepared correctly by a trained chef.

Why is **fugu toxic**?

Fugu fish is toxic because of a molecule called tetrodotoxin. This molecule binds to the sodium ion channels in nerves' cells, shutting down all communication in your nervous system. The toxin is not affected by cooking, so chefs that prepare this dangerous fish have to be highly trained to avoid serving parts that contain tetrodotoxin (the liver, ovaries, and skin can have very high levels of this poison).

How do Pop Rocks® candy work?

Pop Rocks® is a carbonated candy like soda is a carbonated beverage. The fizzing that goes on in your mouth is due to carbon dioxide (CO_2) bubbles escaping from the candy. To make this unique treat, the ingredients are heated, exposed to high pressures of carbon dioxide, and then cooled to trap the CO_2 inside the candy.

What is MSG, and is it really so bad to consume?

MSG stands for monosodium glutamate. It is a natural, but nonessential amino acid. It is used by food manufacturers as a flavor enhancer, and by itself has the taste of "umami"—the fifth taste sensation we mentioned in "Macroscopic Properties: The World We See." There have been countless studies on its safety to eat, and the data overwhelmingly supports MSG as being safe to consume even in absurdly large quantities. But don't try to disprove those studies—everything in moderation—your mother was right about that.

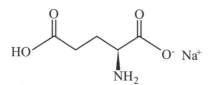

How does popcorn pop?

The shell of a popcorn kernel is a hard, moisture-resistant coating protecting an interior filled with starch and some water. As the kernel gets hot, the starch inside first softens and then the water turns into steam, which raises the pressure inside the kernel. At some temperature the pressure inside the corn kernel is high enough to rupture the hard shell, and the kernel pops. The hot starch quickly cools, trapping tiny air bubbles inside and making a (delicious) solid foam.

What's the difference between saturated and unsaturated fats?

The terms saturated and unsaturated refer directly the to the chemistry definition of unsaturation. Unsaturated fats contain carbon–carbon double bonds (units of unsaturation), while saturated fats have only completely saturated carbon chains. Saturated fats (like butter and lard) stack together well in the solid state, so they are generally solid at room temperature. The double bonds in unsaturated fats disrupt this lattice in the solid state, making these types of fats (like olive oil and vegetable oil) liquid at room temperature.

What about cis and trans fats?

The carbon–carbon double bonds along the hydrocarbon chain of fatty acids can be either the cis or trans isomers. Naturally occurring unsaturated fats have more cis fats, while manmade fats (like margarine) have a higher level of trans fats. Trans unsaturated fats turn out to be particularly bad for humans as they increase cholesterol, leading to heart disease.

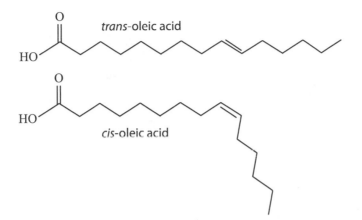

CHEMICALS IN THE NATURAL WORLD

How do **ants know** how to **organize their colonies** so efficiently?

Ants use a set of pheromones to communicate information between individual members of the colony. There are different chemical markers that indicate trails to food, alarm pheromones that are released when an ant is killed, and a number are used in the reproductive process. Some ant colonies even use pheromones to trick enemy colonies into attacking themselves or to convince the opposition into becoming worker ants for their own benefit.

If **chlorophyll** gives plants its **green color**, then **what cells** give me **my skin color**?

Melanin is the molecule that is responsible for skin color in humans. It's also what gives your eyes and hair their color. There isn't one single structure of melanin, rather it refers to a whole class of highly colored molecules that are all derived from the amino acid tyrosine.

Why do **polar bears** have **black skin** and **clear fur**?

There is some debate on the subject, but the combination of black skin and clear fur is probably the best for keeping the polar bear warm while not giving away its position when hunting. The black color of the skin is the best for absorbing energy from the Sun—objects that appear black don't reflect any wavelengths of light. The clear fur allows that light to get to the skin, but still looks white so the bear can blend in with the surrounding ice and snow.

215

Polar bears look white, but they actually have clear fur and black skin.

A few years ago, or maybe you heard this recently because it has achieved urban legend status, a rumor you may have heard is that polar bear fur is actually like a fiber optic network for harnessing or focusing sunlight toward the bear's skin. While this would have been cool, it's unfortunately completely bogus.

If **polar bears** have **black skin** and **clear fur**, why do they **look white**?

Polar bears appear white for the same reason that a pile of snow looks white—reflection. If there are a lot of surfaces that reflect light (either in the polar bear's fur or a snowdrift), then any light that hits the object will be bounced around many times before coming back toward your eye. Most wavelengths of light are scattered equally well, so you end up with the object looking white (in other words, no wavelengths are absorbed, which would give rise to a color).

Why are **snowflakes unique**?

Excellent question! All snowflakes are made up of frozen water (ice), so why don't they look exactly the same? Their unique shape comes from the unique conditions under which each snowflake is made. As a tiny ice crystal is blown into and out of a cloud, or rises or falls in a jet of slightly warmer or colder air, the shape of the growing snowflake can change. In a sense, their shape describes a story of how they were made.

What is a **flame**, really?

From a chemical perspective, a flame is the visible light product of an exothermic oxidation reaction. This reaction takes C–H bonds and oxygen (O_2) in the atmosphere and starts a radical chain reaction that keeps going as long as there is still fuel and oxygen around. Actually, bonds other than just C–H bonds can give rise to burning; that's just

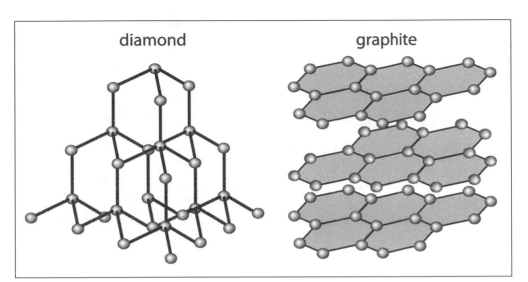

In a diamond crystal, carbon atoms are arranged in a rigid structure, while in graphite the carbon is arranged in layers, making a substance that can easily slough off its layers.

what's happening during the burning of objects you're most familiar with (like candles and wood fires).

Where does all that **wax go** when I **burn a candle**?

Why, it burns! The long hydrocarbon chains that make up candle wax are transformed into carbon dioxide and water. Candle wax is the fuel for the flame; the wick just helps draw it up to the flame by capillary action.

Why is **diamond** so **hard**?

Pure diamond is a crystalline form of only carbon atoms. The crystal lattice is three-dimensional, and all carbon atoms are attached to four other carbon atoms in a perfect tetrahedral geometry. Because the lattice repeats in all three dimensions, there is no easy way to distort the structure, making it a very hard material. The structure of graphite, for comparison, is many stacked-up layers of carbon atoms. These two-dimensional sheets can "slide" relative to one another, making graphite relatively soft in comparison to diamond.

What about the **fake diamonds** at jewelry stores? What are those?

Diamonds themselves cannot be made in a laboratory very well, but crystals with similar properties and appearances can be made. One popular type of "simulated diamond" is a cubic zirconia crystal, which is a crystal with the chemical formula ZrO_2. Cubic zirconia are more dense than diamonds, or, in other words, a cubic zirconia is heavier than a diamond of the same size. They are also a fairly hard material as minerals go, but they are not as hard as diamonds and can be scratched by diamonds. Cubic zirconia are usu-

217

ally colorless, which isn't the case for diamonds. Most diamonds contain some amount of noncarbon impurity that makes them appear colored, and only very pure diamonds appear colorless.

What is **snake venom**?

Most venoms are mixtures of dozens of compounds, and the active toxic ingredients are proteins that wreak havoc on the recipient in a variety of ways. While the exact enzymes vary from species to species, and even geographically within a species, many snake venoms contain some sort of neurotoxins which block signals sent through your nervous system, leading to numbness or even paralysis.

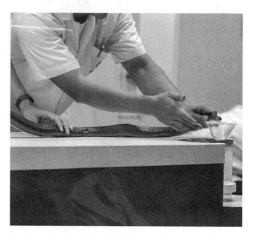

An expert extracts venom from a snake, which will later be used to create antivenom for snake bite victims. Snakes can have different types of venom: neurotoxins or cytotoxins. Neurotoxins work by blocking nerve function, while cytotoxins destroy cells directly.

CHEMICALS IN OUR WORLD

What makes a **rubber band stretchy**?

Rubber bands are made up of long, tangled-up, chainlike molecules called polymers. These long chains prefer to be tangled up because it maximizes their entropy or just the number of ways they can arrange themselves (see "Physical and Theoretical Chemistry" for more on entropy or "Polymer Chemistry" for more on polymers). When we stretch a rubber band, it straightens out these long chain molecules into a state with less entropy, so basically there are fewer stretched-out conformations possible than there are tangled-up ones. When we release the rubber band, it contracts to a more disordered state, primarily just because there are more possible conformations associated with the contracted length.

How does an **air freshener** work?

Although there are many different types of air fresheners, most of them are simply perfume dispensers. The perfume is a strong-smelling substance that has a pleasant odor, which masks the stink.

What's **hair made of**?

Keratin is the major component of hair (and nails). It's a protein that makes up the outer layer of your skin, your hair, fingernails, and toenails. Actually, keratin is everywhere in the animal world: the hooves, claws, and horns of most mammals, the scales of reptiles, the shells of turtles, and the feathers, beaks, and claws of birds. Many copies of this protein molecule assemble into a large helical structure that provides the structure and rigidity to the material. Keratin proteins contain lots of sulfur atoms in the form of cysteine amino acid residues. These sulfur atoms can form linkages between keratin strands and between the larger helices that are formed, leading ultimately to a curling of the hair strand. Yep—you guessed it: The more disulfide bonds there are, the curlier your hair will be. Hair straighteners work by breaking these disulfide bonds, relaxing the hair to a straight conformation.

We spread salt on icy roads to melt the ice, which works because the salt lowers the freezing point of frozen water.

Why does **salt melt ice**?

Adding salt to ice lowers the melting point of water (see freezing-point depression in "Macroscopic Properties"). So when you pour salt on the ice on your driveway, the melting point of the saltwater mixture you made by adding the salt may be lowered sufficiently such that it is below the outside temperature, which means the solid ice will start to melt.

Why do **farts stink**?

Farts are mostly nitrogen (N_2), but that's an odorless gas. The smelly molecules in farts are mostly sulfur-containing compounds that reek like rotten eggs. Other odoriferous compounds include skatole and indole. How did chemists figure this out? They analyzed fart gas by gas chromatography. No, really.

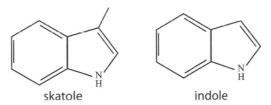

skatole   indole

What is **inside** of a **zit**?

Gross. You really want to know? Okay…the major component of most pimples is a mixture of keratin and sebum. We just talked about what keratin earlier in this chapter. Sebum is an oily mixture that your skin naturally secretes. Earwax is also made up mostly of sebum.

What about **blackheads and whiteheads**? Are they the same?

Almost. If sebum builds up underneath the surface of the skin, it stays white. If sebum collects around a hair follicle or is in some other way exposed to the air, the sebum undergoes oxidation reactions. As sebum oxidizes it darkens, eventually appearing black.

Is it better to use **"chemical-free" products**?

There are no such things! Everything is made up of chemicals! Be careful when choosing which ones you eat or use, but the truth is, they're all made up of chemicals.

How do **mood rings** work?

Mood rings change color based on your body temperature. The piece that changes color is a liquid crystal thermometer, which is the same technology that is used in some disposable medical thermometers as well as the adhesive thermometers commonly used in aquariums.

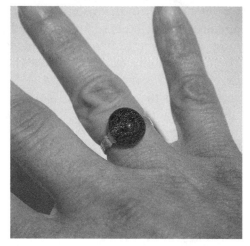

Mood rings were all the rage in the 1970s. They work the same way crystal thermometers work do: thermotropic liquid crystals within a hollow quartz shell respond to temperature changes by twisting, which causes them to reflect different wavelengths of light depending on temperature.

How does **waterproof mascara** work?

Like we talked about in the chapters on "Macroscopic Properties" and "Biochemistry," hydrophobic materials do not dissolve well in water. So it makes sense to make waterproof mascaras from something hydrophobic. Waterproof mascaras contain a combination of waxes to make them resistant to washing off in water.

How do **mirrors reflect** light so well?

Most modern mirrors are made of smooth layers of silver or aluminum along with other chemicals and coatings to aid in their construction. While the chemical properties of silver play some role in reflecting light, the fact that these layers are very, very smooth is more important here. If the surface of the silver layer were rough, the light would be reflected in a variety of different angles, which you would see as a distorted image. When the surface is perfectly smooth, the light bounces straight back into your eye and you see an accurate reflection of the object. This is why you can see clear reflections in very still water or a shiny piece of leather—both are smooth surfaces.

What are **pigments**?

Pigments are molecules that selectively absorb particular (ranges of) wavelengths of light. The remaining reflected wavelengths are the color that you perceive the pigment to be. So pigments can only subtract wavelengths of light from the spectrum, but they cannot generate their own light. For example, a blue pigment absorbs red and green light, but reflects every other color. Alternatively, the blue pigment might only absorb orange light, which is the color complement of blue.

Why was **gold chosen** as a form of **currency** rather than any other element in the periodic table?

What a neat question! There are very basic attributes to money that made gold the obvious element. The element should be a solid at all temperatures it might encounter over the course of its lifetime—no one wants their money to boil away on a hot day. The element needs to be very stable—we wouldn't want our coins rusting, bursting into flame, or slowly giving off radiation. And finally, the element needs to be rare, but not *too* rare. After you remove elements that don't fit these parameters, we are left with rhodium, palladium, silver, platinum, and gold. Rhodium and platinum weren't discovered until the nineteenth century, so they weren't an option for ancient civilizations. The furnaces of the ancient world couldn't reach the temperature required to melt platinum (3200 °F, 1800 °C), so it couldn't be made into coins. That leaves silver and gold. Gold has a lower melting point than silver and it doesn't tarnish in air like silver does, making it the clear winner for a currency element on our planet.

What causes the "glow" in **glowsticks**?

Glowsticks contain three main components—a dye, diphenyl oxalate, and hydrogen peroxide. The hydrogen peroxide is the chemical you release when you "crack" or activate the glowstick. Diphenyl oxalate reacts with the hydrogen peroxide to generate a molecule called dioxetanedione. This particular molecule decomposes to release two molecules of carbon dioxide and energy that excites a dye molecule to a higher energy state. To get back to its stable state the dye releases a photon of light, making your glow stick glow. The particular structure of the dye molecule controls the wavelength (color) of this light, which is how you can get glowsticks in different colors.

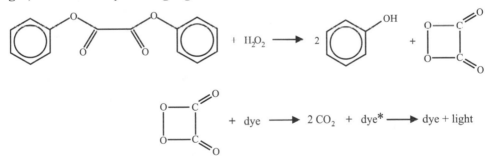

What causes **nonstick pans** to be **nonadhesive**?

Nonstick pans and other surfaces are coated with a polymer that does not interact strongly with other surfaces (hence, no sticking). Teflon®, developed by DuPont® in the 1940s, is most frequently used for this application. Teflon® is a polymer of tetrafluoroethylene, which is very hydrophobic, so water or any other substance (like food) doesn't stick to it. More recently, other polymers have also been developed for this application. Thermolon® is a silicon oxide polymer with some similar properties to Teflon®, and EcoLon® is a nylon-based product that is reinforced with ceramics for toughness.

How does a **microwave work**?

Microwaves work by using a process called dielectic heating. The microwave surrounds your food in a field of electromagnetic radiation that is constantly changing directions. The polar molecules in your food, particularly water, align their dipoles with the direction of this applied field. The constant shifting of the direction of this field causes the polar molecules to tumble around, and this molecular motion warms your food.

What makes **leaves change color** in the fall?

Leaves are normally green because of chlorophyll. Chlorophyll absorbs blue and red wavelengths of light, so it appears green. This molecule is crucial for photosynthesis but when the days begin to get shorter as winter approaches, plants begin to produce less chlorophyll. As the level of this green chemical in leaves falls, we can start to see other highly colored molecules. In particular, carotenoids, which appear yellow, orange, or brown, start to become visible. These molecules are always present in the leaves, it's just that the green color of chlorophyll dominates most of the year.

How does **hand sanitizer work**?

The active ingredient in hand sanitizer is usually an alcohol like isopropanol. This chemical is also used in antiseptic wipes and pads because of its ability to kill bacteria, fungi, and viruses.

How does **soap work**?

Soap molecules have polar end groups and long hydrophobic tails. In the presence of water, they arrange themselves into spheres called micelles. These structures can transport greasy particles in their interiors, helping remove the bits that water can't remove on its own from your hands or clothes.

Why does **bleach kill** everything?

The active ingredient in bleach, sodium hypochlorite (NaOCl), has a few ways of killing off microbes. One method involves causing particular proteins in the microbes to unfold, preventing their normal function and eventually killing the bacteria. Alternatively, bleach can disrupt the membrane that forms the outer shell of a bacteria. Since most bacterial membranes are very similar, bleach is very effective

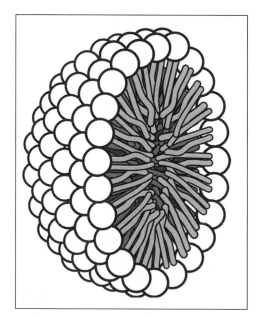

Soap molecules arrange themselves into micelles like this one (shown as a cross-section), which can move grease into the center of the structure.

against a whole host of different types of bacteria. The human body actually produces hypochlorous acid (HOCl) itself to combat bacterial infections.

Why do you put **iodine** on **cuts**?

The iodine you buy at the drugstore is usually an ethanol solution of elemental iodine (I_2). Iodine is a general disinfectant, meaning that it kills all sorts of pathogens, including spores, which are notoriously difficult to kill.

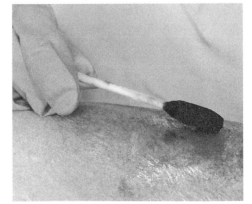

Iodine is useful as a disinfectant that can kill a wide variety of pathogens.

How do **fireworks work**?

Fireworks, those that explode in the sky, technically called pyrotechnic stars or aerial shells, have only a few essential components. Once the shell gets up in the air, the main event is fueled by aluminum metal or sometimes a mix of aluminum and magnesium metal. By themselves, these elements don't burn quickly enough in the atmosphere and the flames they produce are just boring white. So this fuel component is mixed with another chemical that helps the aluminum burn (technically oxidize) faster. Different oxidizing compounds make different-colored flames when they react: purple (KNO_3), blue ($CsNO_3$), green ($BaCl_2$), yellow ($NaNO_3$), or red ($SrCO_3$). There are lots of other components of modern fireworks (like gunpowder to help disperse these chemicals to make big shapes), but all of the chemistry is based on these simple reactions.

Why is **ultraviolet light** from the Sun potentially **dangerous**?

Most of us like being out in the Sun, but we've also all heard to be careful not to get sunburned since too much sun can cause serious problems like skin cancer, along with less serious problems like wrinkles and dry skin. The Sun can damage your skin because the ultraviolet rays from the Sun are relatively high-energy photons that can damage the elastin fibers in your skin. These can then lose their ability to go back to their original position after they are stretched and also lose their ability to heal as quickly when wounded or bruised.

As we discussed briefly in "Biochemistry," the cause of cancers generally involve damage to the genetic material (DNA) of your cells, which interferes with their ability to replicate (or stop replicating) normally. With regard to cancers arising from too much exposure to ultraviolet radiation, DNA can be damaged in two ways. The first, probably more obvious, route is that the ultraviolet radiation could be absorbed by the DNA directly, causing changes in its chemical structure. The second possibility is that ultraviolet radiation can be absorbed by other molecules first, forming reactive, damaging radical species (like hydroxyl radicals or singlet oxygen), which then diffuse through cells and can damage DNA.

223

How does **sunscreen** work?

Sunscreens either reflect or absorb ultraviolet light from the Sun. To reflect the light sunscreens contain either titanium or zinc oxides, which are both very white solids (so all wavelengths of visible light are being reflected). To absorb light, sunscreens can contain organic chemicals that interact with harmful UV wavelengths. While almost all sunscreens use titanium and/or zinc oxides, the particular organic compounds that are used vary widely across brands and countries.

What is a **CD made of**?

All types of optical discs (CDs, DVDs, Blu-Ray® Discs, etc.) have basically the same components. The outer layer of clear polycarbonate plastic protects the inner layers that contain the data from damage. A layer of a highly reflective metal, usually aluminum, is used to reflect the laser that is used to read the data. The data itself is stored on another polycarbonate layer that contains teeny-tiny little pits. The pits are arranged in spiral tracks, just like a vinyl record (if you've ever seen one of those), and are about 100 nm deep and about 500 nm wide. The laser can detect the change in height by measuring changes in how the light is reflected.

Why is **arsenic** so **poisonous**?

Arsenic interrupts some of your body's most basic and common biochemical pathways. Arsenic, particularly As^{3+} and As^{5+} oxides, interferes with the citric acid cycle and respiration (specifically reduction of $NAD+$ and ATP synthesis—see "Biochemistry"). If that weren't enough to kill you, and it probably is, arsenic also boosts the level of hydrogen peroxide in your body, which causes another whole set of problems. Unfortunately, these toxic forms of arsenic are not only water soluble but can be found in well water from natural sources and man-made contamination from mining.

How do our **brains tell time**?

Until very recently, scientists presumed (and it's not clear that there were many experiments to support this) that our brain had a stopwatch of sorts built into its machinery. By stopwatch, scientists meant some biological system that created some signal at regular intervals. If this were true, we would be equally good at estimating short and long periods of time, but that's totally not the case. Humans are pretty awful at guessing how long extended periods of time are.

Instead, Dean Buonomano at UCLA proposed in 2007 that our brains tell time in a different way. To steal his analogy, imagine a rock being tossed into a lake, which creates a series of ripples in the surface of the water. If you were to throw a second rock in, the ripples from the two rocks would interact with each other. The pattern of this interaction depends on the time between the two rocks hitting the lake. The firing of neurons (the rocks hitting the lake) create these unique patterns in their signals—and neurons can use the different patterns of signals interacting to tell time between events. The wonderful piece of this theory is that it explains why we're good at telling time over short durations, but not long ones—over a long time, the ripples in the water just fade away.

What **chemical process** leads to the **formation of fossils**?

There are of course many different types of fossils, but most form through some sort of mineralization process. What's that? In water that has lots of dissolved minerals, after an organism dies, those minerals can slowly deposit in the tiniest of spaces within that organism, even within cell walls. Well-preserved fossils require the organism to be covered with sediment quickly after death (like on the bottom of a lake) so that the body doesn't decay before the slower mineralization process can take place.

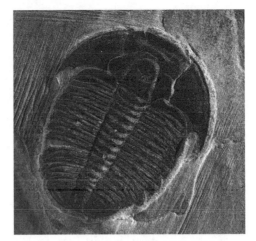

This fossil of an extinct trilobyte was formed millions of years ago when the animal was covered by sediments at the bottom of an ocean or lake. Over time, minerals replaced the decaying flesh and solidified into the shape of the original animal.

Is **sodium laureth sulfate** toxic?

You're probably asking this because you've seen this chemical listed on your shampoo or toothpaste—and at the levels in those products, no, it is not toxic. This molecule is a surfactant, very similar to many soaps. And like most soaps, if you get them in your eyes, it hurts. The chemical is an irritant, but it does not cause cancer, like you might have heard.

What's in **toothpaste** that makes our **teeth cleaner**?

If you get your teeth whitened at the dentist or buy some over-the-counter whitening products, the active ingredient is usually hydrogen peroxide. The hydrogen peroxide reacts with the colored molecules that are staining your teeth to remove them or at least make them colorless. This can be a slow reaction, so at the dentist's office, sometimes they use a bright light to speed up the breakdown of the H_2O_2.

If you're wondering about whitening toothpastes, those contain an abrasive compound that simply rubs the stain molecules off of your teeth—no fancy chemistry at all.

How does **fluoride work** in toothpaste?

Fluoride, usually in the form of sodium fluoride (NaF) in toothpaste, strengthens the enamel in your teeth. But how does it do that? Let's back up a bit first.

Enamel is the outer layer of your teeth, and it's made of a mineral called hydroxyapatite. It's a calcium phosphate structure with one hydroxyl group ($Ca_5(PO_4)_3(OH)$). This mineral dissolves in the presence of acids, which is exactly what bacteria generate when they metabolize sugars in your mouth. This is how cavities form when you drink soda.

Fluoride ions help rebuild your tooth's enamel by replacing the hydroxyl group in the apatite mineral. The new mineral ($(Ca_5(PO_4)_3F)$, or fluorapatite, is more stable in the presence of acids, so your teeth are more resistant to decay.

What is the **hardest material** in your **body**?

Tooth enamel is the hardest material in your body. It's even harder than bone!

Why do **old library books** begin to **smell** after sitting on shelves for years?

The smell of old books is due to hundreds of volatile organic compounds that form from the slow degradation of the book's paper and other materials used in its construction. Acetic acid (vinegar) and furfural (smells like almonds) are two common chemicals ascribed to the smell of old books. Scientists can analyze the volatile compounds to identify the materials used without having to destroy a part of the historical document.

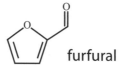

furfural

Why does the **inhalation of helium** make **voices higher**?

After you inhale helium your voice might sound higher, but the pitch (or frequency of the sound waves) is exactly the same. Your vocal cords vibrate at the same frequency because your body doesn't adjust for the presence of a less-dense gas in your throat. What *does* change is the speed of sound in helium versus air—because helium has a lower molecular weight than air, the speed of sound is higher. You've probably heard this is because helium is less dense—that's not technically correct, but let's not go there.

So the speed of sound is faster, but why does that make your voice sound weird? The tone is actually identical; what's different is the timbre. Specifically, the lower frequencies of your voice have less power, so your voice sounds squeaky—like a duck.

What is **ink made of**?

Inks can be very complicated mixtures, but two key ingredients are the pigment used to color the ink and the solvent used to dissolve (or at least suspend) the pigment particles. While modern inks come in every color imaginable for a host of different pen types, historically inks fell into one of two major categories.

The first is carbon-based inks. Residues from burning wood or oil, like soot, were used as coloring in these inks, which were suspended in the sap of the acacia tree (known today as gum arabic).

The other type of ink used historically is called iron gall. The iron was usually added in the form of iron sulfate ($Fe^{2+}SO_4^{2-}$); "gall" refers to gallotannic acid that was extracted from growths, or galls, on oak trees. Iron gall ink slowly darkens as the iron ions undergo oxidation from Fe^{2+} to Fe^{3+}. The acidity of the ink solution can cause damage to the paper it is used on, so preservation of historical documents that were written with iron gall ink is challenging.

Why is **graphite** so good to **write with**?

Graphite is an attractive chemical for writing for a few reasons. Graphite, unlike most inks, is not dissolved by water or affected by moisture, but it's easy to erase.

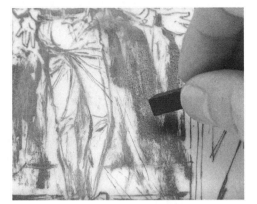

Pencil "lead" is actually made of graphite, a form of carbon that is easy to write and draw with, as well as easy to erase.

Fun fact: We commonly refer to the graphite in pencils as "pencil lead," but there's no lead (Pb) in there. The Romans did use lead for writing, but the practice didn't make it much farther in history than that. The paint on the outside of pencils did, however, contain lead up through the 1900s.

How do **thermometers work**?

There are actually many kinds of thermometers available, but let's talk about the two types that you probably have in your home.

The first type is a glass tube filled with either alcohol or mercury. As the temperature rises the volume of the liquid also increases, so it rises up the tube. The height of the liquid is calibrated with a scale so you can read the temperature value easily.

The second type is known as a bimetallic strip thermometer. While you've probably never heard this name before, you've likely used this type of thermometer. They are the most common models of thermostats (before they went digital), used as meat thermometers, oven thermometers, and the little thermometers that you see baristas using at coffee shops when they're steaming milk for your latte. You can tell from the name bimetallic that there are two metals involved here (usually steel and copper). In order to measure temperatures, these two metals need to expand at different rates when they are heated. If you make a strip of these two metals and wind that strip up into a coil, the difference in their thermal expansion will cause the coil to wind tighter or unwind as the temperature changes (depending on which side of the coil you place the material that expands more). This coil then turns a needle to indicate that the temperature is rising or falling.

Why do we need to **sleep**?

Even though this sounds like a pretty straightforward question, the truth is that scientists still don't know the whole answer! There are several theories, including the idea that sleep promotes restorative functions, that sleep promotes development and structural changes in the brain, that we sleep because it helps us to conserve energy, or simply that it may have been safer for our evolutionary predecessors to remain inactive at night. While there is evidence to support each of these (and other) theories, it remains a very difficult question to answer conclusively.

How do **bees make honey**?

The first step is searching out a flower to pick up some nectar. Nectar is a mixture of sugar and water. Specifically, the sugar in nectar used to make honey is sucrose, a disaccharide (see "Biochemistry"). A honeybee produces enzymes in its body that can break down the sucrose into monosaccharides, fructose, and glucose as well as gluconic acid. These sugars are the primary constituents of honey. Most of the water evaporates, which is what makes honey so viscous and sticky.

What is **testosterone**?

Testosterone is a steroid hormone molecule that is found in human males and females as well as many other species. In humans testosterone serves as the primary male sex hormone, and it plays a crucial role in the development of the male reproductive system. In males it is secreted from the tesisticles, and in females it is secreted from within the ovaries. Males use significantly more testosterone than females, and for this reason males produce testosterone at about twenty times the rate of females. Oddly enough, males are also less sensitive to testosterone than females.

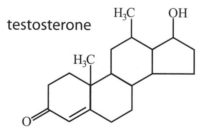

What is **progesterone**?

Progesterone is a steroid hormone that is crucial to regulating the female menstrual cycle and pregnancy cycle in human females and in some other species as well. It is produced in the ovaries and the adrenal glands and is stored in fat tissue.

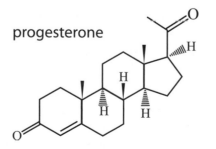

What **causes** the **tides** in the ocean?

The ocean tides are caused by the gravitational forces between the Earth, the Sun, and the Moon, along with the centrifugal force imposed by the Earth's rotation. As the Earth rotates, and as the three bodies move relative to one another, the gradual and recurring shifts in the balance of gravitational forces cause the water in the oceans to tend to move toward one coast or the other.

How does **fertilizer work**?

Fertilizers are used to get elements that plants need into the soil when the natural environment doesn't provide sufficient quantities. The typical elements are nitrogen (N), phosphorus (P), potassium (K), calcium (Ca), magnesium (Mg), and sulfur (S). They're actually labelled by the elements they contain. Next time you're at a garden store, look for "NPK" or "NPKS" on a bag of fertilizer. The numbers after these codes tell you the weight percent of these elements in the bag.

What **nutrients** do **plants** obtain from the **soil**?

There are thirteen mineral nutrients that plants obtain from the soil, and these are divided into the categories of macronutrients and micronutrients. The primary macronutrients include nitrogen, phosphorus, and potassium. Plants require these primary macronutrients in relatively large quantities and deplete them from the soil more rapidly than others. The secondary macronutrients include calcium, magnesium, and sulfur. Micronutrients are required in smaller quantities than the macronutrients, and these include boron, copper, iron, chloride, molybdenum, manganese, and zinc.

What **happens** when something **biodegrades** in a landfill?

Biodegradation describes the process by which microorganisms consume a material and convert it to compounds that are found in nature. This process can happen either aerobically (with oxygen involved in the process), or anaerobically (without oxygen involved in the process). A related term is "compostable"—this specifically indicates that a material will biodegrade/break down when it is placed in a compost pile.

What is **smoke**?

Smoke is a cloud of particles given off by a material that undergoes combustion. Its chemical composition will vary depending on what material is being burned. Smoke may consist of hydrocarbons, haloalkanes, hydrogen fluoride, hydrogen chloride, and a variety of sulfur-containing compounds, among others. These compounds can vary widely in their toxicity, so the severity of the health hazards associated with the smoke from a fire will depend on what is being burned.

How much **salt** (NaCl) is in the average **human body**?

The average adult human body contains about 250 g, or roughly half of a pound, of salt.

What **fraction** of the **oxygen** on Earth is **produced** by the **Amazon rainforest**?

It is estimated that about 20% of the diatomic oxygen (O_2) on Earth at a given time was produced by the Amazon rainforest!

What makes **hot peppers** so **spicy**?

The molecule that makes hot peppers so spicy is called capsaicin (see its chemical structure below). This molecule behaves as an irritant to humans and other mammals, but some of us still really like its flavor!

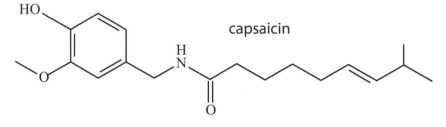

capsaicin

What is **blood doping** in sports?

Blood doping is the act of artificially increasing the number of red blood cells in a person's blood for the purpose of improving athletic performance. This works based on the fact that red blood cells are responsible for carrying oxygen to muscles, and thus more red blood cells can provide more endurance against muscle fatigue. This was originally done by transfusions of red blood cells, either from another person or by collecting and storing a person's own red blood cells to be used later. In the past couple of decades, a new type of blood doping has come about. This is based on the hormone erythropoietin, which stimulates the body to produce red blood cells. Erythropoietin is produced artificially in mass quantities and is commonly used to treat anemia, but is also sometimes used by athletes for the purposes of blood doping.

What makes **carbon monoxide dangerous**?

When carbon monoxide is inhaled, it is readily absorbed through the lungs and into the bloodstream where it binds to the Fe center in hemoglobin (see also "Biochemistry"). Unfortunately, hemoglobin binds to carbon monoxide much more strongly than it does to oxygen, so carbon monoxide rapidly interferes with the ability of hemoglobin to deliver oxygen throughout your body. If this happens, then your muscles and your brain will begin to run out of oxygen, similar to what happens when a person is drowning! Carbon monoxide is a colorless, odorless gas, which makes it difficult to detect unless you have a carbon monoxide detector around. Carbon monoxide levels of 100 parts per million or higher can be hazardous or fatal to humans.

Poisoning from carbon monoxide can result in brain damage, damage to the endocrine system, to the nervous system, and to the heart and other organs. It represents the leading cause of accidental poisoning-related death and poisoning-related injury

> ## Why is the water in the ocean salty?
>
> **O**ne source of salts in the ocean comes from minerals on land dissolving in rainwater and streams, which eventually make their way into the ocean. Since salts do not tend to evaporate along with the water in the ocean, their concentration can build up over time. Another source of salts in the ocean are hydrothermal vents; seawater can flow into these vents, where it becomes warm and dissolves minerals before flowing back out. Underwater volcano eruptions also contribute to the presence of minerals in seawater.
>
> The majority of the salt ions in the ocean are sodium and chloride—these make up about 90% of dissolved ions in seawater. The remainder of the ions present are mainly magnesium, sulfate, and calcium. The concentration of salts in seawater is fairly high, and, on average, seawater is about 3.5% salt by weight.

globally. We should mention that even if a person survives carbon monoxide poisoning, there may be long-lasting effects. Early symptoms of carbon monoxide poisoning can include headaches or nausea, and the treatment for carbon monoxide poisoning typically involves having a person breathe 100% O_2 (recall that air only contains ca. 20% O_2), so that O_2 can more competitively bind hemoglobin to replace carbon monoxide.

What is **permafrost**?

Permafrost is soil that is below the freezing point of water for two consecutive years or more. Most of it is located near the North or South Poles or at very high altitudes.

What chemicals are commonly found in **mosquito repellent**?

A chemical known as N,N-diethyl-meta-toluamide (more commonly known as DEET) is the most commonly used insect/mosquito repellent. This chemical can be applied directly to the skin to prevent mosquito and other bug bites. It is believed to work because mosquitoes dislike the smell of DEET, so they try to stay away from it. In some cases DEET can be an irritant and, in very rare cases, it has been associated with more serious health problems such as seizures.

SUSTAINABLE "GREEN" CHEMISTRY

What is **green chemistry**?

Green chemistry is the practice of designing chemical products with the goal of minimizing the amount of hazardous waste that is generated in the process. This is frequently a complex issue since one needs to consider not just the synthesis of the chemical product but also the sources from which the reagents are obtained, the manufacture of the product, and the end of the product's life cycle. In a seminal book on the topic, Paul Anastas and John Warner defined green chemistry as "the utilization of a set of principles that reduces or eliminates the use or generation of hazardous substances in the design, manufacture, and application of chemical products."

What are the **twelve principles** of green chemistry?

The twelve principles of green chemistry, as originally suggested by Paul Anastas and John Warner (and explained by us), are as follows:

Prevention—The best way to minimize the environmental effects of chemical waste is to prevent waste from ever being generated in the first place. It's easier, and better for the environment, if we can minimize the amount of waste we need to treat, clean up, or store.

Atom Economy—The idea behind this principle is that chemists working on developing synthetic methodologies should generally strive to incorporate as much/many of the reagents as possible into the final product. This helps to prevent the generation of chemical waste and also helps to reduce the quantity of reagents that need to be produced in order to generate the ultimate target.

Less Hazardous Chemical Syntheses—Chemists are encouraged to seek out synthetic routes that avoid or minimize the use of highly toxic chemicals. In addition to trying to minimize the total *quantity* of waste, we also want to minimize the *toxicity* of the waste we generate.

233

Designing Safer Chemicals—In addition to minimizing the toxicity of the waste generated during a chemical synthesis, chemists should strive to select synthetic targets that will also be low in toxicity or have minimal negative impact on the environment.

Safer Solvents and Auxiliaries—Whenever possible, chemists should seek out routes that avoid the use of large quantities of solvents, separation agents, or other chemical auxiliaries. When these cannot be avoided they should be used in minimal quantities, and the choices that are safest for the environment should be made.

Design for Energy Efficiency—The energy costs of synthetic methods should be considered and minimized whenever possible. This may include carrying out reactions and work-up procedures at ambient temperatures and pressures.

Use of Renewable Feedstocks—Reagents and solvents should be obtained from renewable sources whenever possible.

Reduce Derivatives—The number of synthetic steps involving derivatization, such as protection and deprotection steps or use of blocking groups, should be minimized whenever possible. The target of this principle is also to reduce waste and to promote atom economy.

Catalysis—Reagents that behave catalytically, and as selectively as possible, should be chosen over reagents that react stoichiometrically.

Design for Degradation—At the end of their functional life, chemical products should degrade to yield nontoxic products that present minimal threat to the environment.

Real-Time Analysis for Pollution Prevention—Analytical techniques should support in-process monitoring so that the formation of hazardous substances can be prevented.

Inherently Safer Chemistry for Accident Prevention—The use of chemicals should be carried out in a manner that minimizes the likelihood of accidents that may be damaging to the environment or to human health. This includes both the choice of chemicals used as well as the method selected for carrying out a chemical process.

You may notice that the twelve principles are focused closely on the practical synthesis and use of chemical products, and thus they may need to be adapted somewhat depending on the situation (e.g., basic chemical research, large-scale industrial production, nonsynthetic applications of chemistry, etc.).

What are the **goals** of **green chemistry**?

If you have read through the twelve principles of green chemistry, you will likely have realized that the focus of green chemistry rests on minimizing the impact of the development, manufacturing, and use of chemical products on the environment. While these twelve principles may not enumerate every possible method for reducing the impact of chemicals on the environment, they provide a foundation for the outcomes that green chemistry seeks to achieve.

What is **DDT**?

DDT, or dichlorodiphenyltrichloroethane, is a substance that was widely used as an insecticide until its harmful effects on human health and on wildlife became known; DDT was essentially poisoning the humans and wildlife who came into contact with it. In the context of the advent of green chemistry, DDT carries a special significance in that it helped to awaken the public to the fact that the indiscriminate use of chemicals was causing harm to the environment. News of the harmful effects of DDT was spread by Rachel Carson's 1962 book, *Silent Spring*, which explained the numerous negative environmental effects of spraying DDT on crops. Knowledge of the harmful effects of DDT helped to awaken the public to the idea that releasing large amounts of relatively untested chemicals into the environment was potentially causing damage to humans and wildlife. DDT was officially banned in the U.S. in 1972.

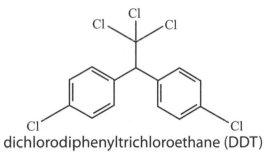

dichlorodiphenyltrichloroethane (DDT)

What is **thalidomide**?

Thalidomide is a drug that was once used to treat the symptoms of morning sickness during pregnancy as well as to help with sleeping problems. A few years after it became widely used, people started to realize that thalidomide was causing birth defects in newborns. These birth defects included phocomelia (abnormal formation of limbs, facial features, nerves, and other parts of the body), problems with sight or hearing resulting from abnormalities in the eyes and ears, gastrointestinal disorders, pasley disorder of the face, underdeveloped lungs, and problems with the digestive tract, heart, and kidneys. The use of thalidomide was then discontinued, though even today there is still some research underway into its possible use in treating cancers. Similar to the situation with DDT, the problems that occurred during the use of thalidomide were particularly influential in motivating the government to tighten regulations on testing drugs and pesticides before their use.

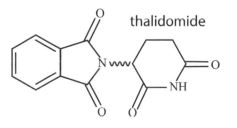

thalidomide

How do you **measure** how **"green"** a **chemical reaction or process** is? What is a **life-cycle analysis**?

Of course, quantifying the question of "How green is a chemical process?" is not always easy to answer! Even in simply comparing a set of alternative processes, it can still be difficult to determine which is better since each may have advantages and disadvantages for different aspects of human health and the environment.

A life-cycle analysis, or LCA, is a tool used to evaluate and compare the effects of a product on the environment. As the name implies, this includes everything that happens between the time the product is created until it is disposed of. Of course, this is no small task! Typically it involves identifying all relevant materials that go into the production of a product, as well as all of the waste produced during the course of using the product, including things like emissions into the atmosphere, soil, and water, as well as the solid waste produced. Then one needs to evaluate the environmental impact of each of those materials and waste products, hopefully in a manner that allows for the results to readily be compared to those from other products or services. The total of these environmental impacts describes the life-cycle impact of the product.

These inputs and outputs are then converted into their effects or impact on the environment. The sum of these environmental impacts represents the overall environmental effect of the Life Cycle of the product or service. Conducting LCAs for alternative products allows comparison of their overall environmental impacts.

What is **bioremediation**?

Bioremediation is an approach to removing pollution from the environment that relies on the use of microorganisms to metabolize pollutants into nontoxic products. Sometimes these microorganisms have been genetically engineered for a specific application. For example, a bacterium called *deinococcus radiodurans* has been genetically engineered to digest ionic mercury compounds and toluene from nuclear waste sites. In such cases, bioremediation can often be accomplished by introducing a microorganism at the site of the pollution, thus avoiding the need for physical cleanup and transportation of the waste to a new location.

What is **phytoremediation**?

Some pollutants, such as heavy metals, are not often readily treated by bioremediation techniques. In such cases, phytoremediation may be useful. Phytoremediation re-

Bioremediation is being used to clean up oil-contaminated soil in the Amazon rainforest.

What is bagasse?

After sugarcane stalks are crushed and pressed to remove their juice, the remaining plant matter is known as bagasse. After the water is removed (and bagasse can contain a lot of water), bagasse is mostly composed of cellulose, hemicellulose, and lignin. Typically this material is burned by the sugar mills directly to generate heat for other processes running at the mill or electricity, which can also be used at the mill or sold back to the electric grid. Bagasse also finds its way into paper production. Recently bagasse has been targeted as a potential source of ethanol.

lies on the introduction of certain plants that are capable of absorbing a pollutant and concentrating it in the above-ground portion of the plant, which can then be removed. The pollutant-containing plants can then be destroyed in an incinerator to concentrate the pollutants even more, or, in some cases, the pollutants can even be recycled for additional use.

How are **plastics sorted** for **recycling**?

The first step in recycling plastics is to sort the plastics by their resin type, or resin identification code. The resin identification code is a number assigned to a plastic product (or container) according to the type of polymers it is made of. While it was once common to directly use this code to identify the types of polymer(s) present, there are now other methods, such as near-infrared spectroscopy or density sorting approaches, that are used to sort mass quantities of plastic samples for recycling. (See "Polymer Chemistry" for more information on resin identification codes.)

What **fraction of plastics** used in the **United States** are **recycled**?

In 2008, about 6.5% of the plastic waste generated was recycled. Roughly another 7.7% was burned to generate energy, while the majority of the remaining waste went into landfills. It is interesting to note that, as plastic production has continued to increase, the fraction of plastics being recycled has decreased. In some cases, this may simply be due to the increasing volume of plastic products or to the fact that it is not easy to build a profitable business around recycling plastics.

What are **alternative solvents**?

Alternative solvents are relatively environmentally benign solvents that can be substituted for the more hazardous choices, which, while they may be traditionally used, may have established precedence without environmental safety or toxicity concerns in mind.

One such example is 2–methyl tetrahydrofuran. As a solvent, it possesses similar characteristics to the widely used dichloromethane and tetrahydrofuran solvents, but with significant environmental advantages in that it is produced from renewable feed-stocks, like corn cobs and bagasse, and is easier to separate and clean up.

You can compare the chemical structures of 2-methyl tetrahydrofuran (left) with dichloromethane (center) and tetrahydrofuran (right) below.

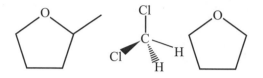

How can **reactions** be run in a **solvent-free environment**?

There are a few common ways to avoid the use of solvents when running a chemical re-action. The simplest situation is when one of the reagents can serve as the solvent for the reaction. This is commonly referred to as running a reaction "neat" (yes, like scotch). Reagents that are not liquids at ambient temperatures can also be used in the molten state so that they can be used as solvents. Some reactions can also be run on solid-supported catalysts that do not require a solvent. By avoiding the use of solvents, each of these approaches cuts down on costs and on the amount of waste generated.

What are **supercritical fluids**, and why can these be **useful** as **green solvents**?

A supercritical fluid is a substance that has reached sufficiently high temperature and pressure to be beyond a "critical point" on the phase diagram (see "Macroscopic Prop-erties"). This "critical" value of temperature and pressure will be different for each sub-stance, and it corresponds to a set of conditions beyond which the distinction between the liquid and gas phases of matter is no longer clear. That is to say, beyond this value, the density and other properties of the substance can be changed continuously, and readily, with changes in the temperature and pressure of the system.

This makes for useful green solvents because having the ability to tune the density, solubility properties, and diffusivity of the supercritical fluid allows for reaction or ex-traction conditions to be sensitively manipulated.

Let's consider one supercritical fluid that has drawn particular interest: carbon diox-ide, or CO_2. Some of the advantages of CO_2 as a supercritical fluid are that it cannot be oxidized, it is aprotic, and it does not tend to participate in reactions involving free rad-icals. This makes carbon dioxide robust toward undergoing chemical reactions itself, and it also means that it is relatively benign as a contaminant (ignoring its role as a greenhouse gas, for the moment). However, CO_2 is a gas at ambient temperature and pressure, and thus, to serve as a good solvent, it must be used at elevated pressures and/or temperatures. At varied temperatures and pressures, supercritical CO_2 is also ca-pable of dissolving a wide range of chemical compounds and is miscible with gases in almost any proportion. Supercritical CO_2 can also often be recycled as a solvent and thus does not tend to generate large amounts of waste. Indeed, carbon dioxide can also be used as a solvent in its liquid form, but it then loses several of the advantages with regard to the tunability of its properties that we mentioned above.

What is an **example** of an **alternative green reagent**?

In a similar spirit to alternative solvents, alternative reagents are relatively environmentally benign reagents that are used to replace more toxic ones. One example of an alternative green reagent is dimethyl carbonate, which can be used to effect methylation and carbonylation reactions. Traditionally phosgene or methyl iodide have been used to carry out this same reaction, but the drawback is that these reagents are significantly more toxic and are thus also more costly to dispose of properly. Dimethyl carbonate is a non-toxic compound and it can be readily produced via an oxidative reaction of methanol with oxygen, thus avoiding any environmentally hazardous synthetic procedures.

Chemical structures of dimethyl carbonate (left), phosgene (middle), and a methyl iodide (right):

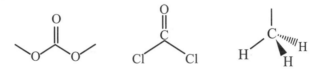

What is an **auxiliary substance**?

Chemical auxiliaries are substances like solvents, separation agents, or dispersing agents that are used in the course of a chemical synthesis but are not reagents because they are not incorporated into the chemical product.

Why does the **fifth principle of green chemistry** seek to **avoid** the use of **auxiliaries**?

Ideally one would minimize the use of auxiliary substances used in a chemical synthesis since doing so would generally be expected to reduce the amount of waste produced, thus minimizing the potential for environmental hazards.

Why is **heating** often used in **chemical synthesis**?

Heat is often used in chemical synthesis to increase the rate of a reaction. In some cases, heating is also used to effect a phase change. To minimize the environmental impact of a chemical synthesis, it is often optimal to seek reactions that proceed readily at ambient temperatures so that energy input in the form of heating is not necessary.

What is an **"E-factor"**?

The "E-factor" is a metric of how environmentally friendly, or harmful, a chemical process is. Specifically, the "E-factor" is the ratio of kilograms of waste generated per kilogram of product synthesized. Thus the lower the E-factor, the more environmentally benign a process should be. Of course, this is only a single metric, and other factors, like the toxicity of the waste produced, should also be taken into account. In pharmaceutical companies, the E-factor for the synthesis of drug products is typically in the range of about 25 to 100.

What is an **example** of how to **"green"** a **chemical process**?

At the pharmaceutical giant Pfizer, the synthesis of Viagra® (see "The World Around Us") originally had an E-factor of 105. However, even before Viagra® was made available to the public, a team of researchers at Pfizer re-examined the entire synthesis step by step. Relatively toxic chlorinated solvents were replaced with less toxic alternatives. The synthesis was also modified to recycle the solvents wherever possible. The use of hydrogen peroxide, which carries some associated health hazards, was removed from the process. Another reagent, oxalyl chloride, is also no longer used; use of this reagent results in the production of carbon monoxide, which is now avoided. In the end, the E-factor for synthesis of Viagra® was reduced to only 8, an over thirteen-fold reduction!

Subsequently, similar changes were made to processes throughout Pfizer. The E-factor for Lyrica®, an anticonvulsant drug, was similarly reduced from an initial value of 86 to now only 9. These sorts of improvements are eliminating millions of tons of chemical waste, while in most cases simultaneously lowering production costs, making safer work conditions, and making products safer for consumers.

How can **microwaves** be used to **promote green chemistry**?

Microwaves are electromagnetic radiation in the frequency range of 0.3 to 300 GHz. When microwaves are absorbed by many substances it causes their temperature to increase, thus heating a sample. This is also the same way the microwave in your kitchen heats and cooks food. Microwaves thus offer chemists the opportunity to use heat to promote reactions in cases where conventional heating methods are not possible. This can be useful for promoting reactions under green conditions, such as when we desire to heat the reagents involved in a reaction in the absence of a solvent.

What is the role of **photochemical reactions** in **green chemistry**?

Photochemical reactions can often serve as excellent choices for green syntheses. One reason for this is that a photon, unlike chemical catalysts or reagents, leaves behind no waste or excess atoms. Photochemically initiated reactions can often proceed rapidly at ambient temperatures, since photoexcitation can be used to generate highly reactive species. In some cases, photochemical syntheses can also reach the synthetic target in fewer steps than those which rely on thermally initiated reactions.

What are **green chemical products**?

Green chemical products are products that were designed with the principles of green chemistry in mind so that they will not have harmful effects on human or animal health or on other aspects of the environment. Since the advent of green chemistry, a tremendous number of green products have been developed, ranging from safer household paints to greener cleaning products to new types of plastic products. You may have heard the phrase "benign by design," which is often used to describe such products.

What are some of the **most harmful organic pollutants** toward the environment?

The EPA provides a list of twelve particularly persistent organic pollutants to watch out for. The list includes aldrin, chlorodane, dichlorophenyl trichloroethane (DDT), dieldrin, endrin, heptachlor, hexachlorobenzene, mirex, toxaphene, polychlorinated biphenyls, polychlorinated dibenzo-p-dioxins, and polychlorinated dibenzofurans. The EPA has colloquially named this group of compounds as the "dirty dozen."

What was **Agent Orange**?

Agent Orange was an herbicide consisting of a mixture of 2,4-dichlorophenoxyacetic acid and 2,4,5-trichlorophenoxyacetic acid used by the U.S. military during the Vietnam War. The intent was to defoliate rural areas of the country, thus removing strategic ground cover and food sources from the rural areas. It was also discovered that the 2,4,5-trichlorophenoxyacetic acid was contaminated with 2,3,7,8-tetrachlorodibenzodioxin, which is an extremely toxic chemical. Agent Orange was sprayed throughout rural areas of southern Vietnam at high concentrations (an average concentration of thirteen times what was recommended by the USDA for domestic use), resulting in roughly 20% of southern Vietnam's forests being sprayed. The use of the Agent Orange herbicide resulted in extremely negative health effects for people in these areas, and the effects still persist today despite the Vietnam War having ended in 1975. It is estimated that one million people are currently disabled or suffer major health problems as a result of the use of Agent Orange.

Chemical structures of 2,4–dichlorophenoxyacetic acid (left), 2,4,5–trichlorophenoxyacetic acid (middle), and 2,3,7,8–tetrachlorodibenzodioxin (right):

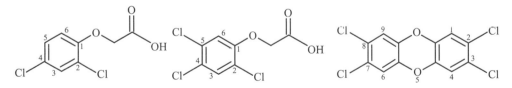

How has the **advent of green chemistry affected the chemical industry** and its role in the public sphere?

Over the past several decades, green chemistry has played a crucial role in turning the focus of large chemical companies toward the environmental impact of their products. There are many reasons for this transformation, some of which include government regulations, changing public opinion, and, of course, the desire to preserve and protect

The Environmental Protection Agency library is located in Washington, D.C. One of the roles of the EPA is to educate the public about green industry.

the environment. Industrial chemists, government agencies, and the general public alike are now constantly considering the effects of chemicals on the environment, which will hopefully serve to prevent recurrences of chemical-related tragedies like those involving DDT.

What is the **role of the EPA** in **promoting green chemistry**?

The United States EPA, or Environmental Protection Agency, has taken great efforts to promote green chemistry. The EPA offers numerous scholarships and awards to promote awareness of green chemistry and the observation of green principles in the chemical industry. It also undertakes efforts to educate the public on green chemistry and, more generally, on the effects of chemical products on human health and on the environment. The EPA also funds research on sustainable technologies and small-business innovation as well as the American Chemical Society Green Chemistry Institute, which promotes partnerships with chemical industry.

What is the role of **aqueous hydrogen peroxide** as a **green reagent**?

Hydrogen peroxide, or H_2O_2, is an ideal choice for a green oxidant because it reacts with high atom efficiency and can react to produce water as the only byproduct. To be a particularly clean oxidant, hydrogen peroxide can be used in aqueous solvents, allowing chemists to avoid the use of any organic solvents. Fortunately there exist catalysts capable of making hydrogen peroxide behave as an efficient oxidant in aqueous conditions, producing products with excellent purity. There are also other reasons that

> ## What is the Warner Babcock Institute for Green Chemistry?
>
> The Warner Babcock Institute for Green Chemistry was founded by John Warner and Jim Babcock to promote the development of environmentally safe and sustainable technologies. The institute offers training in the principles of Green Chemistry for both scientists and nonscientists alike.

hydrogen peroxide is an ideal green reagent, including the fact that it is relatively inexpensive and is produced in mass quantities. In fact, 2.4 million metric tons of hydrogen peroxide are produced each year. One matter of concern is that high concentrations of hydrogen peroxide can be dangerous, so reactions should be run at concentrations of less than about 60% H_2O_2.

Has a **Nobel Prize** ever been **awarded** for a discovery in the field of **green chemistry**?

Yes! Although green chemistry is a relatively young field, one Nobel Prize has already been awarded for work in this area. This prize was awarded in 2005 to three men (Robert Grubbs of the California Institute of Technology, Richard Schrock of the Massachusetts Institute of Technology, and Yves Chauvin of the Institut Francais du Petrole) for their work in the development of olefin metathesis reactions. Olefin metathesis is a type of chemical reaction that involves two carbon–carbon double bonds reacting to form two new carbon–carbon double bonds, effectively exchanging the substituents attached to each carbon. This reaction takes place catalytically under mild reaction conditions, produces little hazardous waste, and has been shown to be effective in a broad range of situations, including the synthesis of new drugs.

Has any **legislation** been passed regarding the implementation of **green chemistry**?

Also yes! A couple of examples are the Registration, Evaluation, Authorisation, and Restriction of Chemicals (REACH) program in Europe and the California Green Chemistry Initiative in California in the United States.

The purpose of the REACH program is to require that companies make data available that demonstrates the safety of their products. This includes the potential chemical hazards during the use of a product, and it also describes means of restricting the use of specific chemicals. A similar piece of legislation, the Toxic Substances Control Act, exists in the United States, but this has received criticism for being far less effective.

The California Green Chemistry Initiative was approved in 2008, requiring the California Department of Toxic Substances Control to place priority on specific "chemicals of concern." This initiative effectively shifted the responsibility for testing chemicals

What is the Presidential Green Chemistry Challenge Award?

These awards were created in 1995 in an effort to promote and recognize innovation in green chemistry in the United States. Five awards are given each year to individuals or companies for work in green chemistry in the following categories: Academic, Small Business, Greener Synthetic Pathways, Green Reaction Conditions, and Designing Greener Chemicals.

away from individual companies and placed it on the government agency. These laws received criticism for not incentivizing research and education regarding green chemistry in the industry. Due to widespread opposition to the initially proposed regulations, the implementation of this initiative had to be postponed at least once due to the need to rewrite the proposal.

What are the **benefits and challenges** of using **water as a solvent**?

The advantages of using water as a solvent are numerous: water is plentiful, environmentally benign, spans a wide range of temperatures while in the liquid phase, and cuts down on waste. Of course, if there weren't also some challenges to making it work, we would just be using it for every reaction. A primary challenge of using water is that many compounds are either unstable or insoluble in water. Additionally, many reactions that were developed in organic solvents do not proceed similarly under aqueous conditions for a variety of reasons, so the majority of existing knowledge surrounding organic synthesis (most of which was developed under non-aqueous conditions) often cannot be directly applied to reactivity under aqueous conditions. Water can also be difficult to remove from reactions relative to many organic solvents due to its higher boiling point. Since the advent of green chemistry the amount of research into aqueous synthesis has skyrocketed, and significant progress is being made every day toward the use of water for a growing number of synthetic applications.

What are some **examples** of **biological feedstocks**?

It is desirable to use biomass, or plant-based materials, as feedstocks for chemical synthesis and energy production. Through photosynthesis plants are able to efficiently capture and store energy from sunlight, and finding ways to use biomass for green chemistry applications is extremely advantageous in advancing the goals of the field. Sources of biomass can be grouped into several categories, including cellulose, lipids, lignin, terpenes, and proteins. Cellulose is often found in structural parts of plants. Lignin is a polymer often found along with cellulose in woodlike parts of plants. Lipids and lipid oils are often extracted from seeds and soybeans. Terpenes are found in pine trees, rubber trees, and a selection of other plants as well. Proteins are found in relatively small quantities in many types of plants and also in larger quantities in animals. Some efforts are also underway to use genetic transplants to create plants that produce increased amounts of proteins. One of the primary challenges in using biological feedstocks to produce chemicals or energy involves separation and purification of the desired materials.

What was the **Bhopal disaster**?

The Bhopal disaster (also referred to as the Bhopal gas tragedy) was a gas leak in Bhopal, India, that happened in December of 1984. At the Union Carbide plant in Bhopal, methyl isocyanate gas was accidentally released during a manufacturing process. The gas poisoned thousands of people in the surrounding city, most of

Protesters in Bhopal, India, rage against the injustice of the Union Carbide plant disaster that poisoned thousands with methyl isocyanate gas.

whom were asleep when the gas leak occurred. The effects of the release of this gas were felt for years to come, with over half a million injuries reported in the nearly three decades since the incident. Following this accident, criminal and civil suits were filed against the company and several of its highest-ranking employees.

Why is the process of **manufacturing ibuprofen** an excellent example of **green synthesis**?

The modern industrial-scale synthesis of ibuprofen has very high atom efficiency, and it has been modified from the original synthesis to be both more environmentally friendly and more cost effective. The original method involved six synthetic steps but used stoichiometric (as opposed to catalytic) quantities of reagents, had lower atom efficiency, and produced undesirable quantities of waste. The modern alternative, on the other hand, requires just three steps, each of which is catalytic in nature. The first step employs a recyclable catalyst (hydrogen fluoride, HF) and produces almost no waste. The second and third steps each achieve 100% atom efficiency (wow!). This process truly represents an ideal benchmark for excellence in green synthesis on the industrial scale.

What is **biocatalysis**?

Biocatalysis, as the name suggests, involves using enzymes or other natural catalysts to carry out chemical reactions. This tends to work well within the context of green chemistry since biological reactions are often catalyzed in water at mild temperatures and pH values. Moreover, enzymes are themselves environmentally benign and obtained from natural sources. In addition to these benefits, biocatalyzed reactions typically proceed with high selectivity and specificity and require relatively few synthetic steps, thus minimizing the amount of unwanted byproducts produced. Some examples of biocatalyzed syntheses are those of penicillins, cephalosporins (another class of antibiotics), and pregabalin (a drug to relieve pain from damaged nerves). The use of biocatalysis is gaining popularity as the necessary technology becomes more readily available and as more people realize the benefits of this approach.

What are the **five environmental spheres**?

In the past environmental science focused on the health of four areas, or spheres, of our world. These are the hydrosphere (dealing with water), the atmosphere (dealing with the air), the geosphere (dealing with the Earth), and the biosphere (dealing with living organisms). Environmental scientists have recently been increasingly recognizing a fifth sphere, the anthrosphere, which deals with the ways that humans modify the overall environment by carrying out their daily activities.

What is the **greenhouse effect**, and how does it affect **Earth's temperature**?

The greenhouse effect involves the thermal radiation from the Earth's surface being absorbed and re-emitted by gases in the atmosphere, termed greenhouse gases. The re-emitted radiation, much of which is in the infrared region of the spectrum, is sent out in all directions, meaning that some of the energy is sent back down toward the lower atmosphere and the Earth's surface. This results in an overall increase in the surface temperature due to the presence of the greenhouse gases. It should be noted that a certain amount of this greenhouse effect is entirely natural, but the effect can be increased when additional greenhouse gases are introduced into the atmosphere as a result of human activity.

As a side note, the name for this effect arises from the fact that greenhouses allow solar radiation to pass through glass and remain inside. In fact, a greenhouse operates on a different principle; the presence of the glass walls in a greenhouse prevent heat from being lost due to convection currents.

How is **rainfall affected** by **pollutants** in the atmosphere?

Air pollution can affect local and global weather patterns, including the amount and frequency of rainfall. At low concentrations particles floating around in the atmosphere may help clouds and thunderstorms to develop, but as their concentration increases these same particles can inhibit the formation of clouds that give rise to rainfall and thunderstorms. This topic is also relevant to issues surrounding climate change, since clouds are understood to generally have a cooling effect on the climate because they reflect incoming sunlight.

What **pollutants** are introduced into the atmosphere by **volcanoes**?

Volcanoes are a major source of sulfur dioxide (SO_2) gas in the atmosphere. This gas is poisonous and is an irritant to the mucous membranes found in your throat, eyes, and nose. Sulfur dioxide also reacts with oxygen, sunlight, dust, and water to create SO_4^{2-} droplets and sulfuric acid (H_2SO_4), which leads to a type of smog referred to as volcanic smog, or "vog." Vog can cause asthma attacks and damage the upper respiratory tract. The sulfuric acid produced can also cause acid rain.

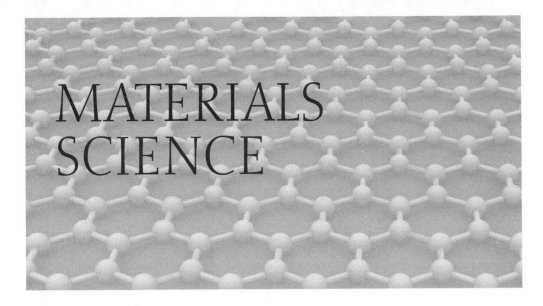

MATERIALS SCIENCE

What is **materials science**?

Materials science is a field at the intersection of the basic sciences and engineering with a focus on the relationship between the microscopic (atomic or molecular) structure of a material and its macroscopic properties. Many techniques relevant to chemistry are used to characterize materials, and the descriptions of the underlying microscopic structure of the material are discussed with regard to solid-state chemistry. Thus many components of materials science represent applications of chemistry. In this section we will provide a brief introduction to materials science with a focus on some of the topics that are more relevant to chemistry.

What are some of the different **classes of materials**?

Biomaterials—materials involving various types of biological molecules

Carbon—materials built from networks of carbon atoms, such as graphite, graphene, diamond, or carbon nanotubes

Ceramics—inorganic (nonmetallic) solids, these are typically prepared by heating and cooling

Composite materials—materials made from two or more components with distinct physical properties

Functionally graded materials—any material that varies gradually in structure or properties throughout its volume

Glass—amorphous solids, typically appearing to have the properties of a solid at the macroscopic level

Metals—composed of metallic elements, good conductors of electricity and heat, typically malleable and ductile

Nanomaterials—materials whose structural features are observable on the nanoscale (typically length scales of less than a tenth of a micrometer)

Polymers—materials/compounds consisting of multiple repeating structural units

Refractory—materials that retain their strength even when they are heated to very high temperatures

Semiconductors—materials with conductivity properties intermediate to those of metals and nonmetals

Thin films—materials that are used in very thin layers, typically ranging from a single layer of molecules to layers that may be several micrometers in thickness

Why do we study materials science?

People study materials science so that others are to be able to choose an appropriate material for an application based on considerations of performance and cost. We want to be able to understand the capabilities and limitations of various materials as well as how, if at all, their properties change after repeated use. By studying materials science, we also become better able to design new materials with the characteristics we desire.

What macroscopic properties of materials are typically studied?

Some of the most commonly studied properties of materials include:

- Thermal conductivity (how well they transmit heat)
- Electrical conductivity (how well they transmit electrons)
- Heat capacity (how their temperature changes with added heat)
- Optical absorption, transmission, and scattering properties
- Stability toward mechanical wear and chemical corrosion

What properties are being targeted for optimization in modern materials design?

Below is a short list of current goals in materials science engineering. This is by no means a comprehensive list, but it is just meant to give you an idea of what is going on in the field of materials science research today.

- Develop structural materials with high temperature stability to increase engine efficiency at high temperatures
- Develop strong, chemically stable, rust- and corrosion-resistant materials for use in construction
- Develop lightweight, mechanically strong materials for high-speed flight
- Develop strong, cost-efficient types of glass to make unbreakable windows increasingly available to the general public
- Develop materials to facilitate the processing of nuclear waste
- Develop fibers with extremely low light absorption for use in optical communication cables

What is an **atomic packing factor**?

The atomic packing factor is the fraction of the volume of a crystal that is filled up by its atoms. In other words, the higher the atomic packing factor, the less empty space there is in the material.

How are **ceramics made**?

Ceramics are nonmetallic materials that are made of a mixture of metallic and non-metallic elements. A ceramic is made by taking an inorganic material, heating it to a high temperature such that the (atomic/molecular) components can rearrange easily, and then allowing the material to cool to room temperature. The resulting materials are typically strong, hard, brittle, and are poor conductors of heat and electricity.

What is **tribology**?

Tribology is a subfield of materials science dedicated to studying the wear of materials. This may include the effects of friction on a material as well as how to better engineer surfaces or lubricate interfaces between surfaces to extend their lifetime.

What is a **fullerene**?

A fullerene is any molecule made up of only carbon atoms that has a shape of a sphere, ellipsoid (a distorted sphere), or a tube. The name fullerene comes from Richard Buckminster Fuller, an architect who designed the geodesic dome (Spaceship Earth at Epcot Center is a geodesic dome). The U.S. Post Office recently commemorated Fuller and and his geodesic dome on a stamp.

What are **buckyballs**?

Buckyballs, or buckminsterfullerenes, are sphere-shaped fullerenes. The most common is C_{60}, which is a sphere composed of alternating five- and six-membered rings of carbon atoms like a soccer ball. This molecule can actually be found in common soot, but don't think you can start selling the remains of your bonfire for cutting-edge fullerene research—it's very, very difficult to purify C_{60}.

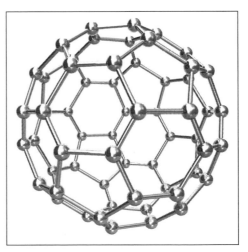

Buckminsterfullerenes are sphere-shaped molecules made only of carbon atoms.

What are **carbon nanotubes**?

Cylindrical fullerenes are known as carbon nanotubes. While C_{60} is a mix of five- and six-membered rings, nanotubes are usually arrays of only six-membered rings. They are just a few nanometers wide, but can be up to several millimeters long. The properties of this form of matter are almost unique in the world, and as a result carbon nanotubes have caught the attention of many chemists. Nanotubes conduct heat and electricity very well, but are also extremely strong (specifically in tests of tensile strength).

What is **scanning electron microscopy**?

Scanning electron microscopy (SEM) is a technique used to capture a picture of a sample by focusing a beam of electrons onto the sample, scanning the beam around the surface of the sample, and then detecting the electrons after they have been scattered off of the sample. The scattered electrons are then analyzed to produce an image of the sample. In general, imaging methods that make use of electrons can offer higher resolution than those based on light due to the shorter wavelengths associated with electrons (as opposed to photons). SEM can be used to obtain very high-resolution images of a sample on length scales as short as one nanometer. The downside to electron-based methods (again, as opposed to using light) is that the electron-based methods are often damaging to the sample (especially to live samples), whereas shining a beam of light on a sample doesn't typically cause a lot of damage. SEM has been useful for characterizing materials as well as a wide range of other kinds of samples.

What is **transmission electron microscopy**?

Transmission electron microscopy (TEM) is similar to SEM in that it uses a beam of electrons to study the sample, but in this case the beam of electrons passes directly through the sample to reach the electron detector. TEM images are able to provide higher resolution than that attainable using a light microscope. The first transmission electron microscope with a resolution greater than that attainable with a light microscope was built in 1933, and there have been commercial TEMs available since 1939. So while it may seem like a very advanced technology, TEM is, in fact, quite an old technique.

What is **graphene**?

In one sense graphene is an unrolled carbon nanotube, or a flattened buckyball. Graphene is a material made of carbon atoms arranged in a hexagonal, "honeycomb" lattice. It is similar to graphite, except that it is only one sheet of atoms thick! A square meter of graphene weighs less than a milligram. Since its discovery, graphene has garnered significant attention for its electronic, thermal, optical, mechanical, and other properties. There is currently a huge amount of research into the properties and applications of graphene,

Graphene is a material made of carbon arranged in a flattened buckyball pattern. It has many potential uses in electronics and other mechanical and engineering applications.

and one Nobel Prize has been awarded (the 2010 Prize in Physics) for research into its properties.

How do **photovoltaics convert light** into **energy**?

Photovoltaic cells are the materials responsible for converting the energy in photons of light from the Sun into energy that can be stored or used. Individual cells usually range in size from areas of roughly one square inch to one square foot, and thousands of cells can be used simultaneously to harvest large amounts of energy. When photons of light strike the photovoltaic material, they excite electrons from a piece of silicon that has been treated such that the excited electrons will gather on one side. This creates a potential difference within the photovoltaic cell such that there is now a positive and negative side (similar to a battery). At this point, the photovoltaic cell has now converted (some of) the energy from the photons into electrical potential energy. This potential difference can be discharged to transfer the energy for immediate use or to be stored while the photovoltaic cell continues to collect additional photons.

How is **hydrogen stored** for use as a fuel?

It would be nice if we could store hydrogen as a liquid; however, it has a very low boiling point (–252.9 °C), which makes this rather inefficient. Due to the strong tendency to evaporate at room temperature, significant energy must actually be expended just to keep hydrogen in its liquid phase. To store hydrogen gas, one possibility is to just compress it inside a metallic container similar to what is typically done with other gases. There have been several other approaches used, however, to attempt to store hydrogen for use as a fuel. These include both chemical and physical storage methods. Some of the chemical storage systems that have been investigated include metal hydrides (like $NaAlH_4$, $LiAlH_4$, or $TiFeH_2$), aqueous carbohydrate solutions (which release H_2 via an enzymatic reaction), synthesized hydrocarbons, ammonia, formic acid, ionic liquids, carbonite compounds, and others as well. These methods generally rely on chemical reactions to make H_2 available for use as an energy source. Physical storage methods include cryogenic compression (involving a combination of low temperatures and high pressures) and a variety of materials, such as metal-organic frameworks, carbon nanotubes, clathrate hydrates, capillary arrays, and others as well. Unfortunately, few of these physical storage methods have thus far been able to demonstrate strongly promising results in working toward a practically useful method of storing hydrogen as a fuel.

What are some **applications** of **functionally graded materials**?

Recall from above that a functionally graded material is one that varies in one or more properties throughout its dimensions. These constitute a relatively young class of materials with promising applications in a variety of areas. For example, the living tissues in your body, including your bones, are classified as (natural) functionally graded materials, so if scientists want to develop materials capable of replacing these, they are looking to develop a functionally graded material. They are also useful in aerospace ap-

Why are scientists so interested in semiconductors?

Recall that semiconductors are a class of materials defined by their conductivity properties. Specifically, they have intermediate conductivity properties between those of things that conduct extremely well (like metals) and things that don't tend to conduct well at all (insulators); this is what makes them so useful. Scientists are able to use semiconductors to control the flow of electricity in circuits, which has been crucial for the development of all of the complicated electronic devices you're familiar with. Semiconductors can be "doped" with materials containing extra electrons, or with materials that are electron deficient, to control the direction of electron flow through the material. Semiconductors have also played a big role in developing solar energy capture devices. The amount of energy a semiconductor needs to absorb to "release" an electron such that electricity can flow can be finely tuned, allowing scientists to develop materials capable of storing solar energy (from photons of light) in the form of electricity.

plications, where materials that can withstand a large thermal (temperature) gradient are needed. Functionally graded materials are commonly found in energy conversion devices and have also been used in gas turbine engines. They can also be good at preventing the propagation of cracks through the volume of a material, which makes them promising candidates for defense applications like developing bullet-resistant materials to create armors for humans and vehicles.

What are some **applications** of **thin films**?

Thin films are layers of material ranging in thickness from nanometers (10^{-9} m) to micrometers (10^{-6} m). They are used commonly for coating optical surfaces and also for coatings on semiconductors. Thin films are used to make mirrors, and these can be finely tuned (in terms of their composition and thickness) to obtain optical surfaces with a wide variety of specific reflection properties, such as wavelength specific mirrors and two-way mirrors (the ones that are transparent from one side but reflective from the other). Thin films are also very useful for coating semiconductors to tune their conductive properties for different applications.

How does **materials science** help to keep your home **warm in the winter**?

By designing effective ways to insulate your home, of course! In total, 48% of the energy used annually in the U.S. is spent on heating buildings during the colder seasons and keeping them cooled during the warmer seasons. Insulation materials help to maintain the temperature differences inside and outside of buildings. In addition to polystyrene and other types of insulation, materials to seal window panes and other potential leaks can significantly reduce the amount of energy we need to spend on heating and cooling.

How does **materials science** help to **improve the fuel efficiency** of your **car**?

One way is through better tires; there are currently tires in development made of rubber that rolls along with less resistance, which has the potential to improve gas mileage by as much as 10 percent just through the tires! Another way involves using special lubricants that work well at a wider range of temperatures, allowing for better fuel efficiency even when the engine is cold as well as when it's warm. This also has the potential to improve fuel efficiency by about 6 percent. Additionally, materials science is also the field responsible for developing lightweight materials to construct all of the components of a vehicle, which further contributes to improving vehicle fuel efficiency.

What are a few of the **big challenges materials science researchers** are working on **now**?

One challenge is to make materials that will allow us to start making smaller, lighter cars that can more easily be powered by electricity. LED lights are another big area in materials science right now. We need to develop materials that can use energy-efficient LEDs to produce the kinds of light we need at low costs. One last area to mention is reducing the amount of waste we produce in general. There are many approaches to reducing waste, and from a materials science perspective we would like to make products out of long-lasting, durable materials that can be repurposed or reused at the end of a product's lifetime.

What is **electrical resistivity**?

Electrical resistivity is a measure of how well a material resists the flow of an electric current. A material with a high electrical resistivity is a poor conductor of electricity, while a material with a low electrical resistivity is a good conductor of electricity. This property is typically expressed in units of ohms × meters ($\Omega \times m$). Recall that ohms (Ω) are a unit of resistance (see "Physical and Theoretical Chemistry").

What is **magnetic permeability**?

The magnetic permeability of a material describes the extent to which it is able to support a magnetic field (inside itself). It describes the amount of magnetization that takes place in a material when an external magnetic field is applied. Materials with a high magnetic permeability are able to support a stronger magnetic field within themselves.

What is **"heat treating"** a material?

Heat treating can be used to achieve different purposes for different materials, but it generally involves heating or cooling a material to a relatively extreme temperature with the goal of changing its properties. Most often, this is done to make a material harder or softer. Heating or cooling the material allows the internal/microscopic structure to change in a way that is preserved when the material returns to ambient temperature.

253

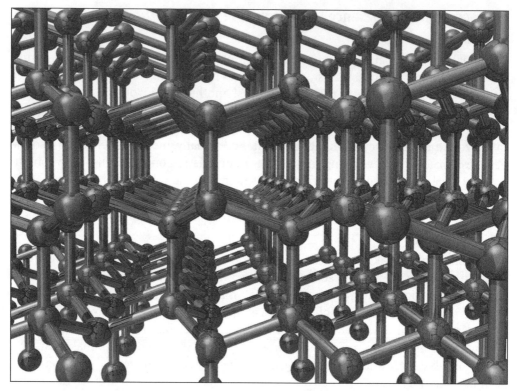

Lonsdaleite is made up of carbon atoms, just like diamonds, but the hexagonal structure makes it even harder than the sparkling gem.

What makes a fabric **"waterproof"**?

One type of waterproof fabric is waterproof simply because the threads are woven so tightly that water cannot easily get inside or through. Other waterproof materials are waterproof because they have been treated with a rubber (or some other) coating that keeps water out. Some coatings will only temporarily waterproof a material such that the coating or treatment wears away over time and needs to be reapplied.

What is the **hardest material** that has been discovered?

At some point you may have heard that diamond is the hardest material known to humankind. While diamond is extremely hard, there are actually a few materials that are even harder still! Years ago, some synthetically produced nanomaterials were discovered that are even harder than diamonds. Even more recently, two additional naturally occurring materials, both of which are even harder yet, were discovered. These materials are wurtzite boron nitride and a mineral called lonsdaleite. Wurtzite boron nitride has its atoms arranged in a very similar structure to the arrangement in diamond, but they are just different atoms (boron and nitrogen, rather than carbon). The other material, lonsdaleite, is actually also made from carbon atoms, but these are arranged differently from those of diamond. Lonsdaleite is also sometimes called hexagonal diamond and can be formed when meteorites, which contain graphite, hit the Earth at very high speeds. Wurtzite boron nitride is produced naturally at high temperatures and pressures during volcanic eruptions. To date, there are only small amounts of either of these materials that have ever been found or synthesized.

What happens when you **"fire"** a **wet clay pot** in a kiln?

Before the clay is placed in the kiln, it is usually dried in the air for at least several days. This first step has already removed the majority of the water, but there will still be some trapped inside the clay. As it is heated in the kiln, the remaining water will turn to steam as it evaporates from the clay. If it is heated too fast, it may turn to steam while still trapped in the clay and cause the pot to explode! As the pot continues to heat some of the organic materials in the clay will burn off, which is necessary for the clay to form a strong final structure.

The next stage is an interesting one, and to understand it we need to consider the chemical composition of clay. Clay consists of a unit of alumina (Al_2O_3) and two units of silica (SiO_2) complexed with two molecules of water. So even after all of the "excess" water has evaporated away, there is still a significant quantity of water that remains chemically bonded within the clay (at this point water accounts for about 14% of the mass of the clay). As the temperature continues to increase those remaining water molecules begin to be released, and they too evaporate away. This is another step where the heating must be done slowly, otherwise the water can create steam pockets within the clay that will expand and eventually explode.

Other changes occur as well, such as changes in the crystalline structure of the silica that will occur multiple times as the pot is heated. Eventually the glass-making oxides within the clay melt, and the clay will fuse into a ceramic material. The materials that melt relatively easily will tend to fill in remaining empty spaces, strengthening the final product. One final note is that changes in the crystalline structure of the silica will also occur upon cooling, and one must take care to cool the pot sufficiently slowly so that these changes don't cause cracks to develop during cooling.

What **kind of glass** is used in your **iPhone** (or other **smartphone**) **screen**?

The glass in most iPhone screens to date, along with that in many other smartphones, is trademarked with the name Gorilla® Glass. This is an alkali-aluminosilicate glass that has been used in over one billion devices! It is lightweight, thin, and resists scratching and cracking significantly better than many other types of everyday glass.

What is a **Pyrex®** **baking dish** made of?

Pyrex® is a type of glass that was originally introduced in the year 1915 to be used in laboratory glassware and home kitchenware as well. It is a borosilicate glass composed of approximately 51% oxygen, 38% silicon, 14% boron, 1% aluminum, 1% potassium, and

0.3% sodium (by mass). Since its original introduction in the early 1900s, a different company has now become responsible for making Pyrex® glassware and they now make it from a soda-lime glass, which is different from the original formula. This new formula is cheaper to produce than the original and is more resistant to breaking when dropped, but has poorer heat resistance.

How do **OLED screens work**, and what are they made of?

OLEDs (organic light-emitting diodes) are a class of LEDs (light-emitting diodes) in which an organic material emits light

Pyrex® baking dishes are a common sight in many kitchens. The material is a soda-lime glass that is resistant to breaking and cracking.

when an electric current is applied. These can be used to create television screens, computer monitors, cellular phone screens, etc. An OLED may employ either small organic molecules or polymers. One advantage of OLED screens is that they do not need a backlight, which allows them to be thin and lightweight and also to display deeper black image levels than backlit screens.

What is the **sticky stuff** that you lick to **seal an envelope**?

The glue that you lick on the seal of an envelope is typically a substance called gum arabic, which is made of polysaccharides and glycoproteins. This gum can be found in the sap of acacia trees.

What is a **gel**?

Gels are solid materials that have flexible properties but do not actually flow in the same way that liquids do. They are made from a crosslinked bonding network of atoms, which actually contains a majority of liquid-like molecules interspersed by weight, but it still behaves as a solid. The crosslinked network within the gel gives it its solid-like properties, while the fluid component gives the gel its stickiness.

What are **metamaterials**?

A metamaterial is a type of artificially engineered material that typically features a pattern or periodic arrangement of a material. Metamaterials are characterized by the fact that they take on specific macroscopic physical properties based on their structure or the pattern in which the material is arranged, but not necessarily the composition of the material. Another way to say this is that the elemental makeup of a metamaterial is not as important as the internal structure of the metamaterial. By relying on the structure of the material to influence its properties, metamaterials have been able to achieve properties that haven't been achieved in other types of materials. One example includes materials with a negative refractive index (see "Physical and Theoretical Chemistry"), which have been able to achieve the first demonstrations of "invisibility cloaking" over certain wavelength ranges, and this technology will hopefully continue to progress as time goes on.

What is **Aerogel**?

Aerogels are materials that are very similar to normal gels, except for the key detail that the liquid component has been replaced with a gas! Since liquids were such a large component of gels by weight, Aerogels are very lightweight. This type of material is translucent and has been given nicknames like "solid smoke" or "solid air." It is produced by extracting the liquid component from a gel using a process called supercritical drying, which allows the liquid to be evaporated away without causing the solid network of chemical bonds to collapse. Aerogels have been produced from a variety of materials including alumina, silica, chromia, and tin dioxide.

What is a **superalloy**?

Superalloys are alloys that display a particularly excellent ability to resist deformation under stress at high temperatures along with good resistance to corrosion and great surface stability. Most often, a superalloy involves nickel, cobalt, or nickel–iron as the base alloying element. Superalloys have been used primarily in turbines and in the aerospace industry.

What are **auxetic materials**?

Auxetic materials have a unique property—when they are stretched, they actually become thicker in the directions perpendicular to the applied force! Think about this in comparison to anything else you stretch; it is really quite a strange phenomenon. This occurs as a result of hinged arrangements within the material that flex apart when a force is applied to stretch it.

ASTROCHEMISTRY

What is **astrochemistry**?

Astrochemistry is chemistry in space! Astrochemists try to do what all chemists do—study molecules and the reactions of those molecules—except that these chemists are looking in outer space, where temperatures are extremely cold and concentrations are exceptionally dilute. These two properties of outer space combined mean that lots of very strange molecules can exist for a relatively long time.

How do chemists **study molecules** so **far away**?

Special kinds of telescopes allow astrochemists to perform spectroscopy on the light (or any type of electromagnetic radiation, not just visible light) coming from a star or other celestial body. Certain features of this radiation allow chemists to measure the quantities of different elements and the surface temperatures of objects like stars and comets.

What was the **first molecule detected in space**?

Hydrogen was probably the first molecule ever detected in outer space, but you're probably thinking of something larger than that. If you exclude diatomic molecules (like H_2, N_2, O_2, etc.), formaldehyde (H_2CO) was the first molecule detected in space.

Can **radio telescopes** detect any kind of **molecule**?

Radio astronomy can only detect molecules with dipole moments, and the stronger the dipole the easier it is to detect. As a result, carbon monoxide (CO) is very easy to detect in space because of both its strong dipole moment and its relative abundance.

So **how** then do astrochemists **detect H_2 in space**?

By looking at other parts of the electromagnetic spectrum, chemists can detect molecules with no net dipole like H_2 (detected by UV radiation) or CH_4 (detected by IR).

What is **atomic emission spectroscopy** (AES)?

AES measures the wavelengths of light that are emitted (or absorbed) when a sample is burned. The wavelengths of the spectral lines are unique to each element because the energy of the photon released from the atom depends on the electronic structure of the particular atom.

What **molecules** have chemists **detected in interstellar space**?

A recent count puts the number of distinct molecules detected in interstellar space at around 150. The list includes small diatomic molecules that are common on Earth (e.g., CO, N_2, O_2), high-energy diatomic radicals that have exceedingly short lifetimes on Earth (e.g., $HO\cdot$, $HC\cdot$), and all the way up to organic compounds like acetone, ethylene glycol, and benzene.

Do **nuclear reactions** take place in **outer space**?

Yes—inside stars! Nuclear reactions occur frequently in stars as hydrogen atoms undergo fusion to produce helium and eventually other heavier elements. Some additional, higher-energy nuclear reactions also take place when two neutron stars merge.

What are **meteorites**?

Meteorites are chunks of material (such as giant rocks) from the solar system that manage to make it to the ground as a larger meteor passes through the atmosphere. Meteors entering from outer space pass through the atmosphere at extremely high rates of speed (from 15 to 70 kilometers per second) and tend to mostly burn up in the process. As they pass through the atmosphere, they may be visible from the ground as a streak of light, and this is what is commonly known as a "shooting star."

What are **comets made of**?

Comets are space snowballs with some dirt sprinkled in. No, really. Comets are typically balls of rock, frozen water, and gases. Sometimes other chemicals are present in much smaller amounts; these include methanol and ethanol, hydrocarbons, and in 2009 NASA's Stardust mission confirmed that glycine (an amino acid) was present in the comet called Wild 2.

Do other **Earthlike planets** exist?

At least one other Earthlike planet has been discovered, and it is possible that many more may exist. In 2011, NASA's Kepler space telescope discovered a planet (named Kepler–22b for now) orbiting a star about 600 light years from Earth (recall that a light year is the distance light travels in one year, so that's extremely far away). This planet is estimated to be about 2.4 times the size of Earth and exists in an area known as a "habitable zone," which means that it is in an area that could potentially serve as a host to life as we know it. Little is known at this time about the atmosphere or composition of this planet, but it is nonetheless interesting that it exists. How many more Earthlike planets might be out there? It's hard to say, but considering the tremendous number of stars out there (see below), it's possible that there are a lot!

How does the **density of matter in outer space compare** with that in **Earth's atmosphere**?

To give a quantitative idea of just how dilute the matter in outer space is, consider that roughly 2.5×10^{19} particles, typically atoms, occupy 1 cm^3 (1 mL) of volume in the Earth's atmosphere. In outer space there is, on average, only one single particle in this same volume. Outer space is a much better vacuum than even the best vacuums ever created on Earth!

How do **space probes** (like the Curiosity rover) **look for molecules** on the Moon or Mars?

The Curiosity rover has a whole suite of chemistry tools on board. The laser-induced breakdown spectroscopy (LIBS) tool is probably the coolest. This instrument breaks down rocks and bits of soil by firing a (freaking) laser at the target. The elements that made up that rock are then detected by atomic emission spectroscopy. Curiosity also contains an alpha particle (He^{2+} ion) X-ray spectrometer (APXS), which is also used to measure what elements make up a sample. If the NASA scientists want to know more

The Hoba meteorite, located in Namibia, Africa, weighs over sixty tons and is the largest piece of naturally occurring iron known. It is so large and heavy that it has never been moved from the spot where it was discovered.

Could comets have been a "seed" for life on Earth?

Interestingly, some scientists are starting to think so. Fairly recently, amino acids (see the "Biochemistry" chapter) were observed in comets, prompting the idea that these molecules, which play a key role in life on Earth, may actually be extraterrestrial in origin. Subsequent research on how these amino acids might have gotten into comets in the first place uncovered that interstellar model ices were able to serve as a breeding ground for dipeptides (a chain of two amino acids). This result further supports the notion that comets from deep space may plausibly have delivered the building blocks for life on Earth today. This contrasts with a long-standing hypothesis that the Earth's early oceans may have served as the source from which the building blocks for life first formed.

than just what elements make up a sample, they can use the quadruple mass spectrometer, which can measure the mass of ions of gases and organic compounds.

What is the **Sun made of**?

The Sun is made up of extremely hot gaseous elements, primarily hydrogen and helium. There are also small amounts of oxygen, nitrogen, carbon, neon, iron, silicon, and magnesium. Because it is so heavy, the Sun produces an extremely strong gravitational pull which leads to very high pressures and temperatures, especially near the Sun's core (roughly 27 million degrees Fahrenheit, or 15 million degrees Celsius). These extreme conditions can cause two hydrogen atoms to undergo fusion (see "Nuclear Chemistry") to create a helium atom. The other two parts of the Sun are the radiative layer (the middle layer) and the convective layer (the outermost layer).

Below is a table listing the relative abundance of elements in the Sun. In total there are at least sixty-seven elements that have been identified as being present in the Sun—this table lists the ten most abundant ones.

Most Abundant Elements in the Sun

Element	% of Atoms	% of Total Mass
Hydrogen	91.2	71.0
Helium	8.7	27.1
Oxygen	0.078	0.97
Carbon	0.043	0.40
Nitrogen	0.0088	0.096
Silicon	0.0045	0.099
Magnesium	0.0038	0.076
Neon	0.0035	0.058
Iron	0.030	0.014
Sulfur	0.015	0.04

Why does the sun glow?

The fusion processes happening near the core of the sun cause energy to be given off in the form of photons (see "Physical and Theoretical Chemistry"). The photons given off in the core of the sun collide with other atoms, which absorb photons and, in turn, give off additional photons. This process repeats, potentially millions of times, before photons at the surface of the sun are emitted off into space.

As a side note, everything tends to emit radiation in this way, at least to an extent—it's just that most things on Earth are not nearly as hot as the sun. Even your own body releases electromagnetic radiation, but the photons coming from your body are in the infrared region of the spectrum, so we cannot see them with our eyes. However, infrared cameras can use the photons given off by a person's body to locate people (or animals) in this way.

What **elements** are **stars other than the Sun** made of?

The Sun is just one of many, many stars that exist. Despite the fact that different stars span a wide range of temperatures and sizes, they are all essentially made up of the same elements as the Sun. Of course, there will be some variations in the relative quantities of the elements present, but, just like the Sun, the main two elements believed to be present in every star are hydrogen and helium.

What happens **chemically** as **stars age**?

As stars age, they continually produce helium from hydrogen by fusion. So as time goes by, the amount of helium in a star increases and the amount of hydrogen decreases. In order to keep the fusion reaction going, stars heat up and get brighter as they age. Stars also continually give off a small portion of their mass, which generates solar (or stellar) wind. For our sun this is an exceedingly tiny amount of material, so don't worry about it vanishing anytime soon. Finally, stars slowly make elements heavier than helium as they age. This is typically quantified by reporting the ratio of iron to hydrogen in a star. Iron is not the most abundant of the heavier elements present in stars, but it is among the easiest of the heavier elements to detect.

Is our **Sun unique**?

Aside from the fact that we're spinning around it, not really. It's a yellow dwarf star of average size (6.960×10^8 m radius, 1.989×10^{30} kg) and surface temperature (5500–6000 K).

Is there **water** on **other planets**?

There is, though in most cases where water exists on other planets it does not exist predominantly in the liquid form (like here on Earth). On some planets there may be trace   263 amounts of water vapor in the atmosphere, beds of ice on a planet's surface, or super-heated, ionized water near a planet's core. There may well be other planets out there with liquid water, but humans have not found many with large amounts of liquid water.

What is a **galaxy**?

A galaxy is a huge system consisting of stars, planets, gas, dust, and lots of other inter-stellar media. The galaxy we live in is called the Milky Way. Galaxies have a lot of stars—the smallest have roughly ten million, while the largest have a hundred trillion! All of the "stuff" in a galaxy orbits around the center of mass of that galaxy.

How **many stars** are in **our galaxy**?

Astronomers currently believe that our galaxy, the Milky Way, contains between two hundred billion and four hundred billion stars.

What is the **hottest planet** in **our solar** system, and why is it so hot?

Venus is the hottest planet, with an average surface temperature of 900 degrees Fahren-heit or 481 degrees Celsius. It's the second closest to the Sun, with only Mercury orbit-ing closer. Interestingly enough, the high temperatures on Venus are largely due to a greenhouse effect due to the very high levels of carbon dioxide (CO_2) in its atmosphere.

What is the **Big Bang theory**?

The Big Bang theory is a model that attempts to explain how the universe was formed. This theory suggests that the universe as we know it came into existence a little less than fourteen billion years ago and that everything started from a very dense, hot, state from which the universe as we know it began to expand. This theory is based on, and is consistent with, all of the current observations we have surrounding the known uni-verse, such as the fact that it is expanding or that there exists a large abundance of light elements in the universe. The one thing left unexplained by this theory, though, which you may likely be wondering about, is how that initial state came to be in the first place. How did the first matter originate, and why was it all packed together in a very dense state? Unfortunately we cannot answer that one, and neither does the Big Bang theory. Rather, the Big Bang theory is only focused on explaining the evolution of the universe from that initial state to what it is today and to what it will become in the future.

Are there **alternate theories** to the **Big Bang theory** out there?

There sure are. While the Big Bang theory is probably the most widely known and ac-cepted theory surrounding the origins of the current universe, there are still other the-ories being explored and proposed. Some of these are much more scientifically feasible than others; take a look around the Web and you can find lots of different ideas out there. One alternative that has received notable attention describes the universe as a continuous cycle of expansion and rebirth; the expansion period, similar to that de-

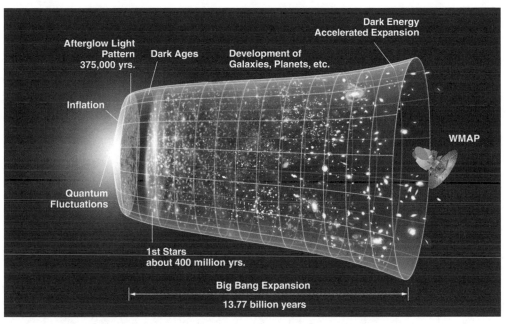

The Big Bang theory postulates that the universe began as a singularity about four billion years ago, expanding rapidly and eventually forming the stars, galaxies, planets and everything else we see.

scribed in the Big Bang, is followed by a period in which the universe once again becomes a dense mass of condensed matter, and then the expansion begins again.

What is a **black hole**?

A black hole is a region of space where the gravitational pull is so strong that nothing, including light, can escape! The size of the black hole is sometimes described by an imaginary surface called the "event horizon," beyond which nothing will be able to escape the gravitational pull.

What **kind of fuel** is used to power a **spacecraft**?

The space shuttle Endeavor, which made its final space voyage in September of 2012, used primarily hydrogen, oxygen, hydrazine (N_2H_4), monomethylhydrazine (CN_2H_6), and nitrogen tetroxide (N_2O_4) as fuels. As it took off for a space voyage, the spacecraft would carry 835,958 gallons of these fuels with a total weight of roughly 1.6 million pounds!

What are the **rings around Saturn**?

You have probably heard about, and looked at pictures of, the rings around the planet Saturn. On average, these rings are about twenty meters thick and they are made up of 93% ice and about 7% carbon. There is actually a pretty large distribution of particle sizes in these rings, ranging from specks the size of dust to chunks of material ten meters in length. Actually, the origin of the rings is not completely understood, and they may be due to either a destroyed moon or to leftover material from when Saturn was formed.

Does **metal rust** in **outer space**?

Sort of. Rusting in outer space doesn't happen in quite the same way it does on Earth due to differences in the amount of available water. On Earth, iron rusts when it interacts with water molecules causing oxidation of some of the metal atoms to metal oxides. Recalling this information, it is clear that some source of oxygen atoms must be present for metal to rust! There is very little oxygen or water floating around in outer space, so the reaction doesn't proceed as quickly or via the same mechanism. Actually, in outer space, the very small amounts of oxygen (O_2) or water (H_2O) that are around are believed to undergo photochemical reactions with metals to produce metal oxides, like rust (Fe_2O_3). Scientists can get a sense of how rapidly metals rust in outer space by looking at iron-containing meteorites that reach the Earth.

What is the **temperature** in **outer space**?

There is not actually a single uniform temperature for all of outer space; it tends to be warmer (at least in a relative sense) in areas that are closer to stars or planets. On average, though, the temperature is only about 3 Kelvin, which is about -270 degrees Celsius. It is extremely cold in outer space, so we humans would not last long floating around in the far reaches of space (even if we could breathe there, which we can't).

How do scientists determine **how far away a star is**?

The answer to this question lies in the application of trigonometry. An astronomer can look at a star at a given point in time and then look at it several months later, after the Earth has moved a substantial distance in its orbit around the Sun. This allows the astronomer to view the star from two different angles. By comparing the images from the two different angles, it is possible to figure out how far away it is.

If a star happens to be too far away the first method described will not be accurate, but fortunately there is an alternative. If an astronomer measures the visible light spectrum of the star, it turns out that one can get a good idea of its actual brightness (by actual, we mean how bright the star is if you are right up close to it). This relationship isn't entirely straightforward and has only been established after looking at data from thousands of stars. Once the astronomer knows the actual brightness, its brightness can be compared by its apparent brightness as viewed from Earth to determine how far away it is.

How **long ago** was the **light** we see **from stars** emitted?

To figure out how long ago the light we see from stars was emitted, we have to know roughly how far away the star is and use the known speed of light (approximately 3×10^8 meters per second). The Sun is roughly 150 million kilometers from the Earth, from

Why is Pluto no longer a planet?

Since Pluto was first discovered in 1930, there has always been some uncertainty about its properties and how they compare to those of the other celestial bodies defined as planets in our solar system. In large part, it is Pluto's small size that led to it being removed from the list of bodies classified as planets.

According to the International Astronomical Union (IAU), a planet is defined in the following way:

A planet is a celestial body that (a) is in orbit around the sun, (b) has sufficient mass for its self-gravity to overcome rigid body forces so that it assumes a hydrostatic equilibrium (nearly round) shape, and (c) has cleared the neighborhood around its orbit.

Pluto meets criteria (a) and (b) mentioned above, but it regularly encounters the orbit of the larger planet, Neptune, which is the technical reason used to remove Pluto's status as one of the planets of our solar system. It is probably worth noting that this resolution was met with some criticism, as even our own planet Earth encounters asteroids in its own orbit on a fairly regular basis.

which we can calculate that sunlight reaching the Earth left the Sun roughly eight minutes ago. The distance to the next nearest star is much farther, roughly 410×10^{11} kilometers from Earth. This translates into a time of over four years between when light leaves this star and when it reaches telescopes on Earth. Keep in mind that this is the next nearest star, so all others are even farther away. This also means that, if we see a star explode, it really happened many years ago.

How many stars exist in total?

This is a difficult question to answer, but we can provide a very rough estimate. In our galaxy alone, there are roughly 10^{11} to 10^{12} stars. Then consider that there are roughly 10^{11} to 10^{12} galaxies in total, and, if we assume the other galaxies are similar to ours, this would put the total number of stars in existence somewhere between 10^{22} to 10^{24} stars. Clearly there are too many to count!

What are sunspots?

A sunspot is a temporary spot on the Sun that appears relatively dark. These spots are caused by magnetic "storms" that prevent convection from distributing heat evenly over the surface of the Sun. This results in relatively cold areas of the Sun. These spots may be as large as fifty thousand miles in diameter, such that they can be visible from Earth even without a telescope (but that doesn't mean you should look at the Sun).

Why do footprints last extra long on the surface of the Moon?

On the Moon, there is no (or extremely little) wind, so the dust will not tend to blow over and fill in your footprints like it does at the beach on Earth. So the footprints left by the earliest humans to visit the Moon should still be there today.

How fast is Earth moving through space?

When taking into account the motion of the Milky Way galaxy, the fact that our solar system rotates within this galaxy, and the motion of the Earth within our solar system, it is estimated that the Earth is moving at about 500 kilometers per second. Of course, since everything around us is moving at the same rate, we don't tend to notice this.

What is astrobiology?

Astrobiology is a branch of science concerned with looking for signs of life, basically, anywhere other than on planet Earth. This isn't really the search for space aliens hanging out in close vicinity to our own planet, but rather astrobiology is focused on looking for any evidence for the current or prior existence of even the smallest (microbial) life forms elsewhere in space.

What is a biosignature?

A biosignature is a chemical signature that can be observed from a distance that signals the presence of living organisms. These could include complex chemical structures associated with life forms or accumulated quantities of biomass or waste.

What is a superbubble?

A superbubble is a cloud of superheated gas that can be formed when multiple stars that are relatively near each other die out at similar times. This situation can lead to an explosion that spans hundreds of light years in distance. Recall that a light year is the distance that light travels through space in a year, so we are talking about an absolutely huge explosion! The light generated from this explosion is often not in the visible range of the spectrum, so they can't always be seen by the naked eye (not to mention the fact that they are also very, very far away).

What is radio astronomy?

Radio astronomers study what's going on in outer space by using radio waves (as opposed to telescopes, which use light in the visible region of the spectrum). This approach has some distinct advantages, perhaps the most obvious being that radio astronomers can work at any time of the day while astronomers who rely on light telescopes can only work at night. Radio astronomy relies on monitoring weak radio wave signals coming in from outer space, and knowing how to interpret these signals, to draw conclusions about the locations of celestial bodies and events that have taken place far, far away.

CHEMISTRY IN THE KITCHEN

Do chemists really **study food chemistry**?

Yes, they really do! There's even a scientific journal called *Food Chemistry* (published by Elsevier) dedicated to reporting new findings regarding the chemistry and biochemistry of food and raw (food) materials.

What is the **Maillard reaction**?

The Maillard reaction is technically "nonenzymatic browning," which basically includes any kind of browning that happens when you're cooking, but excludes what happens to cut apples you leave out on the counter. At the chemical level, it's a reaction between an amino acid and a sugar in the presence of heat. A huge number of chemicals are formed in these processes, so you can't really pin down the Maillard reaction to a single set of chemical steps. These processes are responsible for the browning of meat, the malting of barley for beer, the roasting of coffee, and the browning of the crust of bread.

Is **caramelization** the **same** thing as the **Maillard reaction**?

Caramelization is the breakdown of sugar molecules with heat (pyrolysis), while the Maillard reaction requires amino acids (proteins). Caramelization, like the Maillard reaction, is a term for hundreds of different chemical reactions taking place at the same time.

How does **baking soda** make my **cookies better**?

Baking soda is sodium bicarbonate ($NaHCO_3$) and is used in cooking as a leavening agent (as in, it helps things to rise). It does this by releasing carbon dioxide (CO_2), which it does in the presence of acids, like buttermilk, vinegar, lemon juice, cream of tartar, and so on. The process is much faster at higher temperatures. Once you put your cookies in the oven, the sodium bicarbonate begins to break down and the carbon dioxide that is released makes tiny little bubbles in the batter. These tiny bubbles get trapped as the cookies bake, making them light and fluffy.

What's the **difference** between **baking soda** and **baking powder**?

Baking powder is a mixture of three things: baking soda, an acid, and a filler—usually cornstarch. The acid takes the place of the buttermilk or lemon juice in a recipe (see the previous question) to release CO_2 from $NaHCO_3$. The starch is in the mix to keep the two components from reacting before it gets into your cookies and to keep everything dry.

If **baking soda** and **baking powder** are different, how can I **substitute** one for another?

Since baking powder is diluted baking soda with acid, if your recipe calls for powder but you only have soda, you'll need to use less, but add an acid. Generally the ratio is three parts baking powder equals one part baking soda with two parts cream of tartar (or another acid substitute).

What **chemicals** are used to **preserve food**?

There are two main types of food preservatives, those that prevent oxidation and those that prevent bacteria or fungi from growing. The first are appropriately named antioxidants, and these molecules work by reacting with oxygen themselves. Unsaturated fats are common targets of oxygen, which causes foods to become rancid. Antioxidants provide an even easier target for oxygen to attack, preventing O_2 from wreaking havoc in other ways. Natural antioxidants include molecules like ascorbic acid (vitamin C); and there are also many unnatural antioxidants on the market.

The second class of preservatives are those that stop the growth of bacteria and fungi (like mold) from growing. Many of these preservatives are acidic molecules that can be absorbed into the cells of bacteria. If enough acid gets into the cell, basic biochemical functions (specifically fermentation of glucose) slow down enough that the cell dies, and your food stays fresh.

What is **cream of tartar**?

Cream of tartar is the potassium acid salt of tartaric acid. This means one acidic hydrogen of tartaric acid is replaced with a potassium ion. The structure is shown below. If you've ever seen crystals of something in a bottle of wine (or even fresh grape juice), it was probably this chemical.

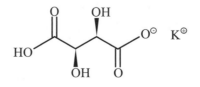

What's in **Jell-O®** that makes it so **jiggly**?

Gelatin, which is a convenient and pleasant term for the mixture of proteins and shorter peptides that are obtained from boiling byproducts of the meat and leather industries (mostly bones and pig skin), makes Jell-O® jiggly. The hydrogen bonds between peptide strands allows gelatin to form a network in the presence of water, which is what the physical structure of the gel is that you are familiar with. These hydrogen bonds can be disrupted with heat, which is why you boil Jell-O® before pouring it into a mold; once it cools back down, the network reforms in whatever shape the liquid is in.

Gelatin, the main ingredient in Jell-O® desserts, is made up of proteins and peptides extracted from boiling leather and meat byproducts. Yummm.

How does **pectin gel make jelly**?

Pectin is a polysaccharide from the cell walls of plants that helps plant cells to grow and also to stick to their neighbors. Commercially, pectin is extracted mostly from citrus peels, but other plants are also used. Pectin is the gelling agent that helps jams and jellies to set, giving them a consistency that is useful for spreading on toast. It does this using a very similar mechanism as the one it uses to keep plant cells together. The chains of sugar molecules (polysaccharides) can form bond between strands, creating an elastic network known as a gel…or jelly.

Why do some **fruits and vegetables turn brown** after you **cut them**?

When you cut or drop an apple, potato, or other fruit, some cells leak their contents, exposing all sorts of cellular machinery to air. The key player in the browning of fruits is tyrosinase, an enzyme that is involved in the oxidation of tyrosine specifically and phenols in general. Tyrosinase plays two key roles in the production of melanin, which is a general term for all sorts of pigments found in plants and animals. So once tyrosinase leaks out of the cell, it is exposed to all the oxygen and phenols it needs to start making brown pigments.

Why does **lemon juice** prevent fruits from **browning**?

If you put lemon juice on your fruit after cutting it to prevent "browning," this is an example of that last category of preservative—one that slows some enzymatic process. The vitamin C lowers the pH (because it is an acid), slowing the enzymes (polyphenol oxidases and tyrosinases) that cause browning of your fruit.

Why do **onions** make you **cry**?

Sulfinylpropene is the main tear-inducing compound (technical term: lachrymator) released when you slice onions. The most interesting part of this story is that this chemical is not present in onions before you slice them. In fact, the generation of this tear-inducing molecule is not a mistake, but part of the defense system that plants have developed to deter animals from continuing to eat them. When you cut, or an animal takes a bite of, an onion, an enzyme, alliinase, that is normally safely stored within the plant's cells is released and starts to wreak havoc. Alliinase converts sulfoxides to sulfinyl groups, which makes you cry like a baby.

So is there any real way to **stop onions** from **making you cry**?

Everyone's mother seems to have their own unique proposal about how to avoid crying over diced onions, but there's only one that makes any chemical sense (excluding wearing a gas mask): put the onion in the fridge before you cut it. Almost every chemical reaction is slower at lower temperature, and since the lachrymator in onions is produced when you cut it (and not naturally present in the onion), you can give yourself more tear-free time to dice if the onion is cold.

How is **refined sugar different** from **raw sugar**?

Refined sugar is raw sugar that has been purified by a series of steps that ends with crystallization of a sugar syrup into white sugar crystals. The debate of whether refined sugar is worse for you than raw sugar is ongoing, but chemically refined sugar is just more pure sugar than raw sugar.

What about **beet and cane sugar**, how are those **different**?

Beet sugar comes from sugar beets. Cane sugar comes from sugarcane. After these two sugars are purified (refined), there is no chemical difference.

So what is **molasses**, then?

Molasses is the byproduct of refining sugarcane. In one of the crystallization steps, the brown liquid that is left behind is concentrated into a syrup known as molasses. The molasses you may have used in cooking or baking is from sugarcane. Sugar beet molasses has a lot of other chemicals and it tastes terrible to us; not everyone seems to mind though—it's a common additive to animal feed.

What is **Splenda®**?

Splenda® is a commercial name for an artificial sweetener, like NutraSweet® or Equal®. Sucralose is the key sweet ingredient in Splenda®, but unlike natural sugar molecules, sucralose is not metabolized, so it effectively has zero calories. It's made from regular table sugar by selectively exchanging three OH groups for chlorine atoms.

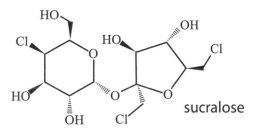

sucralose

What is **Stevia**?

Stevia is a commercial name for an artificial sweetener and is also the common name for the plant that it is extracted from. Steviol is the basic structure of this class of sweetener, but when sugar molecules are attached to steviol (making it steviol glycoside) its sweetness skyrockets to hundreds of times that of regular sugar. This sweetener has been in use for centuries in South and Central America and in Japan since the 1970s. In the United States, it's only been available for a few years as a purified compound (marketed under the name Truvia®); raw plant extracts of the stevia plant are not approved for use in the United States.

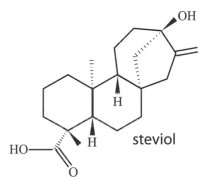

steviol

Okay—last sweetener question—what is **corn syrup**?

Corn syrup is made from cornstarch either by heating up starch in an acidic water solution or by adding an enzyme to break down the long starch molecules into simpler sugars. Chemically it is mostly maltose, which is a disaccharide (two glucose molecules stuck together) along with a small amount of larger chains of sugar molecules (oligosaccharides). The high-fructose variety is made by treating regular corn syrup with a second enzyme that converts glucose into (you guessed it) fructose.

What does **brining** do to **meat**?

Soaking meat in a salt solution (the definition of brining) helps separate the long filaments that make up muscles (myofibril) by dissolving the proteins on their surfaces. With enough salt the actual filaments start to break down, and both effects make the meat seem more tender. Additionally salt allows proteins to hold on to more water, which helps prevent your steak from drying out.

275

What is **removed** from butter to make **clarified butter**?

Proteins and water. Clarified butter is made by melting regular butter at a low temperature. Three layers will form: the top frothy layer contains the proteins from milk (casein, used to make cheese); the middle layer is water with dissolved milk sugars, like lactose; the bottom layer is pure butterfat or milkfat, which is also known as clarified butter. You can instead heat butter at a low temperature for a long time to remove the water by evaporation, and then decant or filter the butterfat. Clarified butter contains almost no proteins, so it has a very long shelf life, and no lactose, so people who are lactose-intolerant can eat it.

How does **nonstick cooking spray** work?

It's not nearly as magical as you might think. Cooking spray is just regular vegetable oil in spray form. To get it to spray out a fine mist, an emulsifier is added, and the can needs something to act as a propellant (usually alcohol, CO_2, or propane).

If **cooking spray** is just oil, then how can it have **zero calories** and **no fat**?

Cooking spray lets you apply a thinner layer of oil than you could probably achieve by pouring normal liquid vegetable oil out of a bottle. The FDA states that any food substance with less than five calories and less than 0.5 g of fat per serving can be labeled calorie-free and fat-free, respectively. So manufacturers of cooking spray adjust the recommended serving size to contain less than those limits. As a result, your can of spray contains hundreds of servings—go check your pantry if you don't believe it!

How do **instant noodles** cook so **fast**?

Because they're already cooked! Instant noodles were invented in Japan in the year 1958 by Momofuku Ando, who was working at Nissin Foods. The noodles are flash fried, creating a dry noodle with a very long shelf life that can be prepared in minutes.

What is **homogenized milk**?

Homogenized milk is milk that won't separate. Normally cream will separate out from milk, forming a layer at the top of the bottle. This is obviously not ideal, so to prevent this separation from happening, milk is treated with pressure to break up the little clusters of fat into much, much tinier pieces. These tiny globules of fat don't recombine at an appreciable rate, so the milk remains a single layer throughout its shelf life.

Instant noodles are pre-cooked and dried by flash frying. In this way, they can be stored for a long time and quickly cooked in hot water.

What makes fish smell fishy?

Fresh fish doesn't smell fishy at all. It's only when proteins and amino acids in fish start to break down, releasing stinky nitrogen and sulfur compounds, that the funk sets in. There are a few reasons that this sort of smelly decay is more common with fish than with chicken, beef, or pork. Fish frequently eat other fish, so their digestive systems contains enzymes that can break down the proteins found in fish. So if some of these enzymes leak out, or if you're slow to gut your fish, those enzymes will go to work…on its own flesh. Fish also generally have higher levels of unsaturated fats, which are less stable than saturated fats to oxidation. Acids, like lemon juice, can slow the enzymes down, and convert the amines into less odorous ammonium salts, which is probably why we're all used to squeezing a lemon wedge on fish.

Why does **elevation matter** for **cooking times** when I'm boiling water?

If you're boiling water in Denver, the temperature of that water will be about 5 °C lower than if you were boiling water in Miami. Because Denver is about a mile above sea level, there is less atmosphere pushing down on that pot of water than there would be at sea level. The decreased temperature that water boils at means that you'll need to increase cooking times the higher up you go.

How does a **pressure cooker speed up cooking**?

If the lower atmospheric pressure in Denver increased cooking times by lowering the boiling point of water, what if we could increase the boiling point of water? That's exactly what a pressure cooker does. Pressure cookers are sealed such that once you start to heat water, the pressure inside the vessel increases. This increase in pressure drives up the boiling point of water because every water molecule that tries to make the transition from water to liquid has a greater force pushing against it. By increasing the pressure inside the pot, pressure cookers can get the boiling point of water up to about 120 °C (250 °F). With water boiling at a higher temperature, your food cooks faster.

Does **hot water** really **freeze faster** than **cold water**?

Sometimes. This observation is known as the "Mpemba effect," named after the Tanzanian student who in 1963 resurrected the idea from Aristotle, Francis Bacon, and René Descartes. Whether or not this effect can be seen seems to depend on so many variables (the size and shape of the container, the initial temperatures of the two liquids, the method of cooling), on how you define freezing (when the first ice crystal forms, when there's a solid layer on the top, or when all of the water has frozen solid), that it's really still unclear if this effect is real or not.

Does all the **alcohol really boil off** when I **cook with wine**?

Not really, no. It's common lore that when you add red wine to pasta sauce that the alcohol evaporates rather quickly. This idea is supported by the fact that alcohol has a lower boiling point than water, so it should evaporate quickly. People who actually studied this, however, have shown that even after an hour, 25% of the alcohol you added is still in the sauce. If you want a truly nonalcoholic marinara, you need to simmer for at least two and a half hours.

When improperly packaged food is put into a freezer, water can be drawn out of the food and crystalize; oxidation can also occur. It is still safe to eat the food, though it is not as appetizing.

What is **freezer burn**?

Freezer burn occurs when frozen food undergoes dehydration due to improper packaging. The humidity level in a freezer is usually quite low, so if food is not stored in airtight packaging, water in the food can be drawn out into the freezer atmosphere by sublimation. Also, because the food is exposed to air, oxidation can occur, though at the lower temperatures in a freezer, these reactions are quite slow. Thankfully, although freezer burn looks nasty it's not a food safety concern, it just causes discoloration.

Why can't I put **raw eggs** in the **freezer**?

You can if you take them out of their shells. Raw eggs expand when frozen, which can break the shell, so don't put whole eggs directly into the freezer.

Are **green, oolong, and black teas** made from **different plants**?

No, all tea is made from the leaves of a single plant, *Camellia sinensis*. That statement excludes herbal teas, though, which are more accurately called infusions. Different categories of tea are prepared using different processes of wilting, bruising, and drying the leaves. Green tea is processed within a day or two of harvest, which preserves the natural chemicals of the fresh leaves. Black tea leaves are prepared by an oxidation process at high temperature and humidity, and then dried. Oolong tea is in between green and black: the leaves are left for a few days to wither, after which a short oxidation process is performed.

What makes a bowl **microwave safe**?

Unfortunately, the only definition is an empirical one: containers that don't get hot in the microwave are microwave safe. Remember (or go look it up now in "The World Around Us") that microwaves heat food by causing molecules with dipole moments to tumble back and forth. If the container has any such molecules, it'll get hot in the microwave. If any water has leached into your ceramic mug, it'll get hot in the microwave. Regardless of whether the container gets hot on its own, the food you are heating up will get hot and transfer that heat to the bowl, so be careful when taking hot things out of the microwave.

What makes **string cheese stringy**?

In the United States, string cheese is usually mozzarella (sometimes with some cheddar thrown in). The production process takes melted cheese and stretches and folds it in a single direction. This stretching aligns the proteins in the cheese, making it possible to peel off long strings of it.

What's the **difference** between **brown and white eggs**?

Besides their color, the only difference is the color of the chicken that laid it. White-feathered chickens lay white eggs; brown-feathered chickens lay brown eggs. That's it. The inside of the eggs are identical in every way, assuming the chickens you're comparing were on identical diets.

Why do **hard-boiled eggs spin**, while **raw eggs don't**?

Hard-boiled eggs are solid all the way through so when you spin the egg, all the energy you apply goes into spinning the whole object. Raw eggs have yolks that are free to move about the interior of the egg, however. So when you spin a raw egg the yolk moves to the outer edge of the inside of the egg, which consumes some of the energy you applied. The other difference you can see is what happens after you stop these two eggs from spinning. Stopping the hard-boiling egg stops the entire system, because the yolk is trapped in the solid, cooked egg white. Inside the raw egg, however, even after you stop it, the yolk is still spinning around. So if you take your finger off of a raw egg that was just spinning, it can start moving again, seemingly on its own.

Why do **hard-boiled egg yolks** turn **green**?

When egg whites cook, a small amount of hydrogen sulfide (H_2S) is released from sulfur-containing amino acids like cystine or methionine. If the H_2S migrates to the yolk, it combines with iron atoms to produce ferrous sulfide (FeS), which is a dark-colored material that looks green mixed with the bright yellow yolk. Ferric sulfide (Fe_2S_3) is also sometimes claimed to be formed in this process, likely because it is itself a yellow-green substance. There is, however, less data to back this up, aside from the convenient color match.

What's in **self-rising flour** that makes it rise?

Baking powder and salt are added to flour to magically transform it into "self-rising flour." Yeah, not so magical after all.

Why is there **lime** in my **tortillas**?

First, let's clarify the question: we're talking about traditional corn tortillas, and by lime we're referring to a calcium hydroxide solution, not the green fruit. Corn tortillas were historically made from what is called nixtamalized corn, which involves soaking corn in a basic solution like calcium hydroxide. The process goes back about three thousand years to the Aztec and Mayan civilizations. Because of its ancient origin, how it was discovered isn't clear, but why it survived is now obvious. Corn lacks one of the essential vitamins required in human diets; it doesn't have niacin, also called vitamin B_3. People who don't get enough niacin in their diet develop a disease known as pellegra (just like if you lack vitamin C, you get scurvy), which has some awful symptoms. Obviously this is a problem if your civilization's staple food is corn. Somehow, the Aztec or Mayan people figured out that cooking corn with strong bases prevented people from getting pellegra. Now we know this is because niacin is not readily available in corn, but can be released by treatment with a strong base.

niacin (vitamin B_3)

Why would anyone **add carbon monoxide** to **tuna**?

Cutting a tuna fish exposes many muscle cells to oxygen, which slowly changes the bright red color of fresh tuna steaks to a darker brown. This is the result of iron-containing enzymes (myoglobin and hemoglobin) being oxidized from Fe(II) to Fe(III). Sushi lovers have come to understand that fresh raw tuna should be bright red and are skeptical of eating any brown-colored fish. The seafood industry figured out at some point that adding CO during the packaging step not only slows the rate of oxidation down, which increases shelf life, but also brightens the red color. It does this because CO binds more strongly to the Fe(II) center than Fe(III). The risk to consumers is not CO exposure here, but that you might be fooled into eating fish that isn't as fresh as you might otherwise think if you judge its freshness based on color.

What is **liquid smoke**?

It's actually exactly what it sounds like, as crazy as that is. Smoke from a wood-burning fire is blown into condensers, which collect many of the volatile chemicals in smoke. This mixture is then diluted with water. It's used to cure bacon and flavor many other foods.

Is **ceviche** really "**cooked**" by lime juice?

The results of cooking with heat and with acid are similar, but of course they are not exactly the same. Both heat and acid serve to denature proteins in food, which is a technical way of saying that the molecules change shape. With access to any number of shapes that it couldn't exist in beforehand, the molecule finds new ways to react both with itself and with other protein molecules. These recently liberated proteins quickly form a solid network, which is why fish gets firmer and whiter when you add lime juice, and it's the same reason that egg whites turn opaque and get harder upon cooking.

Why do **shrimp change color** when they're **cooked**?

Some, but not all shrimp, are grayish when they are raw, but turn pink once they're cooked. It makes sense to guess that this is because some chemical compound with a red color is being produced once you add heat. What's actually happening, though, is that the more intense pigments in the shrimp's shell are decomposing with heat, while the compound responsible for the red color is more stable. That red molecule is called astaxanthin, and it's not only found in shrimp shells but is also the reason that salmon meat is red.

Raw shrimp like this is grayish in color, but when cooked turns a bright pink because a red-colored molecule called astaxanthin remains in the shrimp even after less-stable pigments break up upon being heated.

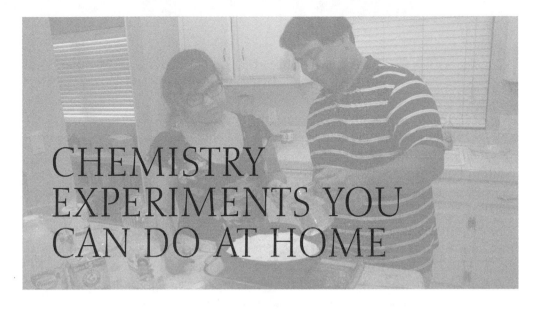

CHEMISTRY EXPERIMENTS YOU CAN DO AT HOME

You'll probably be interested to know that there are a good number of chemistry experiments that you can perform right in your own home! While we can't teach every principle in this book with experiments that you can do with household items, we will do our best to illustrate some of the basic principles (mostly those relevant to basic chemistry lab work) with items you may already have in your kitchen.

How can I carry out a **chromatography experiment** at home?

Chemistry principles encountered in this experiment:

- Extraction
- Filtration
- Chromatography

Materials you'll need:

- Rubbing alcohol (250 mL)
- Small jars with lids (baby food jars work well)
- Tree or plant leaves (5 will do)
- Coffee filters
- Pan containing hot water
- Kitchen utensils

The procedure:

1. Tear a few leaves up into small pieces, and place the pieces from each leaf into their own small jar.
2. Add a small amount of rubbing alcohol to each, such that the leaves are just covered by the rubbing alcohol.

283

3. Place the lids on the jars, and then place them in the hot water for 30 minutes, replacing the water with fresh hot water if it cools. Swirl the alcohol in the jars every 5 to 10 minutes.

4. By the end of 30 minutes, the pigments from the leaves will dissolve, and the alcohol solutions should appear colored. The alcohol serves as a solvent for extracting the pigments in the leaves.

5. Cut long strips of coffee filter paper, remove the tops from the jars, and place one into each jar, with one end in the alcohol, and the other end outside the jar.

6. The pigments will move up the paper different amounts, according to their size. You should be able to see this separation as different-colored regions on the paper. Note that this process is a bit different from some of the separations we described in "The Modern Chemical Lab" based on polarity or other chemical properties. Unlike chromatography on silica gel, the paper is not strongly polar and thus does not interact as strongly with the polar pigment molecules.

7. Remove the strips of paper and allow them to dry. By comparing the relative distances the compounds travelled (technical term: retention factor) of the different compounds with one another, you should be able to identify whether the different leaves contained the same pigments. You can also try to make your alcohol solutions from other pigments, such as those in inks, foods, or drink.

8. Try to design a hypothesis you can test using this experiment, such as whether or not the leaves of different plants contain the same pigments, or whether different markers or pens contain the same inks.

How can I **make slime**?

Chemistry principles encountered in this experiment:

- Hydrogen bonding
- Synthesis
- Polymer chemistry
- Cross-linking

Materials you'll need:

- Water
- Elmer's glue (about 4 oz. or 120 mL)
- Borax powder (4–5 tablespoons)
- Bowl
- Measuring cup
- Small jar (doesn't need to have a cover)
- Spoon or stirring device
- Food coloring (optional)

The key ingredients in making slime are glue and Borax powder. The oxygen atoms in the polymers found in the glue will then link with the hydrogen in the borax. (*Photo by Jim Fordyce.*)

The procedure:

1. Pour about 4 oz. of white glue into the jar. The glue contains several components, including the polymers polyvinyl acetate and polyvinyl alcohol (see "Polymer Chemistry"). Polyvinyl acetate contains oxygen atoms that can serve as hydrogen bond acceptors, and polyvinyl alcohol contains hydroxyl groups that can serve as either hydrogen bond donors or acceptors.

2. Add ½ a cup (4 oz. or 120 mL) of water, and stir it in until it mixes with the glue. Note that there is already some water in the glue, and we are just adding more.

3. (Optional) add food coloring to the mixture to give your slime some color.

4. In the bowl, mix 1 cup (8 oz. or 240 mL) of water with 4 or 5 tablespoons of the borax powder. Stir this well. Here we are preparing a solution of borax so that we can add it to the mixture of glue and water in a more uniform fashion.

5. While mixing, slowly add the glue and water mixture to the solution of borax.

6. As you mix, you should observe the slime forming. Pick it up with your hands and knead it until it seems fairly dry. There will still be extra water left behind in the bowl; that's nothing to worry about. The borax serves to cross-link the polymers

You can play with the amount of borax added to the glue and water mix to give you the consistency you want. Add food coloring, too, to give your slime an even cooler look! (*Photo by Jim Fordyce.*)

by forming hydrogen bonds with the oxygen atoms of the polymers in the glue. These interactions can readily rearrange to form new hydrogen bond donor-acceptor pairs with different oxygen and hydrogen atoms, which is what makes the glue so stretchy and readily deformed.

When you are done with it, you can store the slime in a plastic bag in the refrigerator.

How can I **test the hardness** of **objects** around my house?

Chemistry principles encountered in this experiment:

- Hardness
- Materials science

Materials you'll need:

Collect several materials of known hardness, examples include (numbers are based on the Mohs scale):

- Fingernail (2.5)
- Penny (3)
- Glass (typically 5.5-6.5)
- Quartz (7)

- Steel (typically 6.5-7.5)
- Sapphire (9)

You can also look for additional items from the list provided at the end of this experiment, or search online for additional objects that have been ranked on Mohs scale of hardness. Of course, you can also choose item of unknown hardness and determine their hardness in this experiment!

The procedure:

1. Locate a specimen whose hardness you want to test. Note that you will be attempting to scratch the object, so don't choose anything too valuable or anything you don't want scratched!

2. Select an object of known hardness from those you gathered, and try to scratch the surface of your sample by pressing it with a tip or edge of the object of known hardness. For example, let's say you wanted to test a piece of wood. You could try to scratch it with a piece of quartz by pressing the edge of the quartz into the wood.

3. Inspect your sample to see if you have made a scratch in it. You may need to feel the surface of the object with your finger to check thoroughly. If your object was softer than the sample of known hardness, there will be a scratch. If it was harder, then there will not be a scratch. Repeat the test a couple of times to verify the result.

4. Continue performing the scratch test with various objects of known hardness, until you find two adjacent objects on your list, such as a fingernail (2.5) and a penny (3), in between which the hardness of your sample rests. You will know you've found this pair of objects on your list when the harder object of the pair does scratch your sample, while the softer one does not.

5. Once you have found this place on the list, you can assert that the hardness of your unknown must lie between that of the two objects of known hardness on the list. For example, if a penny (3) scratches your object, and a fingernail (2.5) does not, your object must have a hardness between 2.5 and 3.

Additional Mohs hardness rankings:

- Talc (1)
- Gypsum (2)
- Calcite (3)
- Fluorite (4)
- Platinum or iron (4.5)
- Apatite (5)
- Orthoclase (6)
- Quartz (7)
- Garnet (7.5)
- Hardened steel, topaz, emerald (8)
- Corundum (9)
- Diamond (10)

How can I use chemistry to **make dull pennies shiny**?

Chemistry principles encountered in this experiment:

- Surface chemistry
- Oxidation reactions

Materials you'll need:

- A handful of dull pennies (10 will do)
- 1 teaspoon table salt (sodium chloride)
- ¼ cup white vinegar (acetic acid solution)
- a small, non-metallic bowl
- water
- paper towels or napkins

The procedure:

Copper pennies turn dull because they oxidize over time. A mixture of vinegar and salt reacts with the copper oxide to clean the pennies and make them shiny again. (*Photo by Jim Fordyce.*)

1. Pour ¼ of a cup of vinegar and 1 teaspoon of table salt into the bowl.
2. Stir the mixture until the salt is completely dissolved.
3. Try first dipping one penny into the solution for about 15 seconds and remove it. Do you notice a change in the part that you dipped into the solution?
4. Place the remaining pennies into the solution. You will likely notice a visible reaction as the pennies are placed in the solution. The reason pennies eventually begin to appear dull is that the copper surface reacts with oxygen in the air to create a layer of copper oxide. In this experiment, the vinegar and salt will react with the copper oxide and remove it, which will leave a layer of the original shiny copper exposed on the surface.
5. Allow the pennies to remain in the solution for several minutes. If necessary, try to move the coins around so as to expose both sides of each coin to the solution. If possible, flip the coins over after a couple of minutes.
6. Drain the solution and rinse the coins with clean water. They should now look clean and shiny!

How can I make **black snake fireworks** from items around my house?

(Note: this experiment involves fire and flammable materials, so *adult supervision is required. Also, check your local laws before attempting this experiment.*)

Chemistry principles encountered in this experiment:

- Chemical reactions
- Combustion reactions

Materials you'll need:

- Sand (about 2 cups)
- Lighter fluid (a small bottle of about 100 milliliters)
- Baking soda (1 tablespoon)
- Sugar (4 tablespoons)
- Cup or bowl
- An outdoor location where you can light the black snake firework safely without damaging anything

The procedure:

1. In a cup or bowl, mix 4 tablespoons of sugar with 1 tablespoon of baking soda.
2. Use the sand to form a pile (in your chosen safe outdoor location), and then create a depression in the middle of the sand. This depression is where you will ignite the black snake firework.
3. Pour a small amount of lighter fluid onto the sand to wet it. Try this experiment first with a very small amount of lighter fluid, and, if necessary, repeat the experiment with incrementally larger amounts. It's better to start too small than too big.
4. Pour your mixture of sugar and baking soda into the wetted depression in the sand. You don't have to use it all at once—feel free to experiment with different quantities of the baking soda and sugar mixture.
5. *Carefully* ignite the lighter fluid with a match and stand back. You should see the mixture create long snakes of black ash! The burning sugar and baking soda form sodium carbonate, water vapor, and carbon dioxide gas. The ash in the snake is composed of carbonate and burnt carbon.

How can I make **invisible ink**?

Chemistry principles encountered in this experiment:

- Evaporation
- Combustion reactions
- Acid/base reactions

Materials you'll need:

- Cotton swab or paint brush
- Heat source (can be a light bulb)
- Measuring cup
- Paper
- Baking Soda
- Water
- Grape Juice (optional)

After dissolving baking soda in some water, you can use it like ink to draw a message onto a piece of paper. Once it dries, it will be invisible to the eye. (*Photo by Jim Fordyce.*)

Make your message visible again by carefully heating the paper over a flame. The baking soda will brown before the paper does, making your message visible. You can achieve the same effect by brushing grape juice over the paper. (*Photo by Jim Fordyce.*)

The procedure:

1. To prepare the ink, just mix equal volumes of water and baking soda and stir well.

2. Take a cotton swab, paint brush, toothpick (or something similar) and write a message on a piece of white paper using the water and baking soda mixture you have prepared.

3. Allow the "ink" time to dry. The water will soak into the paper and eventually evaporate, but the baking soda will not evaporate and it will be left behind.

4. To read your invisible message there are a couple of options. One is that you can hold the paper to a heat source, such as a light bulb or gentle flame (don't burn the paper!) This should cause the baking soda on the paper to turn brown, revealing your message! The baking soda burns faster than the paper, which is why it turns brown before the paper does.

5. Another option is to spread purple grape juice, over the paper (you can use a paint brush to do this). The message should appear in a different color/shade compared to the rest of the paper. This works because an acid in the grape juice reacts with the sodium bicarbonate that you used to write your message.

How can I **observe** layers of **immiscible liquids**?

Chemistry principles encountered in this experiment:

- Density
- Miscibility
- Polarity

Materials you'll need*:

- Honey
- Pancake syrup
- Liquid dish soap
- Water
- Vegetable or cooking oil
- Rubbing alcohol
- Lamp oil
- A tall glass of water or other container
- (Optional) food coloring to improve visibility

*Note: for this experiment, it is not necessary to have every material listed.

The procedure:

1. Pour the densest liquid into the glass first. Note that the liquids above are listed from most dense to least dense. *Try to avoid letting the liquid run down the sides of the glass*.

2. Gently pour the second liquid on top of the first. One way to pour it a little more slowly is to pour the liquid over another object such as a butter knife or the back side of a spoon. Allow each layer of liquid to settle for at least a few seconds before adding the next liquid. You'll notice that, instead of mixing, the liquids tend to stay in separate layers. The reason for this is that they are immiscible, which means that it is more thermodynamically favorable for the liquids to stay separated in layers and to form an interface than it is for them to mix together. Whether or not two liquids will be miscible is dictated by the details of the entropic and enthalpic factors associated with the mixing or separation of the two liquids in question. This is often directly related to whether or not the compounds have similar polarity. For example, we know water is a very polar substance, while vegetable oil is composed of primarily long, non-polar hydrocarbon chains. These do not interact favorably with one another, and prefer to stay in separate layers.

3. Continue pouring the third, fourth, etc. liquids on top of each other in order of decreasing density (you can just follow order of your liquids in the list above). As you pour in the successive liquids, they should continue to form separate layers. The densest liquids are affected the most by gravity, so these tend to stay below the less dense liquids. These liquids are not miscible, so they do not mix together to form a single solution. In truth, if you wait long enough, some of these liquids will mix together, but it will take a while.

4. That's it! You should now see a series of separate liquid layers in your container.

How can I **make a volcano** from **vinegar** and **baking soda**?

Chemistry principles encountered in this experiment:

- Chemical reactions
- Gases

Materials you'll need:

- Vinegar
- Baking soda (2 tablespoons)
- Large bowl
- Baking pan
- Flour (6 cups)
- Cooking oil (4 tablespoons)
- Salt (2 cups)
- Plastic bottle
- Dishwashing soap
- Red or orange food coloring (or any color, really)
- Water

A lot of the ingredients in this experiment are mixed together in a large bowl simply to build your volcano structure. If you like, you could achieve the same thing by using clay. (*Photo by Jim Fordyce.*)

After pouring some baking soda, soap, and some coloring into your volcano, slowly add vinegar and stand back to watch the show! (*Photo by Jim Fordyce.*)

The procedure:

1. This is a classic chemistry experiment that you might have done before in school. In a large bowl, first mix 6 cups of flour, 2 cups of salt, 4 tablespoons of cooking oil, and 2 cups of water. Mix these ingredients until they are firm. These ingredients are not involved in the chemical reaction that will make your volcano erupt, but rather this mixture will serve as the "rock" that forms the structure of your volcano.

2. Place the plastic bottle standing vertically in the pan. Use your hands to shape the "rock" material from the first step into a cone shape around the top of the bottle. Be careful not to cover the top of the bottle.

3. Fill the bottle most of the way with water, leaving enough space to add a few ounces of baking soda and vinegar.

4. Add a few drops of dishwashing soap to the bottle. This is not part of the chemical reaction that will take place inside the bottle, but bubbles from the soap will help to catch the gas evolved during the reaction between the vinegar and baking soda.

5. Add 2 tablespoons of baking soda into the bottle.

6. Finally, slowly (or quickly, if you want a really crazy volcano) add vinegar to the bottle, and prepare to witness the eruption! Be careful though—you should avoid getting this mixture in your eyes, or anyone else's, and this combination can make a

spectacular mess in your kitchen. The chemical reaction takes place between the baking soda (sodium bicarbonate, or $NaHCO_3$) and vinegar (dilute acetic acid, or CH_3CO_2H) to release carbon dioxide gas, which is what causes the volcano to erupt. The relevant chemical equation is:

$$NaHCO_3 + CH_3COOH \quad CH_3COONa + CO_2 + H_2O$$

How can I **observe** the **effects** of **electrostatic forces** using **household objects**?

Chemistry principles encountered in this experiment:

- Electric charge
- Electrostatic forces

Materials you'll need:

- Nylon hair comb (or a latex balloon)
- A water faucet

The procedure:

1. Comb your hair with a nylon comb. If you don't have a comb, you can also rub your head with an inflated latex balloon instead. As you rub the comb or balloon to your head, it builds up an electric charge on the object due to the movement of electrons between the object and your head.

2. Go to the faucet and turn it on so that a narrow stream of water flows out. Try to make the stream as thin as possible, while still maintaining a steady, smooth flow of water.

3. With the water running, move the comb or balloon close to the stream of water, but be careful not to actually let the comb or balloon touch the water. As it gets close, the stream of water should be deflected toward the comb or balloon. This is because the charge in the object (comb or balloon) induces an opposite charge in the nearby water, and the object and water then experience an attractive electrostatic interaction (opposites attract).

4. You can experiment with how the amount of deflection varies with the size of the stream of water from the faucet. You can also compare the ability of various objects (different combs, balloons, or different objects altogether), or vary the amount of time you rub the object in your hair.

How can I study the **effects** of **acids and bases** on sliced **fruit** "getting old" and turning brown?

Chemistry principles encountered in this experiment:

- Acids and bases
- Biochemical/enzymatic reactions

Materials you'll need:

- An apple (other fruits like bananas, pears, or peaches will also work)
- Five clear plastic cups
- Vinegar
- Lemon juice
- Baking soda
- Water
- Milk of magnesia
- Measuring cups

The procedure:

1. Prepare aqueous solutions of milk of magnesia and baking soda. The amount of water you use isn't particularly important (feel free to test various concentrations if you'd like). The key aspects are that the baking soda dissolves completely and that the milk of magnesia solution becomes less viscous or thick.

2. Slice your apple (or other fruit of choice) into five pieces. If you have decided to test multiple concentrations of baking soda or milk of magnesia solutions, adjust the number of fruit slices accordingly.

3. Label the cups as follows: vinegar, lemon juice, baking soda solution, milk of magnesia solution, and pure water.

4. Place one slice of fruit in each cup.

5. Add about ° of a cup of the appropriate solution to each of the cups you have labeled. The fruit should not be completely submerged in the solutions, but make sure each slice of fruit gets completely coated with the solution. The vinegar and lemon juice solutions serve as acidic solutions (of acetic and citric acids, respectively). The baking soda and milk of magnesia solutions serve as basic solutions (of sodium bicarbonate and magnesium hydroxide, respectively), while the water serves as a neutral control solution.

6. Write down your observations regarding the physical appearance of each piece of fruit at this time. If you have a camera handy, it might be useful to take a picture of your fruit samples for comparison to the final results.

7. Allow the fruit to sit for one day, and then come back and record your observations again. If you took a picture on the first day, you can compare the current appearance of the fruits to your photograph. Apples and fruits turn brown when an enzyme called tyrosinase (refer back to "Chemistry in the Kitchen") carries out a chemical reaction in the presence of oxygen and phenol containing compounds. How do the acidic or basic solutions affect the browning of the fruit? Since we know that the browning of the fruits is caused by an enzymatic reaction involving tyrosinase, what do the results suggest about the effects of acids and bases on the rate of the reaction involving tyrosinase? Do both acidic solutions affect the rate of browning similarly? How about both basic solutions? Do you think the changes

you observe are due to changes in pH, or the specific chemical involved? Think also about what other experiments you might try to investigate further.

How can I show my friends a **magic trick using pepper**?

Chemistry principles encountered in this experiment:

- Polarity
- Surface tension

Materials you'll need:

- A shaker or small packet of black pepper
- Dishwashing soap (a few drops)
- Bowl
- Water (a bowl full)

The procedure:

1. Begin by filling the bowl with water.

On the left, a finger dipped in water without soap has no effect, but when you put a little dish soap on your finger (right) and put it in the water, the soap spreads out, lowering the water's surface tension and pushing the pepper away. (*Photo by Jim Fordyce.*)

2. Then pour black pepper onto the water to form a thin layer of pepper across its surface.

3. As a control experiment, try dipping your finger below the surface of the water. Nothing too interesting should happen at this time.

4. Now rub a small amount of dishwashing soap on your finger and dip it in the water again. This time, you should see the pepper move away from your finger and toward the edges of the bowl. The non-polar soap molecules don't want to dissolve beneath the surface of the water, and thus they spread across the surface of the water quickly, which lowers its surface tension. Whereas water typically bulges a bit above its surface due to its relatively high surface tension, the water spreads out when its surface tension is lowered. This causes the water to spread out as the soap moves over it, and the pepper is carried away from your finger in the process.

5. Now that you understand the basics of this trick, you can perform it for your friends. Ask a friend to dip their finger in the water and to try to concentrate on trying to make the pepper move away from their finger. When they cannot make it happen, you can step in and use your (already soapy) finger to easily move the pepper away!

How can I make **"hot ice"** (sodium acetate)?

Chemistry principles encountered in this experiment:

- Chemical reactions
- Solubility
- Crystallization and recrystallization
- Colligative properties

Materials you'll need:

- Pan with cover
- Microwave or stovetop
- Vinegar (1 liter)
- Baking soda (4 tablespoons)
- Dish

The procedure:

1. Pour the vinegar into a pan, and very slowly (a little bit at a time) add the baking soda. As you may already know, this reaction will produce large amounts of bubbles (carbon dioxide gas), so you'll need to add this very slowly. This reaction will produce a solution of sodium acetate in water. Sodium acetate is produced according to the chemical equation:

$$Na^+ [HCO_3]^- + CH_3 - COOH \quad CH_3 - COO^-Na^+ + H_2O + CO_2$$

2. Bring the solution to a boil. Allow the solution to continue to boil until a skin or film begins to form on the surface of the solution. This will require heating for a significant amount of time (perhaps, up to an hour), until a large fraction of the water from the vinegar has evaporated. Our goal here is to form a very concentrated hot solution of sodium acetate. You'll recall from our discussion of colligative properties that the solubility of a solute is higher at higher temperatures. As we reduce the volume of the water, the sodium acetate will not evaporate, and we will be left with a concentrated solution at high temperature.

3. When you notice a film start to form on the surface, remove the pan from the heat, and cover it to prevent further evaporation. Place it on the countertop or in the refrigerator to cool. If you see any crystals begin to form, add a small amount of additional vinegar (or, if you are out of vinegar, use water) and stir the solution so that they dissolve.

4. You now have a supercooled solution of sodium acetate that can crystallize out of solution readily if a crystallization nucleus is present.

5. You can now slowly pour the first few drops of the cooled solution onto a dish, and it should begin to crystallize rapidly. If it doesn't, try dragging a fork or knife along the dish to make a tiny scratch, or give the liquid a moment to evaporate a little so that the sodium acetate begins to crystallize before you begin to pour. As you continue to pour, the liquid should continue to crystallize as it contacts the crystals already formed on the dish. This is similar in spirit to how purification takes place during a recrystallization (see "The Modern Chemical Lab"). If you feel the crystals, they will be warm to the touch, since the crystallization is an exothermic process (it gives off heat). This is why it's called "hot ice," but now you know it isn't really ice—it's sodium acetate—and you know how to make it.

How can I **make a pH indicator** at home?

Chemistry principles encountered in this experiment:

- Extractions
- Acid/base chemistry
- Chemical indicators
- Solubility and temperature

Materials you'll need:

- Several leaves of red cabbage
- Blender
- One coffee filter
- Large jar
- Glasses or clear cups
- Water
- Stovetop or microwave
- Pot or pan

You only need some (not necessarily all) of the following ingredients:

- Baking soda (1–2 tsp)
- Lemon juice (1–2 tsp)
- Vinegar (1–2 tsp)
- Ammonia (1 oz of household variety—like what you use for cleaning)
- Antacids (1 tablet; Alka-Seltzer® works)

The procedure:

1. Cut about 2 cups of cabbage and place it in a blender.

2. Boil water in a pot, and then add boiling water to the cabbage in the blender. Turn on the blender and blend for about 10 minutes. The hot water will extract a pigment called an anthocyanin from the red cabbage (along with other components). Recall that solubility tends to increase at higher temperatures. Anthocyanins are molecules that will change color depending on the pH of the solution—this will serve as our indicator.

3. Filter the plant material out by pouring the solution through a coffee filter and into a large jar. The liquid you obtain should be red/blue/purple in appearance. The exact color you observe will depend on the pH of the water you are looking at, which may be influenced by factors like the ion concentration in your tap water and the other plant components that remain in the solution.

4. Pour the solution into various glasses or clear cups. These will be your individual test "beakers" where you can test the pH of various substances.

5. Try adding other substances to your solutions and observe how the color changes as they are added. Note that the amount of solution you add to each glass/cup will influence the amount of each test substance (e.g. lemon juice) you need to add to see a color change. For reference, the list below tells you how the color of your anthocyanin indicator solution should change with pH.

Approx. pH	color
2	red
4	purple
6	violet
8	blue
10	blue-green
12	yellow-green

How can I make a **lava lamp** at home?

Note: OK, so we'll disclose up front that this experiment won't actually make a lamp. It will make a device with bubbles that rise and fall just like a lava lamp, but you'll need a flashlight or other light source if you want it to be illuminated.

Chemistry principles encountered in this experiment:

- Chemical reactions
- Density
- Miscibility
- Gases

Materials you'll need:

- Vegetable oil (20 oz.)
- Plastic soda bottle (20 oz. or 1 liter)
- Water (1 tablespoon)
- Alka-Seltzer®
- Food coloring

The procedure:

1. Begin by filling the plastic bottle almost full with vegetable oil.

2. Add a few drops of food coloring to 1 tablespoon of water, and then add this to the plastic bottle's contents. You'll notice that the oil and water are not miscible, which is what will allow us to make bubbles within the oil. The water sinks to the bottom of the bottle, since it is more dense than the oil.

3. Ground up an Alka-Seltzer® tablet and add the pieces/powder to the bottle, and then seal the cap. When the chemicals in the table dissolve and react, carbon dioxide gas bubbles will be formed (see equation below), which will lower the overall density of the water bubbles, allowing them to rise to the top of the bottle. When they reach the top, they will leave the water bubbles and join the small amount of air at the top of the bottle. At this point, the water bubbles will again be denser than the oil, so they will sink back to the bottom of the bottle. This process will repeat until all of the Alka-Seltzer® has been reacted.

4. You should see colored bubbles move throughout the bottle, similar to a lava lamp. You can also add another Alka-Seltzer® after the reaction has finished.

Blobs of colored water rise and fall within an oily solution after you add Alka-Seltzer® (*Photo by Jim Fordyce.*)

How it works:

The chemical reaction of Alka-Seltzer® with citric acid to release carbon dioxide is described by the following equation:

$$C_6H_8O_7 \text{ (aq)} + 3NaHCO_3 \text{ (aq)} \quad 3H_2O \text{ (l)} + 3CO_2 \text{ (g)} + Na_3C_6H_5O_7 \text{ (aq)}$$

How can I **suck an egg** into a **bottle**?

Chemistry principles encountered in this experiment:

- Combustion
- Pressure
- Ideal gas laws

Materials you'll need:

- Hard-boiled egg
- Bottle or flask with an opening slightly smaller than the egg's diameter
- Paper (a sheet of computer paper or newspaper will do)
- Matches

The procedure:

Note: this experiment involves fire and flammable materials, so *adult supervision is a must!*

1. Peel the hard-boiled egg.
2. Tear off a piece of paper that can easily fit into the bottle, and carefully light it on fire and drop it into the bottle.
3. Quickly place the egg on top of the bottle, covering the opening.
4. The flame will burn, heating the air inside the bottle. This causes the air to expand, and some of it will push past the egg to escape from the bottle. Recall from our discussion of the ideal gas law that, for a fixed number of particles and volume, the pressure inside the bottle should increase linearly with increases in temperature. The increased pressure is what pushes the air out, and you may even see the egg shake a little as the air escapes. Then the egg will come to rest, covering the opening.
5. Eventually, the fire will burn up all of the paper, or all of the oxygen inside the bottle (whichever comes first), and then the air in the bottle will begin to cool. As it cools, the volume it occupies will decrease, lowering the pressure inside the bottle relative to that outside the bottle. The higher pressure outside the bottle is what pushes the egg through the opening and into the bottle.
6. You can get the egg back out by tilting the bottle upside down and blowing air into the bottle, and then allowing the egg to cover the opening before removing your mouth. Thus you can use the same principle regarding equilibration between high and low pressures to force the egg out.

How can I **extract iron** from **oatmeal** or **breakfast cereal**?

Chemistry principles encountered in this experiment:

- Magnetism
- Food chemistry/nutrients
- Extraction

Materials you'll need:

OATMEAL EXPERIMENT

- Iron fortified instant oatmeal packet (check the label to ensure iron content)
- Magnet (it's easiest to see the iron if you can find a magnet that is coated or painted white or another light color)
- Plastic bag or bowl

BREAKFAST CEREAL EXPERIMENT

- Magnet—for this experiment you will want a magnet you can use to stir a liquid (it's easiest to see the iron if you can find a magnet that is coated or painted white or another light color)
- Plastic bag
- Water
- Large glass jar or beaker

The procedure:

OATMEAL EXPERIMENT

1. Open the oatmeal packet and empty it into the plastic bag or bowl.
2. Stir the oatmeal with the magnet. You should see small amounts of grey or brown metal collect on the outside of the magnet. This is iron! Iron is commonly added as a mineral supplement to breakfast cereals and other foods. Now you can see that the iron that goes into your diet is the same element that you find in objects made of iron metal (just in much smaller quantities and pieces). Recall that iron is attracted to magnets due to the fact that it is a ferromagnetic material (see "Atoms and Molecules").

BREAKFAST CEREAL EXPERIMENT

1. Pour 1 or 2 cups of breakfast cereal into a plastic bag.
2. Crush the cereal inside the bag using your hands.
3. Pour about 1 liter of water into the jar or beaker, and add the crushed cereal from the bag to the water. The water will help to extract the iron from the crushed cereal. Whereas the iron bits were looser in the dry oatmeal sample, they tend to be

stuck within pieces of cereal, which is why you need to mechanically crush the cereal and use the water to help extract it; the magnetic interaction wouldn't be strong enough to pull the iron out of the cereal on its own. In a chemistry lab, acids would often be used to help extract metals from a sample, but will work fine for our purposes here.

4. Use the magnet to stir the crushed cereal for about 15 minutes. When you remove the magnet from the water, you should see iron filings collected on the magnet.

How can I **grow crystals** of **rock candy**?

Chemistry principles encountered in this experiment:

- Crystallization and recrystallization
- Solubility

Materials you'll need:

- Sugar (3 cups)
- Water
- Glass jar
- A pencil
- String (cotton; 15 cm will do)

The first step in making rock candy is dissolving sugar in water and boiling it slowly on a stove top. You will then need to wait while it cools. Putting the mixture in the refrigerator can speed up the cooling process. (*Photo by Jim Fordyce.*)

Tie a string or strings to a pencil and dangle the string in the sugar solution. Gradually, crystals will form to make the candy. (*Photo by Jim Fordyce.*)

- Pan (for boiling water)
- Microwave or stovetop
- (Optional) food coloring
- (Optional) flavoring extracts

The procedure:

1. Begin by stirring 3 cups of sugar and 1 cup of water into a pan.
2. While stirring frequently, heat the mixture to a gentle boil. The goal is to just barely get the mixture to its boiling point, and then stop the heating (we don't want to evaporate off too much of the water). Then remove the solution from the heat source.
3. If you want to add food coloring or flavoring, now is a good time to do so. Either way, the candy is made of sugar, so it will still taste fine.
4. Cool the pot containing the solution in the refrigerator until its just below room temperature. In the meantime, tie the cotton string to the pencil, and place the pencil atop the jar, allowing the string to dangle down without touching the bottom. You may need to trim the string such that it doesn't touch the bottom of the jar.
5. You may wish to tie a lifesaver candy or other weight to the end of your string to hold it taut.

6. Wet the string and dip it in a little bit of crystalline sugar (not the sugar you just mixed with water and heated). These sugar crystals will serve as the nucleating sites on which the rock candy from your sugar solution will crystalize.

7. Pour the cool sugar solution into the jar, and hang the pencil and string into the solution.

8. Cover the jar with aluminum foil, a paper towel, or anything else, such that it will not be disturbed.

9. The crystals will take several days, or possibly as long as a week, to grow. Sugar from the solution will continue to crystalize onto the growing crystals on the string. You can check on the crystals occasionally, but you should not bump, tilt, turn, shake, or move the jar, if possible. The crystals will grow larger if you leave them undisturbed. Once they are done growing, remove your string, and your rock candy is ready to eat!

Crystals will slowly form around the string. Be patient, this process can take several days. (*Photo by Jim Fordyce.*)

How can I make **Jell-O®** that **glows** under a **black light**?

Chemistry principles encountered in this experiment:

- Fluorescence
- Gels
- Cross-linking

Materials you'll need:

- Jell-O® or gelatin powder
- 1 cup of tonic water
- 1 cup of water
- Stovetop or microwave
- Large bowl
- Pot (for heating on stovetop)
- A black light source (see "Physical and Theoretical Chemistry" to review how black lights work)

The procedure:

1. Heat one cup of water to a boil.

2. Mix the Jell-O® and the hot water into the bowl, and stir the powder in until it dissolves. The hot water helps the gelatin to dissolve and disperse evenly throughout the solution. Gelatin is a form of collagen (See "Chemistry in the Kitchen"), and when it cools, the Jell-O® will reform cross-links between the collagen strands, which is what traps the water inside to create a gel. You'll recall that a gel is solid material that consists of a bonded network of long strand molecules that contain significant amounts of molecules that would otherwise behave as a liquid trapped within the solid network.

3. Add in one cup of tonic water, stir the solution well, and then place it in the refrigerator for about four hours. The tonic water contains a molecule called quinine that will fluoresce a bright blue color when excited with the appropriate wavelengths of light, which can be provided by the black light. Recall that fluorescence takes place when a molecule absorbs light at one wavelength, relaxes to release a fraction of that energy, and then emits a photon at a longer wavelength (lower energy) than that which was absorbed. Black lights provide light that is at slightly shorter wavelengths (higher energies) than light in the visible spectrum, so it can often excite molecules that will fluoresce in the visible region of the spectrum.

4. When the Jell-O® is finished hardening, take a look at it under the black light. It should glow blue! This is due to the fluorescence of the quinine from the tonic water. This blue glowing color comes from the fluorescence of the quinine molecules, so it will not be affected significantly by the flavor or color of Jell-O® that you decided to use.

How can I **make glue** from **milk**?

Chemistry principles encountered in this experiment:

- Solubility
- Precipitation
- Filtration

Materials you'll need:

- Hot water (from the tap is fine)
- Room temperature water
- 1 tablespoon of vinegar
- ½ teaspoon baking soda
- Coffee filter
- Cup
- Spoon

306

- Small bowl
- tablespoons powdered milk

The procedure:

1. Dissolve 2 tablespoons of powdered milk in ¼ cup of hot tap water.

2. Mix in 1 tablespoon of vinegar and stir the solution well. Recall that household vinegar is a dilute aqueous solution of acetic acid. At this point the milk should begin to form little blobs of insoluble material (curd). These are a substance called casein, which are a class of proteins found in milk. Proteins are typically soluble in water and aqueous environments, but the acetic acid from the vinegar causes the casein to no longer be soluble in this solution.

3. Position a coffee filter on top of a cup and pour the solution through the filter to collect the curd. To further dry the curd, try to use the filter to squeeze out any remaining liquid.

4. Dispose of the liquid in the cup, dry the inside of the cup, and then transfer the curd from the filter into the now dry cup.

5. Break the curd into smaller chunks using a spoon. This will help it to mix more readily with the ingredients you will add next.

6. Add 1 teaspoon of hot water, and ¼ of a teaspoon of baking soda to the cup containing the chopped-up curd. The mixture may foam a little from the reaction of the baking soda with the remaining vinegar, producing carbon dioxide gas.

7. Stir this mixture, which should eventually become smooth and liquid-like. You may need to add a little more water or baking soda to reach a smooth, even consistency.

8. Now you have your glue! You can use it just like you would any other glue, though you should test it out first to make sure it's working well before you use it for that science fair project you're finishing up.

How can I **make a cloud form** in a bottle?

Note: this experiment involves fire and flammable materials, *adult supervision is a must*.

Chemistry principles encountered in this experiment:

- Ideal gas law
- Phase changes
- Droplet formation

Materials you'll need:

- 20 ounce or 1 liter plastic soda bottle
- Warm water (1-2 oz.)
- Matches

Using adult supervision, insert a lit match into a bottle with just a little warm water in it. After some smoke has collected in the bottle, screw on the cap and watch clouds form—the result of water condensing around the tiny particles that make up the smoke. (*Photo by Jim Fordyce.*)

The procedure:

1. Add warm water to the soda bottle until the bottom is just barely covered with water.

2. With the bottle tipped so that you don't burn yourself, ignite a match and insert the burning end of the match into the bottle. Allow the bottle to fill up with smoke.

3. When the bottle is fairly filled with smoke, or when the match goes out, remove the match and screw the cap on to close the bottle. Clouds will form when water vapor forms small-but-visible droplets around particles in the air. In this experiment, the smoke will provide these particles around which the water can form small droplets.

4. With the bottle closed, squeeze it several times. You should see a cloud form! When you squeeze the bottle, the temperature of the gas inside may temporarily increase, but the temperature will quickly equilibrate with the surroundings (recall the ideal gas law we discussed in "Atoms and Molecules," $PV = Nk_bT$). When you release the squeezed bottle, the temperature inside decreases, cooling the water vapor and helping it to liquefy into droplets on the particles provided by the smoke. This is very similar to how real clouds are formed in the atmosphere!

How can I make a **miniature rocket** from a **film canister**?

Chemistry principles encountered in this experiment:

- Gases and pressure
- Chemical reactions

Materials you'll need:

- An empty, plastic 35mm film canister (which are getting more and more rare these days!) If you cannot locate one of these, you could try using any other small, light-weight, plastic container with a lid that easily pops on or off.
- An Alka-Seltzer® or other antacid tablet
- Water

The procedure:

This experiment should be performed outdoors in an open area.

1. Add about one teaspoon of water to the film canister, and leave the lid open.
2. Break the antacid tablet in half and get ready to add it to the container.
3. This step requires you to be quick. Drop the broken antacid tablet into the container, then quickly close the lid and place the canister on the ground with the cap side down. Stand back a good distance, and wait for the rocket to launch. The Alka-Seltzer® will react with the water in a reaction that produces carbon dioxide gas. The pressure of this gas will build and build until it exerts a force so great that it will blast the canister off of the cap, shooting the canister into the air.
4. After about 10 to 15 seconds, the canister will launch into the air!
5. Try repeating this experiment using different ratios of Alka-Seltzer® to water, or using canisters of different types or sizes. You might also compare different brands of antacid tablets, or different methods for crushing the tablet before you add it to the canister. Compete with your friends to see who can make the rocket that shoots the highest!

How can I **inflate a balloon** using **yeast**?

Chemistry principles encountered in this experiment:

- Biochemical/enzymatic reactions
- Gases

Materials you'll need:

- Yeast (5-10 grams powder yeast)
- A small plastic bottle (preferably about 16 oz. or smaller)
- One teaspoon of sugar

- A balloon
- Warm water

The procedure:

1. Add enough water to a small plastic bottle to fill it up about 1 inch. The water needs to be warm for the yeast to do its job.

2. Add about 5-10 grams of yeast (one small packet will do) to the bottle, and mix it around well. Yeast is made of fungal microorganisms that will become active when placed in the warm water.

3. Add a teaspoon of sugar to the water. The sugar serves as the food for the yeast microorganisms. They are going to consume it and produce carbon dioxide gas as a byproduct.

4. Wrap the balloon around the open end of the bottle. You may wish to use tape or a tight rubber band to prevent gas from escaping. Let the yeast do their job for about 20 minutes, and the yeast should soon produce enough carbon dioxide gas to start filling up the balloon. This is the same thing that happens when you use yeast to bake bread! The little holes you see in the bread are from the yeast releasing carbon dioxide as it rises.

5. Try repeating the experiment using different amounts of water, sugar, and yeast, and observe the rate at which the balloon inflates. One hypothesis you might test is whether the concentration of sugar in the water affects the rate of carbon dioxide product. You might also try varying the amount of yeast you use. Pay attention to both the rate at which the balloon inflates, and the final volume of gas it reaches.

How can I make a **chicken bone flexible**?

Chemistry principles encountered in this experiment:

- Solubility in weak acids

Materials you'll need:

- A jar
- Vinegar (enough to fill the jar)
- A chicken bone (one from a drumstick works best)

The procedure:

1. Obtain a chicken bone, and clean it well. Rinse it with water and remove any remaining skin or meat from the bone.

2. Before you soften the bone, try bending it to make sure it is rigid. Don't break it, but test that it is indeed hard.

3. Fill up the jar with vinegar, and drop your clean bone inside.

4. Cover the jar (just so that your whole house doesn't start smelling like vinegar) and let it sit for about 3 days. Recall that vinegar is a dilute aqueous solution of acetic acid, which is a weak acid. Over a few days, the acetic acid helps to dissolve the calcium in the chicken bone. Since the calcium plays a key role in keeping bones hard, the bone will soften once the calcium is dissolved.

5. Remove the bone, rinse it off with water, and now try bending it again. It should be noticeably more flexible than before. Now you can see that without enough calcium, your bones will not stay strong! This might also provide a good indication of why it's a good idea to brush your teeth; if you leave behind any foods capable of dissolving calcium still on your teeth, your teeth may lose some of their calcium and begin to become weak.

How can I **grow** a **large crystal**?

Chemistry principles encountered in this experiment:

- Crystallization
- Solubility

Materials you'll need:

- Hot water (from the tap is fine)
- About 2.5 tablespoons of alum (this is typically found with the spices in the grocery store; it is an ingredient used to make pickles crisp; its chemical formula is $KAl(SO_4)_2 \cdot 12\ H_2O$)
- Nylon fishing line or thread (about 15 cm)
- A pencil or ruler
- 2 jars
- Spoon
- Coffee filter or paper towels

The procedure:

1. Pour ½ a cup of hot water into a jar.
2. Add a little bit of alum to the water and stir it in. Try to do this before the water cools down too much. As you'll recall, we can expect the alum to be more soluble at higher temperatures. We want to create a saturated solution of alum, but we also want to make sure that all of it dissolves. Continue adding alum until it will not dissolve anymore, and then you can add a little bit more hot water to get the last bit of alum to dissolve.
3. Cover the jar with a coffee filter or paper towel and let it sit overnight.
4. The next day, pour the liquid into a second jar. There should be small crystals left in the bottom of the first jar. The process of pouring the liquid off of the top of the solid crystals is known as "decanting." The crystals left behind will serve as the seed crystals to grow a larger crystal of alum.

311

5. Pick out the largest crystal, or one with a nice shape that you like. Tie the fishing line or thread around the crystal, and tie the other end of the line to a pencil, ruler, or other long flat object. Adjust the length of the string such that you can use the pencil or ruler to dangle your crystal from the top of the second jar (the one that now contains the liquid) into the liquid without it touching the bottom.
6. Once your crystal is hanging into the solution, all you have to do is sit back and wait for it to grow. This might take several days. If you notice small crystals growing on the sides or bottom of the jar, it would be a good idea to transfer the solution and your single growing crystal into another jar to prevent the alum in solution from crystallizing into several different crystals (this will ensure that you get a nice big crystal). You can feel free to transfer jars as many times as necessary to continue growing your large crystal.

Physical Constants

Constant	Symbol	Value
Acceleration due to gravity	g	9.806 m/s
Avogadro number	NA	6.0221367×10^{23} particles mol^{-1}
Bohr magneton	μ_β	$9.2740154 \times 10^{-24}$ J/T
Bohr radius	a_0	$5.2917721092(17) \times 10^{-11}$ m
Boltzmann constant	k_B	$6.6260755 \times 10^{-34}$ J $\times$ s
Classical electron radius	r_e	$2.8179403267(27) \times 10^{-15}$ m
Elementary charge	e	$1.60217733 \times 10^{-19}$ C
Faraday constant	F	9.64846×10^4 C mol^{-1}
Free electron g factor	g_e	2.002319304
Gas constant	R	8.31451 $m^2 \times kg/s^2 \times K \times mol^{-1}$
		8.3144621 J K^{-1} mol^{-1}
		5.189×10^{19} eV K^{-1} mol^{-1}
		0.08205746 L atm K^{-1} mol^{-1}
		1.9858775 cal K^{-1} mol^{-1}
Mass—electron	m_e	$9.1093897 \times 10^{-31}$ kg
Mass—neutron	m_n	$1.6749286 \times 10^{-27}$ kg
Mass—proton	m_p	$1.6726231 \times 10^{-27}$ kg
Permeability of vacuum	μ_0	$12.566370614 \times 10^{-7}$ $T^2 \times m^3/J$
Permittivity of vacuum	ϵ_0	$8.854187817 \times 10^{-12}$ C^2 J^{-1} m^{-1}
Planck constant	h	$6.6260755 \times 10^{-34}$ J $\times$ s
Rydberg constant	R_∞	$1.973731568539(55) \times 10^7$ m^{-1}
Speed of light in a vacuum	c	2.99792458×10^8 m/s
Stefan-Boltzmann Constant	Σ	5.670373×10^{-8} W $\times$ m^{-2} $\times$ K^{-4}

Conversion Factors

Mass Conversions

1 g	=	1×10^{-3} kg
1 g	=	1×10^{9} ng
1 g	=	1×10^{12} pg
1 g	=	0.035274 oz
1 mg	=	1×10^{-6} kg
1 mg	=	1×10^{-3} g
1 lb	=	0.453592 kg
1 lb	=	453.592 g
1 oz	=	28.3495 g
1 oz	=	0.0625 lb
1 μ	=	1.66057×10^{-27} kg
1 metric ton	=	1×10^{3} kg
1 metric ton	=	2,204.6 lb

Length Conversions

1 cm	=	1×10^{-2} m
1 mm	=	1×10^{-3} m
1 nm	=	1×10^{-9} m
1 micrometer	=	1×10^{-6} m
1 angstrom	=	1×10^{-10} m
1 angstrom	=	1×10^{-8} cm
1 angstrom	=	100 pm
1 angstrom	=	0.1 nm
1 in	=	2.54 cm
1 in	=	0.0833 ft
1 in	=	0.02778 yd
1 cm	=	10 mm
1 cm	=	1×10^{-2} m
1 cm	=	0.39370 in
1 mi	=	1.609 km
1 mi	=	5,280 ft
1 yd	=	0.9144 m
1 yd	=	36 in

Length Conversions

1 m	=	39.37 in
1 m	=	3.281 ft
1 m	=	1.094 yd

Volume Conversions

1 L	=	1×10^{-3} m^3
1 L	=	1.057 qt
1 L	=	1×10^3 mL
1 L	=	1×10^3 cm^3
1 L	=	1 dm^3
1 L	=	1.0567 qt
1 L	=	0.26417 gal
1 qt	=	0.9463 L
1 qt	=	946.3 mL
1 qt	=	57.75 in^3
1 qt	=	32 fl oz
1 cm^3	=	1 mL
1 cm^3	=	1×10^{-6} m^3
1 cm^3	=	0.001 dm^3
1 cm^3	=	3.531×10^{-5} ft^3
1 cm^3	=	1×10^3 mm^3
1 cm^3	=	1.0567×10^{-3} qt

Energy Conversions

1 J	=	0.23901 cal
1 J	=	0.001 kJ
1 J	=	1×10^7 erg
1 J	=	0.0098692 L atm
1 cal	=	4.184 J
1 cal	=	2.612×10^{19} eV
1 cal	=	4.129×10^{-2} L atm
1 erg	=	1×10^{-7} J
1 erg	=	2.3901×10^{-8} cal
1 L atm	=	24.217 cal
1 L atm	=	101.32 J
1 eV	=	96.485 kJ/mol
1 MeV	=	1.6022×10^{-13} J
1 BTU	=	1,055.06 J
1 BTU	=	252.2 cal

Pressure Conversions

1 atm	=	101,325 Pa
1 atm	=	101.325 kPa
1 atm	=	760 torr
1 atm	=	760 mm Hg

Pressure Conversions

1 atm	=	29.9213 in Hg
1 atm	=	14.70 lb/in^2
1 atm	=	1.01325 bar
1 atm	=	1,013.25 mbar
1 torr	=	1 mm Hg
1 torr	=	133.322 Pa
1 torr	=	1.33322 mbar
1 bar	=	1×10^5 Pa
1 bar	=	1,000 mbar
1 bar	=	0.986923 atm
1 bar	=	750.062 torr

Glossary

absolute zero—lowest theoretical temperature; (0.00 K, −273.15 °C, −459.67 °F)

absorption—capture of one material into another; can be a physical or chemical process

accuracy—closeness of a measurement to the actual or accepted value

acid—a molecule that has easily removable hydrogen ions (Brønsted-Lowry acid), can accept a pair of electrons (Lewis), or releases hydrogen ions in solution (Arrhenius)

actactic polymer—a polymer in which the chiral centers are arranged randomly along the chain

actinide—elements 89–102

activation energy—difference in energy between the reactants and transition state (or activated complete) for a chemical reaction or process

adiabatic—a process that does not absorb or release energy

adsorption—capture of one material onto the surface of another

aerosol—suspension of a solid or a liquid in a gas (e.g., smoke, fog)

aliquot—a sample taken from a larger amount of a material

alkali—a basic substance (i.e., pH > 7)

alkali metal—Group 1 of the periodic table (i.e., Li, Na, K, Rb, Cs, Fr)

alkali earth metal—Group 2 of the periodic table (i.e., Be, Mg, Ca, Sr, Ba, Ra)

alkane—a hydrocarbon with the formula C_nH_{2n+2} (i.e., no double bonds)

alkene—a hydrocarbon at least one double bond

allotrope—different arrangements of atoms of a single element (e.g., diamond and graphite are allotropes of carbon)

alloy—a mixture of metals (e.g., bronze, a mixture of zinc and copper)

alpha particle—a particle consisting of two neutrons and two protons (i.e., a helium nucleus)

amalgam—an alloy of mercury

amorphous—a solid without a repeating, ordered structure

amplitude—the height (or maximum displacement) of a wave

angstrom—a unit of length used often to describe bond lengths; $1\ \text{Å} = 10^{-10}\ \text{m}$

anhydrous—without water

anion—a negatively charged ion

anode—the electrode at which oxidation occurs

antibonding orbital—orbitals in which the component atomic orbitals are out of phase, leading to repulsion or destabilization

atom—smallest unit of a chemical element

atomic number—number of protons in an atom

atomic orbital—an equation that describes the probability of finding an electron around a nucleus

atomic radius—half the distance between nuclei of the same element

atomic weight—the average mass of an atom of a given element

Avogadro's number—the number of particles in one mole, 6.022×10^{23}

azeotrope—a mixture that does not change composition during distillation

band gap—the energy range separating the top of the valence band and the bottom of the conduction band in a semiconductor

barometer—an instrument used to measure pressure

base—a compound that accepts hydrogen ions (Brønsted-Lowry acid), has a pair of available electrons (Lewis), or releases hydroxide ions in solution (Arrhenius)

beta particle—an electron created during nuclear decay reactions

bimolecular reaction—a reaction that involves two molecules in the rate-determining transition state

black body radiation—electromagnetic radiation given off by a black body; at room temperature most of this radiation is in the infrared, but at higher temperatures visible light can be emitted

boiling point—temperature for a given liquid at which its vapor pressure is equal to the external pressure acting on it

boiling point elevation—a colligative property, the increase in boiling point of a liquid as a solute is added

bond angle—the angle relating the orientation of two bonds connecting three atoms

bond length—the distance separating two chemically bonded atoms

bond order—number of pairs of electrons shared by two atoms

bond strength—a measure of the energy required to break a chemical bond

bonding orbital—a molecular orbital that is more stable than the atomic orbitals that were combined to generate it

Boyle's law—law stating that the pressure and volume of gas are inversely proportional

brass—an alloy of copper and zinc, the relative percentages of the two species may vary

bronze—an alloy of copper and tin, with copper as the primary component

buffer—a solution that tends to resist changes in pH upon addition of an acid or base

calorie—a unit of energy equal to 4.184 Joules

calorimeter—a tool used to measure the heat change associated with a chemical reaction

carbanion—an anion in which a carbon atom bears a significant fraction of the negative charge carbocation—a cation in which a carbon atom bears a significant fraction of the positive charge

carbohydrate—organic compounds of carbon, hydrogen, and water, typically with hydrogen and oxygen in a 2:1 ratio, respectively, these are often called sugars

catalyst—any substance that increases the rate of a chemical reaction without being consumed by the reaction

cathode—electrode where reduction occurs

cation—a positively charged ion

Celsius—common temperature scale in which the melting and boiling points of water are defined to be 0 °C and 100 °C, respectively

ceramic—an inorganic crystalline solid typically prepared by heating

chalcogen—a group 16 element (oxygen, sulfur, selenium, tellurium, polonium, or livermorium)

Charles's Law—a law stating that the volume and temperature of a gas are directly proportional

chelation—binding of a ligand to a metal atom through two or more positions

chemical bond—sharing of electrons between two or more atoms

chemical change—a process that alters the arrangement of atoms in a substance

chemiluminescence—emission of light resulting from a chemical reaction

chiral center—an atom whose arrangement of substituents is non-superimposable with its mirror image

chirality—a geometric property of a molecule whose mirror image is non-superimposable with the original molecule

chromatography—a process to separate mixtures, generally by differing affinities of to a solid stationary phase

colligative properties—properties of a solution that depend on the amount of solute dissolved

collision frequency—average number of collision events per second

collision theory—defines reaction rates as a function of the collision frequency

colloid—suspension of particles of one substance in another (e.g., milk)

combustion—a chemical reaction between a fuel and an oxidant to produce heat

compound—a substance composed of more than one element

condensation—conversion of a gas into a liquid

condensation reaction—a reaction in which two molecules combine to form one larger molecule, with the concurrent loss of another small molecule like water or HCl

congener—elements in the same group of the periodic table

coordination number—the number of bonds to an atom

copolymer—a polymer of two or more monomers

coulomb—the standard unit of charge, defined as the amount of charge delivered by 1 amp in 1 second

Coulomb's Law—law describing the force between a pair of charged particles separated by a distance

covalent bond—the equal, or near equal, sharing of electrons between two or more atoms

critical point—a set of conditions at which no boundary exists between two phases of a substance

crystal—a solid with an ordered arrangement of atoms of molecules

crystal field theory—a model used to describe the electronic structure of transition metal molecules, particularly the energy of the d orbitals

crystallization—formation of crystals from a solution of the compound, often used as a purification technique

Curie point—the temperature at or beyond which a ferromagnetic material becomes paramagnetic

d orbital—an atomic orbital with an angular momentum quantum number of 2

Dalton's Law—a law regarding partial pressures of gases that states the total pressure of a mixture of gases is equal to the sum of the partial pressures of each individual component

dative bond—a chemical bond in which one atom is essentially providing both electrons involved in the bond

de Broglie wavelength—also known as matter wavelength; inversely proportional to the momentum of the object or particle; see also wave-particle duality

decant—to pour off a liquid from a solid sediment

decay rate—rate at which a nucleus emits a particle

degenerate orbitals—atomic or molecular orbitals of equal energy

density—mass per unit volume of a given substance

dependent variable—a variable that changes as a function of the independent variable

dextrorotatory—having the property of rotating plane polarized light clockwise

diamagnetic—a material that creates an opposing electric field in response to an applied electric field

diastereomer—a stereoisomer that is not an enantiomer

diffraction—change in the direction of a wave caused by an obstacle (be it a wall, or a nucleus)

diffusion—spreading of something more widely; the movement of a substance from an area of high concentration to an area of lower concentration; the scattering of waves through a space or an object

dilution—process that lowers the concentration of a substance

dipole—a molecule or a property of a molecule that involves the separation of positive charge from negative charge

distillation—a purification technique that separates substances based on their differences in boiling points

DNA—an acronym for deoxyribonucleic acid, which is the biomolecule that stores genetic information in organisms

ductile—pliable, not brittle; a metal that can be drawn into a thin wire

elastic material—a material that deforms when an external force is applied, but returns to its original shape when the external forces are removed

electrochemical cell—a device that either produces an electric current from or uses an electric current to drive a redox reaction

electrolysis—use of an electric current to drive a redox reaction

electrolyte—a substance that forms ions in a solution

electron—fundamental particle with a negative electric charge

electron affinity—the energy change upon adding an electron to a neutral species

electronegativity—a measure describing how strongly atoms attract electrons

electronic wave function—a mathematical description used to describe the electrons in a chemical system

electrophile—a species that is attracted to electron-rich atoms or molecules

element—atoms with the same atomic number

elementary reaction—single step in a chemical reaction

elimination reaction—a reaction in which two ligands or substituents are removed from a molecule

empirical formula—the relative ratios of elements in a substance

emulsion—the suspension of a liquid in another liquid, a type of colloid

enantiomer—non-superimposable mirror images

endothermic—a process or reaction that absorbs heat

enthalpy—for a reaction the heat absorbed or released is defined as the change in enthalpy

entropy—a measure of the disorder, or dispersion of energy, in a system; the 2nd Law of Thermodynamics states that a spontaneous change cannot decrease the entropy of an isolated system

enzyme—a protein-based molecule that acts as a catalyst

equilibrium—a state in which a chemical reaction and its reverse are proceeding at equal rates

evaporation—conversion of a liquid into a gas

excited state—an atom or molecule in any electronic state of higher energy than its lowest energy state

exothermic—a process or reaction that releases heat

extensive property—a property that depends on the amount of a substance present (e.g., size, mass, volume)

extraction—removal of one or more substances from a mixture, typically based on differing solubility of a substance in a solvent

filtration—the process of removing any solid particulates from a solution

fission—a nuclear reaction in which a nucleus splits into smaller parts

flash point—temperature at which the vapor pressure of a liquid is high enough that the vapor can be ignited

formal charge—the amount of charge (typically in integer units of electron charge) assigned to a specific atom in a molecule

freezing point—temperature at which liquid and solid phases coexist in equilibrium

freezing point depression—decrease in freezing point for a solution from that of a pure solvent

frequency—rate of an event per unit time

fuel cell—device that converts chemical energy to electrical energy

fusion—the joining of two (or more) atoms

galvanization—application of a thin layer of zinc metal to steel or iron, which prevents the formation of rust

gamma radiation—high frequency electromagnetic radiation; can be dangerous to living things

gel—a semisolid suspension of a solid in a liquid, a type of colloid

geometric isomer—a molecule with the same molecular formula but different arrangement of those atoms in space

glass—an amorphous solid material

glass transition—transition observed in amorphous materials between a hard material and a more liquid- or rubber-like state

gram—1/1000 of a kilogram, which is the standard unit for mass

ground state—lowest energy state for an atom or molecule

half life—the amount of time required to consume half of the initial amount of a reactant

halogen—elements in group 18 (group VIIA)

hapticity—number of contiguous atoms coordinated to a central atom

heat—transfer of energy between substances with different temperatures

heat capacity—the amount of heat required to raise the temperature of an object by 1 °C

Henderson-Hasselbach equation—equation for pH of a solution as a function of acid strength (pKa); $pH = pK_a + log_{10} ([A^-]/[HA])$

Henry's Law—equation relating the pressure of a gas to its solubility in a liquid

Hess's Law—states that the change in enthalpy for a reaction that occurs in multiple steps is equal to the sum of the change in enthalpy for each step; related to the 1st Law of Thermodynamics

heteroatom—any atom other than carbon or hydrogen

heterogeneous mixture—a sample containing more than one substance

homogeneous mixture—a sample containing only one, pure substance

hybrid orbital—an orbital that is composed of multiple atomic orbitals (e.g., sp3 hybrid orbital)

hydrogen bond—interaction between a hydrogen atom bonded to a highly electronegative atom and a Lewis basic atom

hydrolysis—the breaking of chemical bond(s) by the addition of water to a molecule

hydrophilic—a molecule that interacts favorably with or is attracted to water, typically via hydrogen bonding or other dipolar interactions

hydrophobic—a nonpolar molecular that does not interact with water

hygroscopic—a substance that readily absorbs water from its surroundings ideal gas law—an equation that approximates the properties of gases, commonly written as: $PV = nRT$ (where, P = pressure, V = volume, n = number of moles, R = gas constant, T = temperature)

immiscible—unmixable liquids (e.g., oil and water)

independent variable—a variable that is set to a specific, known value in an experiment

indicator—a substance that undergoes an observable change in response to a chemical input (e.g., pH, redox, presence of a metal ion)

inductive effect—polarization of a chemical bond due to transmission of charge through other chemical bonds

infrared—electromagnetic radiation with wavelengths of 750 nm to 1 mm, just longer than the visible spectrum, but not as long as microwave radiation

insoluble—a substance that does not dissolve in a solvent

intensive property—a property that does not depend on the amount of a substance present (e.g., density, temperature, color)

interference—superposition of two or more waves, resulting in either higher (constructive) or lower (destructive) amplitude

ion—a charged atom or molecule

ionic bond—attraction between two oppositely charged ions

ionization energy—energy required to remove an electron from an atom or ion (i.e., ionization potential)

irreversible reaction—a reaction in which the products cannot be converted back into the reactants

isobaric—with constant pressure

isomer—molecules with different arrangement of atoms in space, but the same molecular formula

isotactic polymer—a polymer in which all substituents are located on the same side of the backbone

isothermal—with constant temperature

isotope—atoms of the same element (same number of protons) with different number of neutrons

kelvin—standard temperature scale in which the triple point of water is defined at 273.16 Kelvin

kilogram—standard base unit of mass

kinase—an enzyme that catalyzes phosphorylation (transfer of a phosphate group)

kinetic energy—energy of an object due to its motion

kinetics—the rate of chemical reactions or processes

lathanide—elements 57–70

lattice—a regular array of atoms or ions

Le Chatelier's Principle—states that any change (concentration, pressure, temperature, volume) to a chemical system at equilibrium will cause the equilibrium to shift in order to counteract that change

levorotatory—having the property of rotating plane polarized light counterclockwise

Lewis acid—a molecule that can accept a pair of electrons

Lewis base—a molecular that can donate a pair of electrons

Lewis structure—a writing convention in which valence electrons are represented as dots and chemical bonds are represented by lines between atom

ligand—a group (an ion or a molecule) that coordinates to a metal atom, forming a coordination complex

lipid—a biochemical molecule that is hydrophobic or amiphilic (e.g., waxes, fats, vitamins A, D, E, K, glycerides, etc.)

liquid—a state of matter with a defined volume, but no fixed shape

London dispersion force—a weak repulsive interaction between molecules resulting from interactions of electron clouds

lone pair—a pair of valence electrons that are localized on a single atom (i.e., not involved in bonding)

magnetic quantum number—the third quantum number, m, which describes the direction of the electron's angular momentum

main group elements—elements of the s and p blocks in the periodic table

malleable—a material that can be pressed into shapes or sheets without breaking or cracking

manometer—an instrument used to measure pressures of gases

mass—the resistance of an object to acceleration; commonly used interchangeably with weight, though the latter depends on gravity while the former does not

matter—any substance that has mass

melting point—temperature at which liquid and solid phases coexist in equilibrium

meniscus—a phase boundary, which is curved due to surface tension

meta—term used to describe two substituents on an aromatic ring that are separated by one position (i.e., 1,3-subsituted)

metal—an element, compound, or alloy that is a good conductor of heat and electricity, also usually reflective, ductile, and malleable

metalloid—an element, compound, or alloy that has metallic and nonmetallic properties

microwave—electromagnetic radiation with wavelengths of 1 millimeter to 1 meter, just longer than infrared radiation, but not as long as radio waves

miscible—liquids that when mixed form a single phase

mixture—a system composed of two or more difference substances

molality—a measure of concentration defined as moles of solute per kilogram of solvent

molarity—a measure of concentration defined as moles of solute per liter of solvent

mole—the SI unit used to describe the amounts of chemical; 6.023×10^{23}

mole fraction—the amount of a substance in moles divided by the total number of moles present

molecular formula—the type and number of atoms in a molecule; unlike empirical formula, the ratios are not reduced

molecular orbital—a mathematical equation describing the position of an electronic in a molecule

monodentate—a ligand that coordinates to a central atom via only one atom (compare with chelate)

monomer—a group of atoms that forms the repeating unit in a polymer

natural abundance—relative abundance of different isotopes of an element found on Earth (i.e., not produced in a lab)

Nernst equation—an equation describing the potential of an electrochemical half-cell

neutron—an uncharged, subatomic particle found in the nucleus of an atom

noble gas—Group 18 of the periodic table, characterized by their general inertness due to their complete valence shell of electrons

noble gas core—used to abbreviate an atom's electron configuration (e.g., the electron configuration of nitrogen is: $1s^2\, 2s^2\, 2p^3$, which can be abbreviated [He] $2s^2\, 2p^3$)

nonmetal—an element that does not possess the properties of a metal

nonpolar—a molecule in which the distribution of charge does not lead to an overall dipole moment

normality—concentration of a solution defined as the molar concentration divided by an equivalence factor (i.e., H_2SO_4 can neutralize two equivalents of base with its two H^+ groups, so a 1 M solution of H_2SO_4 is 2 N; the equivalence factor of H_2SO_4 is 0.5)

nucleation—a process through which a crystal, or a drop of liquid, grows around a small site

nucleic acid—general term for RNA and DNA, composed of a nucleotides, which are in turn each composed of a sugar, a phosphate group, and a nucleobase

nucleobase—nitrogen-containing molecules that form nucleic acids; adenine, cystosine, guanine, thymine, uracil are the major nucleobases; form base pairs via hydrogen bonding between the two helical strands of DNA

nucleophile—a molecule that donates an electron pair (i.e., a Lewis base) to an electrophile (i.e., a Lewis acid), forming a bond

nucleoside—a biochemical molecule consisting of a nucleobase and a sugar molecule

nucleotide—a biochemical molecule consisting of a nucleobase, a sugar molecule, and one phosphate group

nucleus—the center of an atom, consisting of positively charged protons and neutral neutrons

octet rule—a rule for chemical bonding that says atoms "prefer" to have eight electrons in their valence shell

ohm—Ω, the SI unit used to describe electrical resistance

ohmmeter—a device used to measure electrical resistance

olefin—a molecule containing a carbon-carbon double bond, also referred to as an alkene

orbital—a possible arrangement for the density of an electron around the nucleus

ortho—term used to describe two substituents on an aromatic ring that are bonded to adjacent positions

osmometry—the process/study of measuring the osmotic strength of a solution osmosis—the diffusive process of solutes moving from areas of higher concentration to lower concentration

oxidant—an oxidizing agent, or a species that removes electrons from another species

oxidation state—indicates the extent of oxidation for an atom; synonymous with oxidation number

p orbital—an atomic orbital with an angular momentum quantum number of 1, shaped like a peanut

para—term used to describe two substituents on an aromatic ring that are bonded to positions opposite to each other

paramagnetic—a molecule with no net spin; a material that is only attracted in the presence of an external magnetic field

peptide—a polymer of amino acids

pH—a measure of the hydrogen ion concentration of a solution

phase—a state of matter (e.g., solid, liquid, or gas)

phase boundary—the interface between two phases (e.g., solid-liquid, liquid-gas, or solid-gas boundary)

phase diagram—a plot that shows the phase in which a material is expected to exist as a function of variables like temperature and pressure

photon—a quantum (or particle) of electromagnetic radiation

physical change—a change in the macroscopic properties of a substance without a change in its chemical composition

pi bond—a chemical bond in which two lobes of two orbitals on adjacent atoms overlap favorably

pKa—the logarithm of an acid dissociation constant

plasma—a phase of matter in which a significant fraction of electrons have been ionized from their nuclei

polar—a compound in which the charge distribution leads to an overall dipole moment

polydentate—a ligand is described as polydentate if it can bind to a central atom multiple times

polymer—a "chain molecule" consisting of a repeating series of a single unit, or possibly multiple repeating units

polymorph—one of the crystalline forms of a substance that can crystallize in multiple forms

potential energy—the energy contained within an object as a result of its current state (its charge distribution, its position relative to other objects, etc.)

precipitation—the process through which a substance comes out of solution, usually due to insolubility

precision—the extent to which a measurement is repeatable (note that this is distinct from accuracy)

product—substance produced by a chemical reaction

proton—a charged subatomic particle found in the nucleus of an atom with a charge equal and opposite to that of an electron

pyrophoric—a substance that tends to ignite readily and spontaneously upon exposure to air

qualitative analysis—analysis of the identity of the constituent species present in a sample

quantitative analysis—analysis of the specific amounts of the constituent species present in a sample

quantum—a discrete amount of energy, electric charge, or another physical quantity

quantum mechanics—the branch of physics that describes the motion and interactions of subatomic particles

quantum number—values of conserved quantities in a quantum mechanical system; electrons are described using four quantum numbers: n, the principal quantum number; l, the azimuthal quantum number; ml, the magnetic quantum number; and, ms, the spin quantum number

racemic—a mixture of equal amounts of the two enantiomers of a chiral molecule

radiation—the emission of energy as an electromagnetic wave

radical—a species containing unpaired electron density, typically involving an odd number of valence electrons

radioactive—emission of radiation or particles

Raoult's Law—states that the vapor pressure of a solution depends on the amount (mole fraction) of a solute added to it

rare earth element—term for the lanthanides and actinides (including scandium and yttrium)

rate constant—a numerical value that characterizes the rate at which a chemical reaction takes place

rate determining step—the slowest step in a multiple step chemical process; usually the step with the highest energy transition state

reactant—a compound that is consumed or transformed by a chemical reaction

reaction—the process of changing chemical bonds in a molecular or ionic compound

reaction order—the number of chemical species simultaneously involved in a chemical reaction

reaction rate—how fast the reaction takes place, typically measured in terms of a change in concentration of a chemical species with respect to time

reductant—a reducing agent, or a species that gives electrons to another species

resonance—a way of describing the delocalization of electron density in a molecule

reversible reaction—a reaction where the products can be converted back into the reactants

RNA—ribonucleic acid; composed of a chain of nucleotides; encodes genetic information

s orbital—an atomic orbital with an angular momentum quantum number of 0, shaped like a sphere

salt—an ionic compound that can be thought of to form from the neutralization of an acid or base

salt bridge—a method used to place the two sides (solutions) of an electrochemical cell in electrical contact

saponification—technically the hydrolysis of triglycerides (i.e., triesters) with sodium hydroxide; more generally refers to hydrolysis of any ester

saturated compound—a molecule with no pi bonds or ring structures

saturated solution—a solution that contains the maximum amount of dissolvable solution (i.e., the addition of any more solution will not dissolve, and will remain in a separate phase)

scientific notation—a way of expressing a number a the product of a decimal and a power of 10 (for example, $1,050 = 1.05 \times 10^3$)

semi-metal—an element, compound, or alloy that has metallic and nonmetallic properties; also known as "metalloid"

shell—a way of classifying the orbitals in which electrons can reside based on their principle quantum number

sigma bond—a chemical bond involving direct overlap between orbitals on adjacent atoms, sigma bonds are symmetric with respect to rotation around the bond axis

significant figure—a digit in a number that is known with some amount of accuracy

soluble—a property describing a species that can be dissolved in a particular solvent

solute—a substance that is dissolved in a solution, present in a much smaller quantity than the solvent itself

solution—a liquid mixture containing multiple components, typically with one major component and one or more minor components

solvent—the major component of a solution

sorption—attachment of one substance to another; see absorption and adsorption

specific gravity—the ratio of a compounds density to the density of a reference material (usually water for liquids)

specific heat—the quantity of heat energy needed to increase the temperature of a substance by a fixed amount

specific volume—the volume occupied by a compound per unit mass

spectrum—a plot of the absorption intensity of light by a sample as a function of wavelength

spin—a type of angular momentum possessed by electrons and some other particles

stereochemistry—relative arrangement of atoms of a molecule in 3D space (see also stereoisomer, enantiomer, diastereomer)

stereoisomer—molecules that differ only in the spatial arrangement of their atoms

stoichiometry—the ratio of quantities in which reactants are consumed and products are formed in a chemical reaction

sublimation—phase transition from solid to gas without passing through the liquid phase (e.g., the "smoke" that dry ice gives off)

substituent—an atom or group of atoms occupying a designated position in a molecule

substitution reaction—a chemical reaction that replaces one substituent with another

substrate—a molecule that reacts with a reagent; a subjective label as the substrate is also a reagent

sugar—a carbohydrate (a molecule composed of only carbon, hydrogen, and oxygen) that tastes sweet; monosaccharides (or simple sugars) include fructose, galactose, and glucose

surfactant—a solute that reduces the surface tension of a liquid

syndiotactic polymer—a polymer in which substituents are located on alternating sides of the backbone

tautomer—structural isomers that differ only in the position hydrogen atoms

temperature—a physical property of matter that describes the average kinetic energy of molecules

thermoplastic—a type of plastic that hardens when heated and then cooled

titration—the process of measuring the concentration of an analyte by reacting a second solution of known concentration to cause a reaction

transition metal—typically, this refers to any element in the d-block (groups 3 through 12) of the periodic table

triple point—the combination of temperature and pressure values at which solid, liquid, and gas phases of a substance are in equilibrium

ultraviolet—electromagnetic radiation with wavelengths of 10 nm to 400 mm, just shorter than the visible spectrum, but not as short as X-rays

unimolecular reaction—a reaction that takes place involving only a single reactant molecule (there can be one or more product molecules formed)

unit cell—the smallest group of atoms, in a crystal, required to represent the symmetry and overall three-dimensional structure of the crystal

unsaturated compound—a molecule with at least one pi bond or ring structure

valence band—highest energy range in which electrons are normally present at 0 K

valence electron—an electron in the valence shell of an atom

valence shell—the highest partially occupied shell of electrons in an atom

van der Walls force—intermolecular forces between molecules resulting from the interactions of dipoles or induced dipoles

van der Walls radius—the radius of an imaginary sphere that can be used to represent the effective size of an atom

vapor pressure—partial pressure in the gas phase of a substance in equilibrium with its solid or liquid phase

viscosity—a property describing the consistency of a fluid by its ability to be deformed by stress, this is related to its ability to flow

volatile—a substance that easily evaporates at normal temperatures and pressures

voltage—the difference in electrical potential between two locations

voltmeter—a device used to measure an amount of electricity

volume—the quantity of space that a substance occupies

vulcanization—chemical reaction that creates crosslinks between individual polymer chains in rubber

wavenumber—inverse of wavelength, typically reported in units of m^{-1} or cm^{-1}

weight—the force applied by gravity on an object

work—a force acting over a distance, such as a person lifting a box

X-ray—a form of short wavelength electromagnetic radiation

History of Chemistry Timeline

Year	Event
c. 465 B.C.E.	Democritus first proposes that all matter is made up of small particles. He also first proposes the term "atoms" to describe these particles.
c. 450 B.C.E.	Empedocles proposes the four basic elements: air, fire, earth, and water.
c. 360 B.C.E.	Plato first uses the term "elements" to describe the basic components of matter.
c. 350 B.C.E.	Aristotle expands on the theories of Empedocles to include a fifth basic element: aether.
c. 300 C.E.	Some of the earliest known books on the subject of alchemy are written.
c. 770	Persian alchemist Abu Musa Jabir ibn Hayyan, also commonly known as Geber, develops some of the first experimental methods for isolating chemical compounds.
c. 1000–1650	Alchemists searched for a method to transform cheap metals into gold, an elixir to make people live longer, and a universal solvent. While none of these goals were accomplished, the alchemists did make progress in learning how to use plant extracts and metals to treat illnesses. It's also worth noting that as early as c. 1000, some chemists were also already beginning to speak out against alchemy, suggesting that some of its goals were impossible to achieve.
c. 1167	The first recorded references to wine distillation were made by Magister Salernus.
c. 1220	An early version of the scientific method was described by Robert Grosseteste.
c. 1250	Fractional distillation was developed.
c. 1260	Arsenic was discovered by Saint Albertus Magnus.
c. 1310	Geber publishes books establishing a theory that all metals are made of sulfur and mercury (which is today known to be incorrect). Several strong acid solutions, still used today, were also described for the first time.
c. 1530	Paracelsus initiates the development of a subfield of chemistry dedicated to extending life, which may be credited as forming the foundations of the study of pharmacy.
1597	An early chemistry textbook, titled *Alchemia,* is published.
1605	The first description of the scientific method is published by Sir Francis Bacon.
1643	A mercury-based barometer is first invented by Evangelista Torricelli

Year	Event
1661	Robert Boyle publishes a piece of writing titled "The Sceptical Chymist," explaining the difference between alchemy and chemistry, laying the foundation for the modern field of chemistry.
1662	Boyle's Law, which relates the pressure and volume occupied by a gas, is first proposed.
1728	The speed of light is determined by James Bradley.
1752	Benjamin Franklin discovers that lightning is electricity.
1754	Carbon dioxide is first isolated.
1772–1777	Oxygen is first isolated independently by both Joseph Priestly and Carl Wilhelm Scheele, and correctly identified by Antoine Lavoisier.
1787	Charles' Law, which relates the volume and temperature of a gas, is first proposed by Jacques Charles.
1797	The law of definite proportions, which states that elements combine to form compounds in whole number ratios, is proposed by Joseph Proust.
1798	Count Rumford proposes that heat may be a form of energy.
1800	The first chemical battery is created by Alessandro Volta.
1801	Thomas Young performs experiments to demonstrate the wavelike behavior of light by using interference patterns.
1801	Dalton's Law, describing the relationship between the quantities of components present in a gas mixture and their partial pressures, is introduced by John Dalton.
1805	The fact that water is made of hydrogen and oxygen in a 2-to-1 ratio is first discovered.
1811	Avogadro's law is proposed. This states that identical amounts of different species occupy the same volume in the gas phase.
1825	The existence of isomers is discovered.
1826	Ohm's law is introduced, describing the concept of electrical resistance and its relationship to the current through a conductor.
1827	Biomolecules are first classified as carbohydrates, proteins, and lipids (DNA was not yet discovered).
1840	Hess' law is first proposed, stating that the energy change for a chemical reaction depends only on the identities of the reactants and products, and not on the pathway through which they interconvert.
1843	Heat is demonstrated to be a form of energy.
1848	The concept of an absolute-zero temperature, at which all molecular motion stops, is proposed by Lord Kelvin.
1852	Beer's Law, which relates the concentration of a sample to its optical absorption at a given wavelength, is introduced by August Beer.
1857	Carbon is proposed to form four bonds with neighboring atoms in molecules.
1859	James Maxwell proposes a description for the velocity distribution of molecules in a gas.

Year	Event
1859–1860	The early foundations of spectroscopy are established by Gustav Kirchhoff and Robert Bunsen.
1864	The law of octaves, an important step leading to development of the modern periodic table, is first proposed.
1865	The number of molecules in one mole of a substance is first determined.
1869	Mendeleev publishes the first version of the modern periodic table, containing 66 elements, and leaving space for yet-to-be-discovered elements.
1869	DNA is first discovered.
1874	The second law of thermodynamics is proposed by Lord Kelvin.
1874	Electric current is proposed to be caused by the motion of electrons.
1876	The concept of free energy is introduced, by Josiah Willard Gibbs to explain chemical equilibria.
1877	Ludwig Boltzmann provides a definition of entropy, along with statistical derivations of several other physical concepts.
1884	Le Chatelier's principle is first introduced, explaining changes in chemical equilibria upon the introduction of changes in a chemical system.
1887	The photoelectric effect is first discovered.
1888	Radio waves are discovered by Heinrich Hertz.
1893	Alfred Werner establishes that some cobalt complexes involve a central cobalt atom bonded to an octahedral arrangement of six ligands, laying the groundwork for the field of coordination chemistry.
1894	Noble gases are first discovered.
1895	X-ray radiation is discovered.
1897	The electron is discovered.
1900	Planck's constant is first introduced by Max Planck.
1900	Ernest Rutherford establishes that radioactivity is caused by the decay of atoms.
1901	The Nobel Prizes are awarded for the first time.

The Nobel Prize
in Chemistry

1901—Jacobus Henricus van 't Hoff (the Netherlands) wins for "discovery of the laws of chemical dynamics and osmotic pressure in solutions." The first Nobel Prize ever awarded went to van't Hoff for his discoveries surrounding osmotic pressure and chemical equilibrium. If a solution of sugar water was separated from a volume of pure water by a membrane that allowed water, but not sugar, to pass, van't Hoff discovered that additional water would force its way across the membrane until an equilibrium was established. This results in a greater pressure on the side to which the water is moving, and this pressure is known as osmotic pressure.

1902—Hermann Emil Fischer (Germany) for "his work on sugar and purine syntheses." Fischer showed that various molecules from plants and animals, which were initially not believed to be related, were in fact structurally similar. He went on to show that various sugars could be made in the lab from simple chemicals like urea. With the ability to make sugar molecules, he proceeded to establish the three dimensional structure of every known sugar.

1903—Svante August Arrhenius (Sweden) for "his electrolytic theory of dissociation" Arrhenius began studying as a graduate student at the University of Uppsala how much electricity solutions of various molecules could conduct. He correctly proposed that solution of salts were conductive because the salts formed charged particles (later named ions), even in the absence of an electric current. His advisors weren't very impressed, and he earned a third class degree in 1884. Arrhenius continued to develop the work, and was soon awarded the Nobel Prize for his insights.

1904—Sir William Ramsay (United Kingdom) for the "discovery of the inert gaseous elements in air, and his determination of their place in the periodic system." Lord Rayleigh noted in a lecture that he measured a difference in the density of nitrogen made synthetically in the lab, and the same substance isolated from air. Ramsay followed-up on this observation, quickly isolating argon for the first time. He went on to discover neon, krypton, and xenon.

1905—Johann Friedrich Wilhelm Adolf von Baeyer (Germany) for "the advancement of organic chemistry and the chemical industry, through his work on organic dyes and hydroaromatic compounds." Baeyer was the first to synthesize indigo, which is still a widely used dye (although a different method is used to prepare it today).

1906—Henri Moissan (France) for the "investigation and isolation of the element fluorine, and for electric furnace called after him." Moissan was the first to make fluorine gas (F_2), which he did by passing an electric current through a solution of KHF_2 in liquid HF. The second part of the Committee's statement relates to his work on an electric arc furnace, which he designed to try to make synthetic diamonds.

1907—Eduard Buchner (Germany) "for his biochemical researches and his discovery of cell-free fermentation." Buchner ground up dry yeast cells, quartz, and a form of silica known as diatomaceous earth, to release the contents of the yeast cells. He then filtered this solution and added sugar. Buchner observed that fermentation of the sugar still occurred in the absence of the whole yeast cells.

1908—Ernest Rutherford (United Kingdom, New Zealand) "for his investigations into the disintegration of the elements, and the chemistry of radioactive substances." Rutherford's most famous experiment, the "Gold Foil Experiment," was actually done after he received the Nobel Prize. He was awarded the Nobel for discovering that there were multiple kinds of radioactivity (he named them alpha and beta rays—later he added gamma rays to the list), and for proposing that radioactivity was a result of the atom actually breaking apart. Until Rutherford's work, chemists had assumed that atoms were indestructible.

1909—Wilhelm Ostwald (Germany) for "his work on catalysis and for his investigations into the fundamental principles governing chemical equilibria and rates of reaction." Ostwald performed many experiments surrounding the strengths of acids and bases and was the first to carefully explore how reaction rates are affected by acid/base catalysis. He was actually the first to coin the term "catalysis" in the context of chemistry, and his definition of catalysis is still very much the same as the one used today. Ostwald's work corroborated earlier studies into acid/base chemistry by Arrhenius, and his measurement of chemical reaction rates under various acidic and basic conditions placed the understanding of Brønsted-Lowry acids and bases on a more sound footing. Ostwald also expanded the scope of his study beyond acid/base catalysis, and his work led to the understanding of chemical reaction rates as we know them today. He showed that chemical reaction rates—and the factors that influenced them—were quantitatively measurable parameters.

1910—Otto Wallach (Germany) for "his services to organic chemistry and the chemical industry by his pioneer work in the field of alicyclic compounds." Alicyclic compounds are aliphatic (meaning they have no aromatic rings) and cyclic (meaning they do have other rings). Cyclohexane is a simple example of an alicyclic compound. Wallach specifically studied terpenes, which are found in the essential oils of various plants. He was able through chemical transformation to make these liquid molecules into crystalline solids, which made them easier to study with the methods available at the time. Wallach is also known for a number of different organic reactions that all still bear his name today (Wallach rearrangement, Wallach's rule, Wallach degradation, Leuckart-Wallach reaction).

1911—Marie Curie, née Sklodowska (Poland/France), for "the discovery of the elements radium and polonium, by the isolation of radium and the study of the nature and compounds of this remarkable element." Curie was the first woman to win a Nobel Prize, and she is still the only person to win in multiple sciences. Curie was a professor at the University of Paris (another first—she was the first woman on the faculty), beginning in 1900, and then at the Sorbonne in 1906 (again—she was the first female professor there). In 1903 she was awarded, along with her husband, Pierre Curie, and Antoine Henri Becquerel, the Nobel Prize in Physics for her work on radiation. In 1910, Marie isolated radium for the first time by electrolysis of radium chloride in the presence of hydrogen gas. This contribution was not enough for the French Academy of Sciences to elect her to their ranks, but a year later this accomplishment was recognized with Curie's second Nobel Prize.

1912 (two winners)—Victor Grignard (France) "for the discovery of the Grignard reagent" and Paul Sabatier (France) "for his method of hydrogenating organic compounds in the presence of finely disintegrated metals." Grignard was recognized for his preparation and study of organic magnesium compounds, (R-MgX) which he prepared from organic halides (R-X) and magnesium metal. He discovered that these magnesium reagents can react with carbonyl groups to form new carbon-carbon bonds. Sabatier shared the Prize in 1912 with Grignard for his work on hydrogenation, or the addition of a molecule of H_2 to an organic molecule. Sabatier observed that nickel acted as a catalyst for these reactions. The Sabatier Process (the reaction of CO_2 and H_2 to form CH_4 and H_2O) continues to be a useful reaction, and NASA is investigating it as a means to generate water from CO_2 waste (from exhaling astronauts).

1913—Alfred Werner (Switzerland) for "his work on the linkage of atoms in molecules especially in inorganic chemistry." Werner was the first to propose the correct structure of inorganic compounds like $[Co(NH_3)_4Cl_2]^+$. He proposed that the cobalt ion was in the center, surrounded by the ammonia and chloride ligands, in an octahedral arrangement. This proposal was consistent with the fact that two isomers of this compound were observed, as the two chloride ligands can be arranged either $180°$ (trans) or $90°$ (cis) apart from each other.

1914—Theodore William Richards (United States) for "his accurate determinations of the atomic weight of a large number of chemical elements." Richards was the first American to receive the Nobel Prize in Chemistry. He and his students accurately measured the atomic weight of 55 different elements and also showed that some crystalline solids can contain gases or other solutes within their lattices (technical term: occlude).

1915—Richard Martin Willstätter (Germany) "for his researches on plant pigments, especially chlorophyll." Willstätter was a German organic chemist who studied a variety of pigments isolated from flowers and fruits. He was the first to show that chlorophyll was a mixture of two different compounds, known today as chlorophyll a and chlorophyll b. These two molecules absorb slightly different wavelengths of light, so that the plant can capture more of the Sun's energy.

1916–1917—No prizes were awarded.

1918—Fritz Haber (Germany) "for the synthesis of ammonia from its elements." Haber was a German chemist, who developed a synthesis of ammonia (NH_3) from nitrogen (N_2) and hydrogen (H_2), along with Carl Bosch, while working at the University of Karlsruhe. Ammonia is critical to many applications, including fertilizers, explosives, and as a feedstock for other chemicals. It is estimated that because of the widespread use of chemical fertilizers about half of the nitrogen atoms in your body probably passed through the Haber-Bosch process. A significant percentage of our planet's population couldn't exist without this chemical process.

1919—No prize was awarded.

1920—Walther Hermann Nernst (Germany) for "his work in thermochemistry." Nernst made several somewhat related contributions to chemistry surrounding the specific heats of compounds at very low temperatures, the use of Galvanic cells and the ability to calculate chemical affinity from thermochemical properties, and surrounding variations in chemical equilibria at different temperatures. Nernst's most significant gift to chemistry can be stated succinctly by saying that he (and his co-workers) helped to make possible the calculation of whether a reaction will take place to a significant extent under a set of known conditions. Nernst also was the first to formulate the third law of thermodynamics.

1921—Frederick Soddy (United Kingdom) "for his contributions to our knowledge of the chemistry of radioactive substances, and his investigations into the origin and nature of isotopes." Soddy, an English chemist, correctly explained that radioactivity of elements was due to their transmutation, or the changing of one element into another. Specifically, he demonstrated that uranium changes into radium. He also revealed the difference between alpha emission (loss of a helium nucleus, so atomic number decreases by 2) and beta emission (electron emission from the nucleus, so atomic number increases by 1). Finally, the Soddy demonstrated that radioactive elements can have more that one atomic weight, and the idea of isotopes was born.

1922—Francis William Aston (United Kingdom) "for his discovery, by means of his mass spectrograph, of isotopes in a large number of non-radioactive elements, and for his enunciation of the whole-number rule." Following the prize in 1921 for Soddy, the Nobel Committee again recognized the importance of isotopes by selecting Aston for the prize. Aston build the first instrument capable of separating individual isotopes of a given element, which he used to identify over 200 elemental isotopes. This work led him to conclude that all isotopes of all elements have whole number masses (if one defines the major isotope of oxygen to be 16).

1923—Fritz Pregl (Austria) "for his invention of the method of micro-analysis of organic substances." Pregl, a chemist and physician, was award the prize for his work on ways to quantitatively characterize organic molecules. In particular, he made great improvements to elemental analysis, which reveals the amount of various elements in a substance by measuring the combustion products (or, more simply: burn the stuff and see how much CO_2, H_2O, and NO are released).

1924—No prize was awarded.

1925— Richard Adolf Zsigmondy (Germany/Hungary) "for his demonstration of the heterogeneous nature of colloid solutions and for the methods he used." Although the Nobel committee didn't mention it by name, the "methods he used" refers to Zsigmondy's invention of the ultramicroscope, which allowed a visible light microscope to see objects that are smaller than the wavelength of light. Zsigmondy accomplished this (literally) physically impossible task by looking at the light that scattered off of a sample, and not the light that it reflects. Using this new tool, he showed that the red color in so-called "Cranberry glass" was due to small (4 nm) particles of gold, which was of interest to his employer at the time, Schott Glass.

1926—The (Theodor) Svedberg (Sweden) "for his work on disperse systems." Svedberg's disperse systems were, like the previous year's winner, colloids. Svedberg studied their absorption, diffusion, and sedimentation. He was able to produce colloidal particles to validate Einstein's theory of Brownian motion. To accomplish this work he developed the ultracentrifuge, which he used to purify proteins. The unit used to describe the rate at which particles undergo sedimentation is now known as the svedberg (1 svedberg = 10^{-13} seconds = 100 femtoseconds).

1927—Heinrich Otto Wieland (Germany) "for his investigations of the constitution of the bile acids and related substances." The bile acids are a set of steroid acids whose synthesis begins in the liver with the production of chloic acid chenodeoxycholic acid (all of which derive from cholesterol). Wieland isolated and determined the structure of a number of these biochemically significant compounds. During his career he also isolated toxins from poisonous frogs and mushrooms.

1928—Adolf Otto Reinhold Windaus (Germany) for "his research into the constitution of the sterols and their connection with the vitamins." Windaus was awarded the Nobel Prize for discovering that cholesterol (a sterol) was removed to cholecalciferol (vitamin D3) through a series of several steps. One of Windaus' doctoral students, Adolf Butenandt, was himself a Nobel laureate.

1929—Arthur Harden (United Kingdom) and Hans Karl August Simon von Euler-Chelpin (Germany) "for their investigations on the fermentation of sugar and fermentative enzymes." Harden and Euler-Chelpin were biochemists who independently investigated fermentation processes. Harden discovered that phosphate was required for alcohol fermentation. Euler-Chelpin studied the means by which living cells produced energy by degrading sugar molecules, and the machinery that cells used to perform these reactions.

1930—Hans Fischer (Germany) "for his researches into the constitution of haemin and chlorophyll and especially for his synthesis of haemin." Fischer was interested in biologically relevant pigments, specifically those in human bodily fluids like blood and bile, and the green color of plants. He was the first to determine the correct structure for heme B and heme S, which are iron porphyrin molecules (a large ring with four nitrogen ligands coordinated to an iron center). These red-colored molecules help transport O_2, and other important functions. He also determined the structure of chlorophyll a, which has a structure similar to that of a heme, but instead of an iron, a magnesium ion sits in the middle of the large porphyrin ring.

1931—Carl Bosch (Germany) and Friedrich Bergius (Germany) for "their contributions to the invention and development of chemical high pressure methods." Bosch worked with Fritz Haber on the synthesis of ammonia that today bears both their names, but was not recognized in the 1918 prize with Haber. Bergius developed a process for producing liquid hydrocarbon fuel from coal. Bosch also worked on the so-called Bergius process, after Bergius sold his patents to BASF, where Bosch was working at the time. The Bergius process and the Haber-Bosch process both operate at high pressures, and both methods had significant impacts on human history.

1932—Irving Langmuir (United States) "for his discoveries and investigations in surface chemistry." As a graduate student, Langmuir studied light bulbs, and then went on to improve vacuum pump designs. These two interests merged, in some respects, and Langmuir was able to invent the incandescent light bulb. These led to his interest in surface chemistry, after observing that a tungsten-filament (like those in a light bulb) could split H_2 on its surface, forming a single atomic layer of hydrogen atoms. The work of direct relevance to his receiving the prize, however, was his study of thin films of oil and surfactants on the surface of water. Langmuir postulated that the molecules of surfactants would orient themselves into a layer that was a single molecule thick. He went on to develop the physics to describe such thin layers, which would eventually be known as monolayers.

1933—No prize was awarded.

1934—Harold Clayton Urey (United States) "for his discovery of heavy hydrogen." Urey received the Nobel Prize for isolation of deuterium (D_2) by distilling a sample of liquid hydrogen multiple times. He is perhaps better known (amazing that he's better known for something other than what he received the Nobel Prize for) for his work on the Manhattan Project. Working at Columbia, Urey and his team developed a method for enriching uranium by gaseous diffusion. Finally, after WWII ended, at the University of Chicago, Urey and Stanley Miller, one of his graduate students, showed that a mixture of water, ammonia, methane, and hydrogen could produce amino acids by exposing the mixture to electricity. The experiment was designed to simulate conditions in the early days of our planet, and clearly showed that organic molecules that form the basis of all life can be made from basic inorganic building blocks … and a little spark. After Urey's death, it was shown that over twenty different amino acids were present in the mixture, an even more remarkable result than Urey and Miller claimed in their initial publication.

1935—Frédéric Joliot (France) and Irène Curie (France) for "their synthesis of new radioactive elements." Frédéric was an assistant of Marie Curie, who ended up marrying Marie's daughter, Irène. The husband and wife team of Joliot and Curie collaborated on experiments investigated the effect of bombarding atoms with other particles. Specifically, they struck boron, magnesium, and aluminum atoms with alpha particles (He^{2+} ions), creating new short-lived radioactive particles.

1936—Petrus (Peter) Josephus Wilhelmus Debye (the Netherlands) for analyzing "molecular structure through his investigations on dipole moments and the diffraction of X-rays and electrons in gases." Debye developed the theory of how electric fields affect molecules, and figured out a method to determine their dipole moments by measuring how their insulating properties and density vary with temperature. He also measured interferences of X-rays and electrons with molecules in the gas phase, along with other investigations into molecular structure, to determine the chemical structures of molecules. This work provided some of the first detailed structural characterization that allowed chemists to definitively determine the different structures of isomers (compounds of the same chemical composition but different geometrical arrangements of the atoms).

1937—Walter Norman Haworth (United Kingdom) "for his investigations on carbohydrates and vitamin C," and Paul Karrer (Switzerland) "for his investigations on carotenoids, flavins and vitamins A and B2." Haworth's work on carbohydrates expanded on that done by Fischer, making significant further progress into understanding the structures of various isomers of monosaccharides and disaccharides. Prior to this work, very little was known about vitamins, carotenoids, and flavins, and these men were the first to uncover the chemical composition of these molecules. In addition to his work with sugars, Haworth also studied the composition of vitamin C. Karrer shared the prize for his work characterizing the chemical compositions of several carotenoids and flavins, as well as vitamin A and vitamin B2. Knowledge of the chemical compositions of these species provided some of the earliest insights into how they are formed and the roles they play as nutrients in the human body. Prior to their work, little was known regarding the chemistry or reactivity of these species, because their composition was not known.

1938—Richard Kuhn (Germany) "for his work on carotenoids and vitamins." Originally the 1938 Nobel Prize nominations did not yield a suitable candidate, so the 1938 award was held until 1939, when it was awarded to Kuhn. Kuhn received the award for his work studying vitamins and carotenoids. He isolated and characterized the composition of numerous complexes, and also studied the optical properties of some of these species to differentiate those with different chemical structures. He also made significant advances in the understanding of the chemistry of vitamin B complexes, including vitamin B2 (lactoflavin or riboflavin) and vitamin B6.

1939—Adolf Friedrich Johann Butenandt (Germany) "for his work on sex hormones," and Leopold Ruzicka (Croatia) "for his work on polymethylenes and higher terpenes." Butenandt was the first to isolate and crystalize a compound with the characteristics of a male sex hormone by extraction from male urine. Butenandt characterized the chemical formula of this complex, and called it andosterone. It was then discovered that this compound was actually slightly different from testosterone, but Butendant and Ruzicka were both able to synthesize testosterone from andosterone. Ruzicka synthesized the same compound, andosterone, that Butenandt isolated from male urine. He too was able to convert it into the male sex hormone testosterone. Ruzicka also worked to synthesize and characterize polyterpene complexes that are related to physically and biologically important compounds, including sex hormones. Ruzicka's work contributed greatly to the knowledge surrounding sex hormones, which, of course, are very important physiologically, and his work thus laid the ground for future investigations into the roles of these complexes.

1940–1942—No prize was awarded.

1943—George de Hevesy (Hungary) "for his work on the use of isotopes as tracers in the study of chemical processes." de Hevesy was a pioneer in using radioactive isotopes to carry out chemical studies. By using these isotopes as labels, he was able to track what was happening to the isotopes he introduced into a sample, providing unique new insights in a variety of areas. For example, by introducing radioactive sodium into a human body, he was able to track its spatial movement throughout the body, as well as its excretion from the body. He found that it after one day, blood corpuscles had lost/replaced roughly half of their sodium content. Of course, other scientists later picked up on this method and applied it themselves.

1944—Otto Hahn (Germany) "for his discovery of the fission of heavy nuclei." Hahn's and his colleague's work discovered nuclear fission, and in particular that uranium could be split in a chain reaction by nuclear fission. This discovery was perhaps recognized as much for its importance as it was for its potential danger to society if not properly used and controlled, and Hahn himself was keenly aware of the potential for danger. Nonetheless, this discovery paved the way for much future research into nuclear chemistry, as well as for the development of modern nuclear reactors.

1945—Artturi Ilmari Virtanen (Finland) "for his research and inventions in agricultural and nutrition chemistry, especially for his fodder preservation method." Virtanen was quite an interesting chemist, and he was also a farmer! His fodder preservation method made use of hydrochloric and sulphuric acids to slow the processes that normally cause fodder to ferment. He also made several other achievements in the area of nutrition/agricultural chemistry, helping to better provide nutrition to animals raised on farms.

1946—James Batcheller Sumner (United States) "for his discovery that enzymes can be crystallized," and John Howard Northrop (United States) and Wendell Meredith Stanley (United States) "for their preparation of enzymes and virus proteins in a pure form." Summer was the first to provide concrete evidence that enzymes and proteins could be crystallized, paving the way for further research in this area. Northrop and his co-workers explored the conditions that led to formation of crystalline proteins, and mastered this "art," pioneering the way for modern scientists working to crystallize proteins and viruses. Stanley demonstrated that viruses can be crystallized in the same way as proteins, and then showed that viruses are, in fact, proteins themselves.

1947—Sir Robert Robinson (United Kingdom) "for his investigations on plant products of biological importance, especially the alkaloids." Robinson made significant advances in medicinal chemistry, both in synthesis and in gaining a better understanding of the mechanisms by which drugs work. His Nobel Prize was awarded for his work with alkaloids in particular, including quinine, cocaine, and atropine; Robinson studied and helped to understand how these types of molecules have their effects on the human body and mind.

1948—Arne Wilhelm Kaurin Tiselius (Sweden) "for his research on electrophoresis and adsorption analysis, especially for his discoveries concerning the complex nature of the serum proteins."

1949—William Francis Giauque (United States) "for his contributions in the field of chemical thermodynamics, particularly concerning the behavior of substances at extremely low temperatures." Giauque's work did much to prove the third law of thermodynamics (originally proposed by Nernst) and helped make it possible to calculate the free energies of formation of molecules. Much of this was made possible by experiments performed at extremely low temperatures, and Giauque can be credited for developing the methods necessary to perform these experiments at temperatures close to absolute zero. Giauque's work explored the entropy, or disorder, associated with various forms of various substances, and he did much of the pioneering work in studying molecules at low temperatures.

1950—Otto Paul Hermann Diels and Kurt Alder (both Federal Republic of Germany) "for their discovery and development of the diene synthesis." Diels and Alder won the prize for the developing chemical reactions that are broadly relevant in organic synthesis, and they even have a reaction named after them called the "Diels-Alder" reaction. Dienes are complexes containing a pair of conjugated carbon-carbon double bonds, which can react to form a cyclic product in a wide variety of cases. Further research into this class of reactions has found a tremendous number of applications throughout organic chemistry.

1951—Edwin Mattison McMillan and Glenn Theodore Seaborg (both United States) "for their discoveries in the chemistry of transuranium elements." All the way back in 1934, Fermi discovered that heavier elements could be created by bombarding heavy elements with neutrons. McMillan was the first to succeed in demonstrating the existence of the transuranium elements, which are the heaviest elements on the periodic table. Seaborg expanded on this work, discovering a whole additional row of the heaviest elements on the periodic table!

1952—Archer John Porter Martin and Richard Laurence Millington Synge (both United Kingdom) "for their invention of partition chromatography." Martin and Synge won the prize for their development of the basic principles of chromatography, or the separation of chemical substances based on differences in their chemical properties (often their polarity). In their separation method, a drop of a mixture of compounds is placed on a strip of paper, and a solvent, perhaps water or an alcohol (or a mixture), is drawn up the strip of water, resulting in a separation of the compounds in the mixture. This separation occurs because each component in the mixture interacts differently with the solvent. This method has been expanded upon by a large number of later researchers, and chromatographic methods play a very important role in laboratory chemistry to this day.

1953—Hermann Staudinger (Federal Republic of Germany) "for his discoveries in the field of macromolecular chemistry." Staudinger was among the first to propose that macromolecules (polymers) are of significant importance and play an important role in chemistry. These views were not initially well-received by many members of his field, but he was able to demonstrate experimental proof of the existence of macromolecules. Today the importance of macromolecules is widely recognized in polymer chemistry, biochemistry, and numerous other fields.

1954—Linus Carl Pauling (United States) "for his research into the nature of the chemical bond and its application to the elucidation of the structure of complex substances." Pauling made significant progress toward understanding the nature of chemical bonds, which are (obviously) relevant to virtually all chemical processes. He proposed the property of electronegativity as a means for characterizing chemical bonds, and he developed a scale for electronegativity values. Pauling's work with X-rays also uncovered the structures of numerous molecules, and paved the road for X-ray diffraction as a characterization method for even more complex molecules. In 1963, Pauling also was awarded a Nobel Peace Prize. Throughout his career, Pauling was always known for working, and living, at the forefront of scientific discovery.

1955—Vincent du Vigneaud (United States) "for his work on biochemically important sulfur compounds, especially for the first synthesis of a polypeptide hormone." Du Vigneaud was known for bridging the fields of organic and biochemistry by synthesizing biochemically relevant compounds. He synthesized the first polypeptide hormone and developed knowledge of the chemistry surrounding peptides, especially those containing sulfur. In early experiments on the posterior lobe of the brain, du Vigneaud had noted the high percentage of sulfur present, prompting his career-long investigation into the correlation of sulfur with biological activity.

1956—Sir Cyril Norman Hinshelwood (United Kingdom) and Nikolay Nikolaevich Semenov (USSR) "for their researches into the mechanism of chemical reactions." These men made numerous important discoveries regarding chemical reaction mechanisms, in particular with regard to proposing and demonstrating the importance of chain reaction mechanisms. In various instances, these two men showed that observations could be explained by invoking a chain reaction mechanism, which is a mechanism that "self-propagates" for several cycles before terminating. Among other examples, chain mechanisms were shown to be key to chemical reactions resulting in explosions.

1957—Lord (Alexander R.) Todd (United Kingdom) "for his work on nucleotides and nucleotide co-enzymes." Todd laid the foundation for future research in biochemistry, genetics, and biology surrounding the structures and roles of nucleotides in living organisms. He established the chemical structures of nucleotides, and established the (extremely important) role of phosphorylation in biochemical processes.

1958—Frederick Sanger (United Kingdom) "for his work on the structure of proteins, especially that of insulin." Sanger established the structure of insulin by first recognizing that it was made up of two chains of amino acids, one 31 and the other 20 residues long. He proceeded to establish the sequence of amino acids (all 51) in the two chains, thus determining the chemical composition of insulin. Insulin is a key peptide hormone that regulates glucose levels in the human body. In addition to characterizing the structure of this important hormone, Sanger in fact did much more. The methodology he applied to characterizing its structure went on to be used by numerous other scientists in many, many other situations to characterize protein structures.

1959—Jaroslav Heyrovský (Czechoslovakia) "for his discovery and development of the polarographic methods of analysis." Heyrovský was an innovator in the field of chemical analysis, developing a polarographic method of analysis capable of analyzing the presence of virtually any species dissolved in water, and could determine the relative percentages of the species present. He did this using a method that we won't get into here, but it relied on a fairly simple procedure involving the application of a current and measuring drops of mercury.

1960—Willard Frank Libby (United States) "for his method to use carbon-14 for age determination in archaeology, geology, geophysics, and other branches of science." Libby developed a procedure that uses the relative abundance of carbon isotopes in a sample to determine how old it is. This procedure has been extremely useful to scientists from many fields and has played a major role in establishing the ordering of historical events, including the prehistory of humankind.

1961—Melvin Calvin (United States) "for his research on the carbon dioxide assimilation in plants." Calvin received the prize for his work on what is today known as the Calvin Cycle, which is the process by which green plants affect the fixation of carbon dioxide, or in other words the incorporation of carbon dioxide molecules from the atmosphere into other molecules. Calvin identified that there is a close connection between the metabolism of carbohydrates and photosynthesis. This process is quite complex, involving ten different intermediates and eleven different enzymes to catalyze each step of the reaction.

1962—Max Ferdinand Perutz and John Cowdery Kendrew (both United Kingdom) "for their studies of the structures of globular proteins." Perutz and Kendrew sought to elucidate the structures of large proteins using X-ray diffraction techniques, with a particular focus on hemoglobin and myoglobin. They took a variety of innovative approaches, including recording a huge number of X-ray diffractions (about a quarter million), incorporating heavier gold or mercury atoms at well-defined locations into the molecule, and using a computer (one that was advanced for the time) to process the large amounts of data they collected. This was quite a challenging task, as even myoglobin (the smaller molecule of the pair) contains roughly 2,600 atoms. Their work was the first to help understand the principles behind the structure of globular proteins.

1963—Karl Ziegler (Federal Republic of Germany) and Giulio Natta (Italy) "for their discoveries in the field of the chemistry and technology of high polymers." These two great polymer scientists developed several classes of polymers, while simplifying and clarifying the mechanisms of polymerization processes. Ziegler discovered titanium complexes that can catalyze olefin polymerization reactions, and Natta developed a method for the preparation of stereoregular polymers from propylene. At the time, the Nobel committee recognized that the full implications of their work were likely not yet realized, and indeed polymer chemistry was still a relatively young field at that date. Today the work of Ziegler and Natta underpins the technology used to generate many of the plastics you encounter.

1964—Dorothy Crowfoot Hodgkin (United Kingdom) "for her determinations by X-ray techniques of the structures of important biochemical substances." Hodgkin used X-ray crystallography to determine the structure of penicillin, vitamin B12, and a large number of other biologically relevant molecules. As the importance of computers for processing X-ray crystallographic data was increasingly being recognized, Hodgkin was noted for her exceptional ability to process the data, and the Nobel committee recognized that this talent likely played a vital role in her ability to achieve so much during her career.

1965—Robert Burns Woodward (United States) "for his outstanding achievements in the art of organic synthesis." Woodward made tremendous research achievements in a broad range of areas in the field of organic synthesis. He established the structures of aureomycin and terramycin (which are antibiotics), and made possible new synthetic work in this area. He also synthesized quinine, which was considered a great challenge and was used to fight malaria, and later synthesized cholesterol and cortisone. The list of synthetic achievements goes on and on, and he truly ranks among the most successful synthetic chemists ever to have graced the field. In addition to these synthetic achievements, Woodward also established the structures of a large number of important compounds.

1966—Robert S. Mulliken (United States) "for his fundamental work concerning chemical bonds and the electronic structure of molecules by the molecular orbital method." Mulliken received the Nobel Prize in recognition of his work studying that nature of how electrons behaving in molecules, in particular for the molecular orbital approach that he developed. Molecular orbitals are formed by the overlap of the orbitals on individual atoms, and these can be used to rationalize whether bonds will exist between pairs of atoms, how strongly the pairs will be bonded, and what type of reactivity the molecule may be expected to undergo.

1967—Manfred Eigen (Federal Republic of Germany), Ronald George Wreyford Norrish (United Kingdom), and George Porter (United Kingdom) "for their studies of extremely fast chemical reactions, effected by disturbing the equilibrium by means of very short pulses of energy." Norrish and Porter were credited with the prize for their development of flash photolysis methods, which applied short pulses of light to initiate photochemical reactions, allowing them to study very fast chemical reactions that could not previously be observed. Eigen's work differed in that he used sound as his form of energy to probe the chemical reactions, which is far less invasive of an approach in the sense that sound does not cause drastic changes in the behavior of the molecules being studied. Of the two methods, the flash photolysis method is much more akin to the modern spectroscopic approaches used today, while sound-based approaches did not gain as much traction.

1968—Lars Onsager (United States) "for the discovery of the reciprocal relations bearing his name, which are fundamental for the thermodynamics of irreversible processes." Onsager received the prize for his brilliant mathematical work that allowed for a theoretical description of irreversible processes. He additionally made a large number of other contributions to physics and chemistry during his career, including developments surrounding the conductivity of solutions and flow of electrolytes, and a solution to the Ising model.

1969—Derek H. R. Barton (United Kingdom) and Odd Hassel (Norway) "for their contributions to the development of the concept of conformation and its application in chemistry." Barton and Hassel earned the prize for their work surrounding the conformational analysis of molecules. While we draw a molecule on paper in a specific orientation and conformation, there are, in truth, typically many accessible conformations to a given molecule. This is especially true for "floppy" molecules, and these are where conformational analysis can be particularly insightful and important. Their work drew attention to the importance of rotations and other conformational changes in chemistry, particularly with regard to organic molecules. They showed that reactivity can actually be significantly influenced by the conformation of a molecule, and that conformational changes may be necessary to promote reactivity or allow for a reaction to take place.

1970—Luis F. Leloir (Argentina) "for his discovery of sugar nucleotides and their role in the biosynthesis of carbohydrates." Leloir discovered a substance essential for the transformation of one sugar into another, and this turned out to be a sugar bound to a nucleotide molecule (a sugar nucleotide). He soon realized that he had opened the door to a vast number of unsolved problems surrounding carbohydrate synthesis, and proceeded to pursue them with fervor. During his career, Leloir revolutionized the understanding of the synthesis and biosynthesis of sugars.

1971—Gerhard Herzberg (Canada) "for his contributions to the knowledge of electronic structure and geometry of molecules, particularly free radicals." Herzberg was a famous physicist and astrophysicist, and he was awarded the Nobel Prize in chemistry for his great achievements in molecular spectroscopy. A particular demonstration of his skill in spectroscopy was his investigations into the role of free radicals in chemical reactions. Free radicals had long been a difficult target to study, owing to their relatively short lifetimes (on the order of millionths of a second). Herzberg's talent with spectroscopy allowed him to address this, and other similarly challenging (and interesting) problems.

1972—Christian B. Anfinsen (United States) "for his work on ribonuclease, especially concerning the connection between the amino acid sequence and the biologically active conformation," and Stanford Moore and William H. Stein (both United States) "for their contribution to the understanding of the connection between chemical structure and catalytic activity of the active center of the ribonuclease molecule." Together, these three scientists uncovered the structure of the ribonuclease enzyme, and they were able to gain insight into how its structure related to its reactivity. In particular, Moore and Stein were able to correlate the structure of the enzyme's active site to its reactivity. Not only was ribonuclease an important enzyme to study, but their approach paved the way for similar studies to come.

1973—Ernst Otto Fischer (Federal Republic of Germany) and Geoffrey Wilkinson (United Kingdom) "for their pioneering work, performed independently, on the chemistry of the organometallic, so called sandwich compounds." Fischer and Wilkinson received the prize for their role in elucidating the basic properties surrounding bonding and reactivity in organometallic complexes, with a particular focus on their work with organometallic sandwich compounds, which involve a metal center "sandwiched" between two ligands. This prize was somewhat special in that it was openly recognized that the practical implications of their work were not readily obvious, but that the knowledge surrounding organometallic chemistry gained from their work would be invaluable to chemists in the future.

1974—Paul J. Flory (United States) "for his fundamental work, both theoretical and experimental, in the physical chemistry of macromolecules." Flory contributed greatly to polymer science by developing metrics for characterizing polymers and for comparing different polymers against one another. This was a task that had often proved difficult due to the different compositions and conformational arrangements found in different polymers. He made great strides toward putting polymer science on a firm theoretical footing, which was lacking when he entered the field.

1975—John Warcup Cornforth (Australia, United Kingdom) "for his work on the stereochemistry of enzyme-catalyzed reactions," and Vladimir Prelog (Yugoslavia/Switzerland) "for his research into the stereochemistry of organic molecules and reactions." Cornforth explored the stereochemistry of enzyme catalyzed reactions by using isotopically labeled hydrogen atoms to study the geometry of their arrangement in the enzymes active site. Prelog studied how the stereochemistry of organic molecules affected their reactivity, leading to important discoveries. He also explored the stereochemistry of enzyme catalyzed reactions observing how they reacted with simple organic molecules

1976—William N. Lipscomb (United States) "for his studies on the structure of boranes illuminating problems of chemical bonding." Boron hydride (borane) complexes participate in many interesting examples of chemical bonding, and this is due to the fact that boron possesses one less electron than carbon to donate to chemical bonds, but still often forms four-coordinate complexes. Lipscomb won the prize for his pioneering explorations of the chemistry and bonding in borane complexes using X-ray diffraction and quantum chemical calculations. He reached a level of understanding of borane complexes at which he could predict the properties and reactivity of borane complexes reasonably well, and his work led to deeper insight into the nature of the chemical bond.

1977—Ilya Prigogine (Belgium) "for his contributions to non-equilibrium thermodynamics, particularly the theory of dissipative structures." Non-equilibrium thermodynamics is traditionally considered a particularly hard topic to address, because the assumptions one can usually make regarding the behavior of molecules are thrown out the window. Prigogine's work expanded on existing thermodynamic theories to address systems far from equilibrium, such as when a liquid is heated rapidly from below. Prigogine demonstrate that structures called "dissipative structures" could exist under conditions far from equilibrium, and that these structures could only exist in conjunction with the surrounding environment.

1978—Peter D. Mitchell (United Kingdom) "for his contribution to the understanding of biological energy transfer through the formulation of the chemiosmotic theory." Mitchell received the prize for developing a theory of how electron transfer was coupled to ATP synthesis during oxidative phosphorylation and photophosphorylation. He proposed that a gradient of proton concentration (and thus also charge) was built up across the mitochondrial membrane, and that the reverse flow of protons down the concentration gradient provided the driving force for ATP synthesis. This is what is known as the chemiosmotic theory.

1979—Herbert C. Brown (United States) and Georg Wittig (Federal Republic of Germany) "for their development of the use of boron- and phosphorus-containing compounds, respectively, into important reagents in organic synthesis." Brown received the prize for his work with boron reagents in organic synthesis, which also led to the development of organoboranes as a class of molecules. Wittig worked with phosphorus, developing a reaction through which a carbonyl could be converted to an olefin. This reaction is today known as the Wittig reaction. The work of both Wittig and Brown resulted in the development of useful reagents that find wide application in organic synthesis even today.

1980—Paul Berg (United States) "for his fundamental studies of the biochemistry of nucleic acids, with particular regard to recombinant-DNA" and Walter Gilbert (United States) and Frederick Sanger (United Kingdom) "for their contributions concerning the determination of base sequences in nucleic acids." Berg earned the prize for having been the first to design a recombinant-DNA molecule, which is to say he constructed a DNA molecule containing parts of DNA from different species. Gilbert and Sanger shared the other half of the award for their work toward the sequencing of DNA. The sequencing of DNA was initially an extremely time-consuming and expensive task, but today it is becoming more and more feasible to sequence DNA efficiently and affordably.

1981—Kenichi Fukui (Japan) and Roald Hoffmann (United States) "for their theories, developed independently, concerning the course of chemical reactions." Fukui can be credited for the development of frontier molecular orbital theory, which provides predictions surrounding the reactivity of compounds based on the nature of their most loosely bound electrons (or highest energy occupied orbitals), along with the nature of their lowest energy unoccupied orbitals. Hoffman had completed some significant theoretical work while working with Woodward years earlier, and had reached conclusions surrounding the relationship between chemical reactions and the symmetry of the orbitals involved in those reactions. Both of these great scientists succeeded in approaching rather difficult problems by attempting to look for generalizations, simplifications, and basic patterns that were common among their observations.

1982—Aaron Klug (United Kingdom) "for his development of crystallographic electron microscopy and his structural elucidation of biologically important nucleic acid-protein complexes." Klug expanded on the existing technology of electron microscopy to develop an approach based on mathematical processing of the images that gave higher contrast images at relatively low radiation doses and without having to apply heavy metal stains to the samples. His approach made it possible to determine the structures of important aggregates that would have previously been difficult or impossible to accurately characterize.

1983—Henry Taube (United States) "for his work on the mechanisms of electron transfer reactions, especially in metal complexes." Taube investigated a series of ions of cobalt and chromium, finding that certain species reached chemical equilibria in solution, while others did not. A careful series of experiments demonstrated that, in some cases, a bridge was required to form between a pair of metal ions (or one of their ligands) before electron transfer could take place. In other cases, electron transfer could take place at a distance. These observations turned out to have important applications for many chemical processes, particularly in the field of biochemistry. It is also noteworthy that, at the time he won the prize, Taube had already established a strong history of being the first in the entire field of chemistry to report significant discoveries that were of great importance to the field. His importance to the field of coordination chemistry was recognized by the Nobel committee, and the following quote is from one of their reports: "There is no doubt that Henry Taube is one of the most creative research workers of our age in the field of coordination chemistry throughout its extent. He has for thirty years been at the leading edge of research in several fields and has had a decisive influence on developments."

1984—Robert Bruce Merrifield (United States) "for his development of methodology for chemical synthesis on a solid matrix." Merrifield developed a simple and very clever way of synthesizing peptides or chains of nucleic acids. The method involves binding the first of the chain of (let's say) amino acids to a polymer. The subsequent residues can then be added synthetically, and this approach turns out to be faster and yield larger quantities of the desired final product than earlier methods.

1985—Herbert A. Hauptman and Jerome Karle (both United States) "for their outstanding achievements in developing direct methods for the determination of crystal structures." These two men won the prize for developing improved methods of analyzing crystal structure diffraction patterns to yield molecular structures. Their method was based on developing probabilistic equations relating the chemical structure to the observed diffraction patterns, and it relied on measuring many diffraction patterns to determine the molecular structure.

1986—Dudley R. Herschbach (United States), Yuan T. Lee (United States), and John C. Polanyi (Canada / Hungary) "for their contributions concerning the dynamics of chemical elementary processes." Herschbach was recognized for having developed the method of using crossed molecular beams to carry out detailed studies of chemical dynamics. Lee was initially working with Herschbach, and independently further developed the method of using crossed molecular beams to study important reactions of relatively large molecules. Polanyi developed the method of using chemiluminescence in the infrared where weak infrared emissions from recently formed molecules is measured to learn how energy is released during chemical reactions.

1987—Donald J. Cram (United States), Jean-Marie Lehn (France), and Charles J. Pedersen (United States) "for their development and use of molecules with structure-specific interactions of high selectivity." These chemists won the prize for developing/finding molecules that can "recognize" one another so that they react to form complexes in a highly specific way. They investigated the key features of these molecules that allow for such highly specific molecular recognition. In turn, they produced molecules that can mimic the highly specific recognition features of enzymes. This gave way to the area of research that is today known as host-guest chemistry or supramolecular chemistry.

1988—Johann Deisenhofer, Robert Huber, and Hartmut Michel (all Federal Republic of Germany) "for their determination of the three-dimensional structure of a photosynthetic reaction center." This prize was awarded for the atom-by-atom characterization of a membrane bound protein, and, in particular, one that is responsible for carrying out photosynthesis. These proteins are quite difficult to crystallize, and Michel can be credited with achieving this task. He then collaborated with Huber and Deisenhofer to unravel the details surrounding the structure of the crystallized membrane bound photosynthetic reaction center.

1989—Sidney Altman (Canada, United States) and Thomas Cech (United States) "for their discovery of catalytic properties of RNA." While RNA was already known for its role as a molecule involved in heredity and the transport of genetic information, Altman and Cech discovered that RNA can also serve biocatalytic functions. This was a completely surprising result to the scientific community. The discovery was made when RNA was put into a test tube in the absence of any protein enzymes, and the RNA began to cut itself into pieces and rejoin the pieces back together again. This observation led to the discovery of the first RNA enzymes. By the time the prize was awarded, nearly a hundred RNA enzymes had already been discovered.

1990—Elias James Corey (United States) "for his development of the theory and methodology of organic synthesis." Corey was awarded the Nobel Prize for his numerous important contributions to organic synthesis, including the development of theories and methods that allow for the production of biologically active natural products. This led to many pharmaceuticals becoming commercially available, and thus had a clear impact on the general public health. His research was likely so successful because he developed the principles of a methodology known as "retrosynthetic analysis," in which one starts with the structure of the target molecule and analyzes in reverse order what bonds must be broken to generate simpler structures that one can already synthesize readily.

1991—Richard R. Ernst (Switzerland) "for his contributions to the development of the methodology of high resolution nuclear magnetic resonance (NMR) spectroscopy." Nuclear Magnetic Resonance (NMR) spectroscopy has proven to be a powerful tool for physical and synthetic chemists alike. It has been used to characterize the structures of simple organic complexes, complex biomolecules, and to dynamically track the progress of chemical reactions, among other applications. Ernst won the prize for his substantial contributions to increasing the sensitivity and resolution of NMR instruments. Ernst contributed to improving a wide variety of aspects of NMR spectroscopy, including both one- and two-dimensional approaches, and he also proposed a method for obtaining NMR-tomographic images, which was eventually realized.

1992—Rudolph A. Marcus (United States) "for his contributions to the theory of electron transfer reactions in chemical systems." Marcus received the prize for his work studying electron transfer reactions, and developing the theory surrounding this elementary chemical process. The transfer of an electron between two molecules is a ubiquitous process in chemistry, and it is important to a wide range of chemical phenomenon, ranging from the conduction properties of materials, to chemical synthesis, to the capture of light by plants for the purpose of harvesting energy. Marcus completed much of the work surrounding the theory of electron transfer in the 1950s and 1960s, and he was able to explain the greatly varying rates of electron transfer observed in different chemical systems. Some aspects of his theories were difficult to confirm experimentally, and were not proven until the 1980s, which contributed to why it took so long for him to receive the Nobel Prize for his work.

1993—Kary B. Mullis (United States) "for contributions to the developments of methods within DNA-based chemistry for his invention of the polymerase chain reaction (PCR) method," and Michael Smith (Canada) "for contributions to the developments of methods within DNA-based chemistry for his fundamental contributions to the establishment of oligonucleotide-based, site-directed mutagenesis and its development for protein studies." Mullis was awarded the prize for developing a widely used laboratory method called the polymerase chain reaction (PCR) method. PCR allows the use of relatively simple lab equipment to generate millions of copies of a sample of DNA in the timespan of only a few hours. This has found diverse applications in a large number of laboratories. It can be used, for example, to generate copies of DNA from fossils of organisms that are long-since extinct. Smith shared the prize with Mullis for his work "manipulating" the genetic code. In cells, proteins are produced based on sequences of nucleotides contained within the organism's DNA. These sequences dictate the sequences of amino acids that will be incorporated into a protein. Smith developed a method to selectively manipulate DNA to alter the amino acid sequences of the proteins that are produced.

1994—George A. Olah (United States / Hungary) "for his contribution to carbocation chemistry." Carbocations are species that contain one or more positively charged carbon atoms. While these types of intermediates were postulated for years to play key roles in organic reactions, it was never believed that they could be isolated, due to the fact that they should be highly reactive. Olah managed to prepare stable carbocations through the use of extremely acidic compounds, which are actually called "superacids." Olah's work thus revolutionized research into carbocation chemistry, as these species could finally be characterized for the first time. This work has been very influential in better understanding the role of these important intermediates in organic chemistry.

1995—Paul J. Crutzen (the Netherlands), Mario J. Molina (Mexico / United States), and F. Sherwood Rowland (United States) "for their work in atmospheric chemistry, particularly concerning the formation and decomposition of ozone." Each of these researchers made important contributions toward understanding how atmospheric ozone is depleted through atmospheric reactions. Importantly, each demonstrated ways in which pollution from humans was responsible for depleting the ozone layer, and they did this by learning how atmospheric pollutants caused the breakdown of ozone. This information will hopefully continue to help us protect the ozone layer and the stability of the Earth's climate.

1996—Robert F. Curl Jr. (United States), Sir Harold W. Kroto (United Kingdom), and Richard E. Smalley (United States) "for their discovery of fullerenes." This Nobel Prize was awarded for the discovery of a new form of elemental carbon in which the atoms are arranged in a closed shell to form a ball. These three scientists worked together to make this discovery, and the resulting new form of carbon was known as a fullerene. Fullerenes are generated when carbon vapor, obtained by intense pulsed laser irradiation, condenses in the presence of an inert gas. Fullerenes of a variety of sizes are now known.

1997—Paul D. Boyer (United States) and John E. Walker (United Kingdom) "for their elucidation of the enzymatic mechanism underlying the synthesis of adenosine triphosphate (ATP)" and Jens C. Skou (Denmark) "for the first discovery of an ion-transporting enzyme, Na^+, K^+-ATPase." Boyer and Walker shared half of this prize for their work to understand how ATP synthase (an enzyme) catalyzes the formation of ATP. This information bears a crucial relationship to understanding how energy is stored, transferred, and used in cells. Skou received the other half of the prize for his work to discover the first ion transporting enzyme, which is responsible for keeping the balance of sodium and potassium concentrations in cells.

1998—Walter Kohn (United States) "for his development of the density-functional theory" and John A. Pople (United Kingdom) "for his development of computational methods in quantum chemistry." Kohn and Pople represent two of the founding fathers of modern computational chemistry methods. Density functional theory is a method that relies on the a function describing the spatial electron density in a molecule to calculate molecular properties, and Kohn was awarded the prize for his substantial contributions to the development of this now widely used approach. Pople also made major strides toward making computational methods reliable and accessible to a large number of chemists. In addition to his more technical achievements, Pople is responsible for developing the GAUSSIAN computational chemistry software that is probably the most widely used computational chemistry tool available today.

1999—Ahmed Zewail (Egypt / United States) "for his studies of the transition states of chemical reactions using femtosecond spectroscopy." Zewail was awarded the prize for his work using ultrafast femtosecond spectroscopy to study chemical reactions in real time (that is to say, on the timescale on which they actually take place). Zewail's earliest experiments in this area looked at iodocyanide and sodium iodide. He was the first to be able to observe the dissociation and recombination of a chemical species as it actually took place! His great advances in spectroscopic techniques revolutionized the field, and continue to do so today.

2000—Alan J. Heeger (United States), Alan G. MacDiarmid (United States / New Zealand), and Hideki Shirakawa (Japan) "for their discovery and development of conductive polymers." This prize was awarded for the discovery of plastics/polymers that can, under certain circumstances, be made conductive. This was quite a surprising result, since nobody ever expected plastics to be conductive materials. These plastics/polymers consist of chains of conjugated carbon-carbon double bonds (in other words, carbon-carbon double bonds that alternate with carbon-carbon single bonds). These polymers were also doped, which means that electrons were either removed or added artificially by oxidation or reduction.

2001—William S. Knowles (United States) and Ryōji Noyori (Japan) "for their work on chirally catalyzed hydrogenation reactions" and K. Barry Sharpless (United States) "for his work on chirally catalyzed oxidation reactions." Half of this prize was awarded to Knowles and Noyori for their work with chiral catalysts that chirally catalyzed hydrogenation reactions, or reactions in which two hydrogen atoms are added. This work found applications in pharmaceutical production soon after its discovery. Sharpless was awarded the other half of this prize for developing chiral catalysts to carry out oxidation reactions, which represent another important class of reactions in organic synthesis.

2002—John B. Fenn (United States) and Koichi Tanaka (Japan) "for the development of methods for identification and structure analyses of biological macromolecules for their development of soft desorption ionization methods for mass spectrometric analyses of biological macromolecules" and Kurt Wüthrich (Switzerland) "for the development of methods for identification and structure analyses of biological macromolecules for his development of nuclear magnetic resonance spectroscopy for determining the three-dimensional structure of biological macromolecules in solution." In the past, mass spectrometry was limited to the study of relatively small or lightweight molecules. Fenn and Tanaka share half of this prize for their work in extending the technique of mass spectrometry to significantly larger biomolecules. Fenn and Takana independently developed two different methods for obtaining freely hovering protein samples appropriate for mass spectrometric analysis. Wüthrich received the other half of this prize for the use of NMR spectroscopy to determine the three-dimensional structure of biomolecules in solution. This represents a significant advance beyond crystal structure analyses, as the structure of biomolecules may differ in the crystalline and solution phases, with the solution phase structure being more relevant to their true biological function.

2003—Peter Agre (United States) "for discoveries concerning channels in cell membranes for the discovery of water channels" and Roderick MacKinnon (United States) "for discoveries concerning channels in cell membranes for structural and mechanistic studies of ion channels." Agre received half of this prize for isolating a membrane protein that serves as a water channel in cells. For roughly 200 years it had been postulated that cells contained channels capable of moving water in and out of cells, but Agre was the first to finally discover a concrete example of a water channel. MacKinnon received the other half of this prize for his work on ion channels, and in particular for his characterization of the spatial structure of a potassium channel. For the first time, this allowed chemists to "see" how potassium ions flowed in and out of cells.

2004—Aaron Ciechanover (Israel), Avram Hershko (Israel), and Irwin Rose (United States) "for the discovery of ubiquitin-mediated protein degradation." While the overwhelming majority of biochemical research on proteins has focused on how they are synthesized and how they function, these three researchers took a unique path and studied how cells regulate their breakdown. They discovered that proteins are broken down (and rebuilt) at a much faster rate than anyone probably would have expected. Proteins are labeled when it is time for them to be broken down, and then they are chopped up into small pieces so that they no longer function.

2005—Yves Chauvin (France), Robert H. Grubbs (United States), and Richard R. Schrock (United States) "for the development of the metathesis method in organic synthesis." Chauvin, Grubbs, and Schrock shared the Prize for their work on a metal-catalyzed reaction that exchanges (metathesizes) carbon-carbon double bonds. Many of the double bonds that react with these catalysts are very strong, so a reaction that breaks both bonds of a double bond is impressive indeed. Chauvin was the first to propose a detailed mechanism for this catalyzed reaction. Grubbs and Schrock independently developed extremely active catalysts for this reaction using primarily molybdenum and ruthenium metal centers, respectively.

2006—Roger D. Kornberg (United States) "for his studies of the molecular basis of eukaryotic transcription." In order for DNA to be "read" for the purpose of producing a protein, there are a couple of key steps that must take place. Kornberg received the prize for his work to elucidate the details surrounding the first key step: transcription. Transcription involves making a copy of the DNA, which is to be transferred outside the nucleus of the cell. Interestingly, Roger Kornberg's father, Arthur Kornberg, also received a Nobel Prize in Physiology or Medicine in 1959 for his work studying how information is transferred from one DNA molecule to another. Who would have thought that his son would go on to win a Nobel Prize for a very related piece of research?

2007—Gerhard Ertl (Germany) "for his studies of chemical processes on solid surfaces." Surface chemistry is a topic of broad importance to many applied areas of chemistry. These include atmospheric chemistry (as some atmospherically relevant reactions take place at the surface of ice crystals), semiconductors, and solar energy capture, to name a few. Ertl was one of the founding fathers of modern surface science, and he played a role in developing many of the techniques used to study surfaces today. The study of well-characterized surfaces typically requires studies to be carried out under ultra-high vacuum conditions so as to avoid contamination of the surface to be studied by any gas-phase molecules that may tend to adsorb onto the surface.

2008—Osamu Shimomura (Japan), Martin Chalfie (United States), and Roger Y. Tsien (United States) "for the discovery and development of the green fluorescent protein, GFP." These three scientists received the prize for their work with a protein simply called green fluorescent protein, or GFP. This protein was first observed way back in 1962 in a jellyfish. It eventually became a very important molecule in the biological sciences, as clever researchers have found ways to use GFP to study otherwise invisible chemical processes. Shimomura, Chalfie, and Tsien were at the forefront of the major discoveries that led to the understanding of GFP that researchers possess today.

2009—Venkatraman Ramakrishnan (United States), Thomas A. Steitz (United States), and Ada E. Yonath (Israel) "for studies of the structure and function of the ribosome." Ramakrishnan, Steitz, and Yonath are responsible for uncovering the structure and function of the ribosome at the atomic level, and it is for that work that they share this Nobel Prize. They used X-ray crystallography to individually map each atom that makes up the ribosome; it consists of hundreds of thousands of atoms! The ribosome plays a crucial role in each cell, as it is responsible for the synthesis of proteins. On the basis of their results, these researchers have also developed models showing how antibiotics bind to the ribosome.

2010—Richard F. Heck (United States), Ei-ichi Negishi (Japan), and Akira Suzuki (Japan) "for palladium-catalyzed cross couplings in organic synthesis." Heck, Negishi, and Suzuki shared this prize for their development of a class of reactions called palladium-catalyzed cross couplings. These reactions are a powerful tool in organic synthesis, allowing chemists to form carbon-carbon bonds (which is generally not an easy task at all). The reaction is accomplished by having the carbon atoms first join to a common palladium center, from which point carbon-carbon bond formation can proceed.

2011—Dan Shechtman (Israel) "for the discovery of quasicrystals." In discovering quasicrystals, Shechtman initially made some people think he was foolish and just plain wrong. Using a microscope, he found patterns of molecules that appeared to be packed in repeating patterns, but these were patterns that should not be allowed to repeat in a periodic way by simple arguments of geometry. This controversial discovery, and Shechtman's defense of what he observed, led to his being asked to leave his research group at the time of he made the initial discovery. After fighting a long battle, he eventually reached the point where his discovery was recognized. To date, quasicrystals have not found too many important applications, but they have made scientists rethink the fundamental properties of solid matter, which is certainly a pretty big deal.

2012—Robert J. Lefkowitz and Brian K. Kobilka (both United States) "for studies of G-protein-coupled receptors." Lefkowitz and Kobilka shared this prize for their work studying a class of receptors called G-protein-coupled receptors. These play a role in how cells "sense" their environment, and especially in how cells respond to medicines, hormones, and other signaling molecules. Lefkowitz had been working in this area since the late 1960s, and Kobilka since the 1980s, but it wasn't until 2011 that they were able to get an image of the receptor at the exact moment that a hormone activates its signaling mechanism. The prize was awarded for their collective body of work elucidating how these receptors function.

Bibliography

Books and Articles

Anastas, P. T., and J. C. Warner, *Green Chemistry: Theory and Practice,* New York: Oxford University Press, 1998, p. 30. ISBN 978-0198506980.

Arnett, David. *Supernovae and Nucleosynthesis.* First edition. Princeton, NJ: Princeton University Press, 1996. ISBN 0-691-01147-8.

Burke, J. *Connections.* Boston: Little, Brown & Company, 1978. ISBN 0-316-11681-5.

Cline, D. B. "On the Physical Origin of the Homochirality of Life." *European Review* 13 (2005): 49.

DeKock, R. L., and Harry B. Gray. *Chemical Structure and Bonding.* Second edition. Sausalito, CA: University Science Books, 1989. ISBN 978-0935702613.

D'Ettorre, P. Heinze, J. "Sociobiology of Slave-making Ants" *Acta Ethologica* 3 (2001): 67.

Ellinger, Steve. *Twinkie, Deconstructed: My Journey to Discover How the Ingredients Found in Processed Foods Are Grown, Mined (Yes, Mined), and Manipulated into What America Eats.* New York: Hudson Street Press, 2007. ISBN 978-1594630187.

Gilbert, Avery. *What the Nose Knows: The Science of Scent in Everyday Life.* Crown Publishers, 2008. p. 28. ISBN 140008234X.

Karmarkar, U. R., and D. V. Buonomano. "Timing in the Absence of Clocks: Encoding Time in Neural Network States." *Neuron* 53 (2007): 427.

McQuarrie, Donald A., and John D. Simon. *Physical Chemistry: A Molecular Approach.* University Science Books, 1997. ISBN 978-0935702996.

Strlič, M., J. Thomas, T. Trafela, L. Cséfalvayová, I. K. Cigič, J. Kolar, and M. Cassar. "Material Degradomics: On the Smell of Old Books." *Analytical Chemistry* 81 (2009): 8617.

Online Resources

Bellasugar—What Makes Mascara Waterproof?: http://www.bellasugar.com/What-Makes-Mascara-Waterproof-2804912.

Business Briefing, Pharmatech 2002. "Pharmaceutical Quality Control—Today and Tomorrow" a report by Dr. Michael Hildebrand: http://www.touchbriefings.com/pdf/17/pt031_r_17_hildebrand.pdf.

CNN Tech—Can World's Largest Laser Zap Earth's Energy Woes?: http://articles.cnn.com/2010-04-28/tech/laser.fusion.nif_1_largest-laser-national-ignition-facility-energy?_s=PM:TECH.

Discover Magazine—How Your Brain Can Control Time: http://discovermagazine.com/2008/aug/11-how-your-brain-can-control-time.

Energy Quest: http://www.energyquest.ca.gov/.

European Nuclear Society—Nuclear Power Plants World-wide: http://www.euronuclear.org/info/encyclopedia/n/nuclear-power-plant-world-wide.htm.

Green Econometrics—Understanding the Cost of Solar Energy: http://greenecon.net/understanding-the-cost-of-solar-energy/energy_economics.html.

The Guardian—World Carbon Dioxide Emissions by Country: China Speeds Ahead of the Rest: http://www.guardian.co.uk/news/datablog/2011/jan/31/world-carbon-dioxide-emissions-country-data-co2#data.

Harvard Medical School, Healthy Sleep—Why Do We Sleep, Anyway?: http://healthysleep.med.harvard.edu/healthy/matters/benefits-of-sleep/why-do-we-sleep.

Hillman's Hyperlinked and Searchable Chambers' Book of Days: http://www.thebookofdays.com.

International Energy Agency—Key World Energy Statistics 2007: http://www.iea.org/textbase/nppdf/free/2007/key_stats_2007.pdf.

International Union of Pure and Applied Chemistry Gold Book: http://goldbook.iupac.org.

National Historic Chemical Landmarks—Joseph Priestly, Discoverer of Oxygen: http://acswebcontent.acs.org/landmarks/landmarks/priestley.

National Public Radio—A Chemist Explains Why Gold Beat Out Lithium, Osmium, Einsteinium … http://www.npr.org/blogs/money/2011/02/15/131430755/a-chemist-explains-why-gold-beat-out-lithium-osmium-einsteinium.

Nobel Prize Official Website: http://www.nobelprize.org.

Nuclear Energy Institute—Where Does Uranium Come From?: http://www.nei.org/resourcesandstats/publicationsandmedia/insight/insightmay2009/where-does-uranium-come-from.

Science Bob—Experiments: http://www.sciencebob.com/experiments.

Scientific American—Does Turkey Make You Sleepy?: http://www.scientificamerican.com/article.cfm?id=fact-or-fiction-does-turkey-make-you-sleepy.

Scientific American—Our Planet's Leaky Atmosphere: http://www.scientificamerican.com/article.cfm?id=how-planets-lose-their-atmospheres.

United States Energy Information Administration: http://www.eia.gov.

United States Environmental Protection Agency: http://www.epa.gov.

University of Maryland Medical Center—Herbal Medicine:
http://www.umm.edu/altmed/articles/herbal-medicine-000351.htm.

West Coast Analytical Service—Metals Analysis by ICPMS:
http://www.wcaslab.com/tech/tbicpms.htm.

Further Reading

CHAPTER 1

Burke, J. *Connections.* Boston: Little, Brown & Company, 1978.

CHAPTERS 2, 3, AND 4

Oxtoby, David, H. Pat Gillis, and Alan Campion. *Principles of Modern Chemistry,* 7th edition. Belmont, CA: Brooks/Cole, 2012. ISBN 978-0840049315.

Tro, Nivaldo J. *Chemistry: A Molecular Approach,* 2nd U.S. edition. Upper Saddle River, NJ: Prentice Hall, 2011. ISBN 978-0321651785.

CHAPTER 5

Eğe, Seyhan. *Organic Chemistry: Structure and Reactivity,* 5th Edition. *New York: Houghton Mifflin Harcourt, 2003. ISBN 978-0618318094.*

Vollhardt, K. Peter C., and Neil E. Schore. *Organic Chemistry: Structure and Function, Fifth Edition.* W. H. Freeman, 2005. ISBN 978-0716799498.

CHAPTER 6

Cotton, F. Albert. *Chemical Applications of Group Theory,* 3rd edition. *New York: Wiley, 1990. ISBN 978-0471510949.*

Cotton, F. Albert F., Geoffrey Wilkinson, and Paul L. Gaus. *Basic Inorganic Chemistry,* 3rd edition. New York: Wiley, 1994. ISBN 978-0471505327.

Miessler, G. L., and Donald A. Tarr. *Inorganic Chemistry,* 3rd edition. *Upper Saddle River, NJ: Prentice Hall, 2003. ISBN 978-0130354716.*

DeKock, R. L., and Harry B. Gray. *Chemical Structure and Bonding,* 2nd edition. *Sausalito, CA: University Science Books, 1989. ISBN 978-0935702613.*

CHAPTER 7

Harris, Daniel C. *Quantitative Chemical Analysis,* 8th edition. Wiley-Interscience, 1990. ISBN 978-0471510949.

Harvey, David T. *Modern Analytical Chemistry*. New York: McGraw-Hill, 1999. ISBN 978-0072375473.

Skoog, Douglas A., Donald M. West, F. James Holler, and Stanley R. Crouch. *Fundamentals of Analytical Chemistry,* 8th edition. Brooks Cole, 2003. ISBN 978-0030355233.

CHAPTER 8

Lehninger, A., David L. Nelson, and Michael M. Cox. *Principles of Biochemistry,* 5th edition. W. H. Freeman, 2008. ISBN 978-1429224161.

Voet, D., and Judith G. Voet. *Biochemistry,* 4th edition. Wiley, 2010. ISBN 978-0470570951.

CHAPTER 9

Atkins, Peter, and Julio De Paula. *Physical Chemistry,* 8th edition. Oxford University Press, 2006. ISBN 978-0198700722.

Chandler, David. *Introduction to Modern Statistical Mechanics*. Oxford University Press, 1987. ISBN 978-0195042771.

McQuarrie, Donald A., and John D. Simon. *Physical Chemistry: A Molecular Approach*. University Science Books, 1997. ISBN 978-0935702996.

CHAPTER 10

Jaffe, Bernard. *Crucibles: The Story of Chemistry from Ancient Alchemy to Nuclear Fission,* Dover Publications, 1976. ISBN 978-0486233420.

CHAPTER 11

Odian, George. *Principles of Polymerization*. Wiley-Interscience, 2004. ISBN 978-0471274001.

Stevens, Malcolm P. *Polymer Chemistry: An Introduction*. Oxford University Press, 1998. ISBN 978-0195124446.

CHAPTER 12

Yergin, Daviel. *The Quest: Energy, Security, and the Remaking of the Modern World*. New York: Penguin Books, 2012. ISBN 978-0143121947.

CHAPTER 13

Pavia, Donald L., Gary M. Lampman, George S. Kriz, Randall G. Engel. *Microscale and Macroscale Techniques in the Organic Laboratory*. Brooks Cole, 2001. ISBN 978-0030343117.

CHAPTER 14

Fetterolf, Monty L. *The Joy of Chemistry: The Amazing Science of Familiar Things.* Prometheus Books, 2010. ISBN 978-1591027713.

Field, Simon Q. *Why There's Antifreeze in Your Toothpaste: The Chemistry of Household Ingredients.* Chicago Review Press, 2008. ISBN 978-1556526978.

Le Couteur, Penny. *Napoleon's Buttons: How 17 Molecules Changed History.* Jeremy P. Tarcher, 2004. ISBN 978-1585423316.

CHAPTER 15

Anastas, P. T., and J. C. Warner. *Green Chemistry: Theory and Practice,* New York: Oxford University Press, 1998, p.30.

CHAPTER 16

Callister Jr., William D., and David G. Rethwisch. *Materials Science and Engineering: An Introduction.* John Wiley and Sons, 2010. ISBN 978-0470419977.

CHAPTER 17

Miller, Steve. *The Chemical Cosmos: A Guided Tour.* Springer, 2011. ISBN 978-1441984432.

CHAPTER 18

Corriher, Shirley O. *CookWise: The Hows & Whys of Successful Cooking, The Secrets of Cooking Revealed.* William Morrow Cookbooks, 1997. ISBN 978-0688102296.

Ettlinger, Steve Twinkie, *Deconstructed: My Journey to Discover How the Ingredients Found in Processed Foods Are Grown, Mined (Yes, Mined), and Manipulated into What America Eats.* Hudson Street Press, 2007. ISBN 978-1594630187.

Wolke, Robert L. *What Einstein Told His Cook: Kitchen Science Explained.* W. W. Norton & Company, 2008. ISBN 978-0393329421.

CHAPTER 19

Science Bob—Experiments: http://www.sciencebob.com/experiments.

Thompson, Robert B. *Illustrated Guide to Home Chemistry Experiments: All Lab, No Lecture.* O'Reilly Media, 2008. ISBN 978-0596514921.

Index

Note: (ill.) indicates photos and illustrations.

B

Babcock, Jim, 243
Bacon, Francis, 277
bagasse, 237
baking powder, 272
baking soda, 56, 271–72, 292–94, 293 (ills.)
balloon, 309–10
band gap, 85, 85 (ill.)
bases, 200–201
battery, 107
Beadle, Clayton, 167
Becher, Johann Joachim, 9
Beer's law, 104, 146
bees, 228
beet sugar, 274
benzaldehyde, 131, 131 (ill.)
benzene, 70, 70 (ill.)
beta particle, 153
Bevan, Edward John, 167
Bhopal disaster, 245, 245 (ill.)
Big Bang theory, 264–65, 265 (ill.)
bimetallic strip thermometer, 227
bimolecular elimination, 75, 75 (ill.)
bimolecular reaction, 52, 59
bimolecular substitution reaction, 73, 73 (ill.)
binder, 174
binding affinity, 125
biocatalysis, 245–46
biochemical pathway, 124
biochemistry. *See also* brain; genetics; metabolism
    active site, 116
    amino acids, 115, 115 (ill.)
    biomolecules, 113, 114–15
    cell, 114, 114 (ill.)
    cellular organelle, 114
    chirality, 115
    definition, 113
    enzyme, 116
    glycosidic linkage, 118, 118 (ill.)
    native state of protein, 116–17
    peptide bond, 116, 116 (ill.)
    professions, 113
    protein, 116–17
    saponification, 118
    sugar, 117 (ill.), 117–18
biodegradation, 229
biodiesel, 184
biological chemistry, 92
biological feedstocks, 244–45
biological systems, 93
biomass, 244
biomaterials, 247
biomolecules, 113, 114–15
bioremediation, 236, 236 (ill.)
biosignature, 268
Biot, Jean-Baptiste, 70
black hole, 265
black lights, 147, 305–6
black snake fireworks, 288–89
black tea, 278
blackheads, 219
bleach, 222–23
Bloch, Felix, 195
blood, 102–3, 110
blood clotting, 126
blood doping, 230
blood test, 205–6
Bohr, Niels, 14, 20
boiling point, 37, 38, 193
boiling water, 211
Boltzmann, Ludwig, 138
Boltzmann distribution, 41
bond enthalpy, 51
bonds, 64
bones, 93, 310
books, 226
Boyle, Robert, 5–6, 8
brain
    cerebellum, 130
    cerebrum, 129
    frontal lobe, 130
    medulla oblongata, 130
    naturally occurring molecules, 130–34 (ills.), 131–34
    neurons, 129
    neuroscience, 129
    occipital lobe, 130
    parietal lobe, 130
    parts of, 129, 129 (ill.)
    pons, 130
    prion disease, 130–31
    temporal lobe, 130
breathalyzer, 110
breeder reactor, 162
brining, 275
Bronsted acids and bases, 55
bronze, 3–4
brown eggs, 279
browning of fruits and vegetables, 273, 294–95
buckminsterfullerenes, 249, 249 (ill.)
buckyballs, 249–50
buffer solution, 102–3
Buonomano, Dean, 224
Bussy, Antoine, 64
butter, 276

C

caffeine, 131, 132 (ill.), 207
calanolide, 132, 132 (ill.)
calcium, 93
California Chemistry Initiative, 243–44
calorie, 51
calorimetry, 107–8, 110 (ill.), 110–11
cancer, 89
candle wax, 217
candy, 303 (ill.), 303–5, 304 (ill.)
cane sugar, 274
capsaicin, 230, 230 (ill.)
caramelization, 271
carbanion, 72
carbene ligand, 87–88
carbocation, 72

371

F